RARE EARTH ELEMENTS IN GROUNDWATER FLOW SYSTEMS

Water Science and Technology Library

VOLUME 51

The titles published in this series are listed at the end of this volume.

Rare Earth Elements in Groundwater Flow Systems

edited by

KAREN H. JOHANNESSON
University of Texas at Arlington, TX, U.S.A.

 Springer

A C.I.P. Catalogue record for this book is available from the Library of Congress.

ISBN-10 1-4020-3233-1 (HB) Springer Dordrecht, Berlin, Heidelberg, New York
ISBN-10 1-4020-3234-X (e-book) Springer Dordrecht, Berlin, Heidelberg, New York
ISBN-13 978-1-4020-3233-2 (HB) Springer Dordrecht, Berlin, Heidelberg, New York
ISBN-13 978-1-4020-3234-9 (e-book) Springer Dordrecht, Berlin, Heidelberg, New York

Published by Springer,
P.O. Box 17, 3300 AA Dordrecht, The Netherlands.

Printed on acid-free paper

Cover image:
Maps of the Carizo Sand aquifer of southern Texas, U.S.A., also presented in
chapter 9. Made by PhD students S. Haque and J. Tang.

Printed in the Netherlands.

Table of Contents

Preface

Geochemists recognize the utility of the rare earth elements (REE) as especially powerful tools for tracing geochemical processes within the earth. Interest in the REEs reflects, in large part, their unique and chemically coherent properties that arise as a result of their trivalent charge and the similar ionic radii of all 14 naturally occurring rare earths. For example, owing to their trivalent charge, the REEs are chemically fractionated from divalent Ba and tetravalent Hf, their nearest neighbors in the Periodic Table of the Elements. Furthermore, because the ionic radii of the REEs vary systematically across the REE series (i.e., ionic radius decreases with increasing atomic number, which is referred to as the "lanthanide contraction"), individual REEs are also subject to chemical fractionation within the series itself. As the ionic radii decrease systematically across the series with increasing atomic number, geochemical processes that fractionate REEs (e.g., liquid-melt partitioning, aqueous complexation, adsorption/desorption) produce differences in their relative abundances that varies with atomic number. Thus, by understanding how physico-chemical processes fractionate REEs, geochemists can employ the measured differences in REE concentrations in geologic samples to sort out the important geochemical processes that have acted on the samples.

The literature addressing the geochemistry of REEs is vast, and continues to grow at a remarkable rate. Much of the earliest literature concerned the study of REEs in high temperature magmatic systems, including the partitioning of these heavy metals between melts and mineral crystals, the petrogenesis of igneous rocks, and the study of planetary interiors and origins. Subsequently, and as analytical techniques and instrumentation improved, investigation of the REEs expanded to the marine environment to include their quantification in sediments and seawater samples. In addition to scientific journals, a number of particularly useful reviews or compendia focused on the lanthanides exist, including, but not limited to, *Rare Earth Element Geochemistry*, edited by Paul Henderson (published in 1984 by Elsevier as part of the *Developments in Geochemistry* series) and *Geochemistry and Mineralogy of Rare Earth Elements*, edited by Bruce Lipin and Gordon McKay (published in 1989 by the Mineralogical Society of America as part of the *Reviews in Mineralogy* series). Unfortunately, both of these volumes deal only sparingly with the REE geochemistry of low-temperature aqueous environments, containing two and one chapters, respectively.

Beginning in the 1980's, and continuing throughout the 1990's, the numbers of papers published and manuscripts submitted concerning REE geochemistry in natural waters, and particularly, groundwaters, have ballooned. In part, the ever increasing interest in these trace elements in natural waters reflects improvements in the capability of analytical instrumentation to quantify the picomolal and sub-picomolal concentrations that characterize REE abundances in aqueous systems. The idea for this book originated as a direct consequence of the increasing interest in REEs within groundwater-aquifer systems. This interest led to the convening of a session at the annual Geological Society of America held in Reno, Nevada in November 2000 that focused on REEs in groundwater flow systems. Participants in this GSA session recognized the timely nature of their research and the special GSA session. We agreed that a compilation of papers from the session would be a valuable contribution to the literature. Furthermore, our vision was that such a compilation would expand upon previous books, such as those edited by Henderson (1984) and Lipin and McKay (1989), becoming a useful reference to geochemists currently involved in similar research as well as providing a base from which interested students and other researcher could initiate their own investigations of these fascinating trace elements. The results of the participants' research are presented within this volume.

The first two chapters present analytical methods for quantification of REEs in groundwaters at sub-picomolal concentrations (Chapter 1) and for aquifer rocks (Chapter 2). These are followed by a number of site specific studies of REEs in groundwater flow systems. Chapters 3 and 4 examine REE geochemistry of geothermal fluids from New Zealand and acid rock drainage in Idaho, respectively. Chapter 5 focuses on secondary processes leading to REE transport in the vadose zone in tuffaceous rock system, whereas Chapter 6 investigates linkages between the readily exchangeable REE fraction of aquifer rocks and their concentrations in associated groundwaters. Chapters 7 through 9 examine process that occur within aquifers and along groundwater flow paths, including solution and surface complexation reactions, that exert controls on REE concentrations and fractionation patterns. Finally, Chapter 10 presents the case of including Sm-Nd isotopic analysis, as well as Rb-Sr, along with REE concentrations and fractionation patterns in understanding water-rock reactions in aquifers.

Karen Johannesson

Dallas, Texas   September 2004

Acknowledgments

The fact that this book was completed and published is entirely due to the dedication, perseverance, and hard work of each of the contributing authors. As a consequence, I am eternally grateful to them for their efforts, and especially for believing in this project and sticking with it during my numerous, and time consuming transitions. Thanks are also in order to those who agreed and participated in reviewing the various chapters, many of whom reviewed multiple chapters: Michael Bau, Chris Benedict, James Carr, Zhongxing Chen, Eric DeCarlo, Irene Farnham, Robyn Hannigan, Vern Hodge, Matt Leybourne, Berry Lyons, Annie Michard, Rick Murray, Pauline Smedley, Jianwu Tang, and Xiaoping Zhou. The editor is also indebted to her acquisitions editor, colleague, and friend, Petra van Steenbergen at Kluwer Academic Publishers for continuing to be supportive throughout the many years of this project. We have come a long ways from our days in Tempe. Thanks are also in order to Ingrid Aelbers and Martine van Bezooijen, also at Kluwer, for their assistance and hard work, which was instrumental to completing this book. The editor is especially grateful to her two PhD students and research colleagues, Jianwu Tang and Shama Haque, for their constant and willing assistance with this project, and especially for their willingness to donate numerous hours instructing me in the formatting details required to prepare each chapter (i.e., camera-ready) for publication. The editor also wishes to thank L. Douglas James, Hydrologic Science director at the National Science Foundation, for funds that helped support the research presented in at least two of the contributed chapters. Finally, the editor thanks her parents, Harold, Patricia, and Gordon, and siblings, Mark, Kristen, and Eric, for everything else that really matters.

List of Contributors

LESLIE BAKER

Department of Geological Sciences, University of Idaho, Box 44302, Moscow, Idaho, 83844-3022, USA

DAVID BISH

Group EES-6, Geology, Geochemistry, and Hydrology, MS D462, Los Alamos National Laboratory, Los Alamos, New Mexico 87545, USA

STEVE CHIPERA

Group EES-6, Geology, Geochemistry, and Hydrology, MS D462, Los Alamos National Laboratory, Los Alamos, New Mexico 87545, USA

OLEG CHUDAEV

Far East Geological Institute, Russian Academy of Sciences, Vladivostok, Russia

VALENTINA CHUDAEVA

Pacific Institute of Geography, Russian Academy of Sciences, Vladivostok, Russia

ALEJANDRA CORTÉS

Instituto de Geofísica, Universidad Nacional Autónoma de México, 04510 México, D. F., México

BRIAN L. COUSENS

Department of Earth Sciences, Carleton University, Ottawa, Ontario, Canada K1S 5B6

JAIME DURAZO

Instituto de Geofísica, Universidad Nacional Autónoma de México, 04510 México, D. F., México

W. MIKE EDMUNDS

British Geological Survey, Crowmarsh Gifford, Wallingford, Oxfordshire OX10 8BB, United Kingdom

CAIXIA GUO

Harry Reid Center for Environmental Studies, University of Nevada, Las Vegas, Las Vegas, Nevada 89154-4009, USA

ROBYN E. HANNIGAN

Department of Chemistry and Physics, Arkansas State University, State University, Arkansas 72467, USA

VERNON F. HODGE — *Department of Chemistry, University of Nevada, Las Vegas, Las Vegas, Nevada 89154-4003, USA*

KAREN H. JOHANNESSON — *Department of Earth and Environmental Sciences, The University of Texas at Arlington, Arlington, Texas 76019-0049, USA*

JOSE A. R. LEAL — *Instituto de Geofísica, Universidad Nacional Autónoma de México, 04510 México, D. F., México*

MATTHEW I. LEYBOURNE — *Department of Geosciences, University of Texas at Dallas, Richardson, Texas 75083-0688, USA*

ALEJANDRO G. RAMÍREZ — *Instituto de Geofísica, Universidad Nacional Autónoma de México, 04510 México, D. F., México*

PAUL SHAND — *British Geological Survey, Crowmarsh Gifford, Wallingford, Oxfordshire OX10 8BB, United Kingdom*

WILLIAM M. SHANNON — *Department of Geological Sciences, University of Idaho, Box 44302, Moscow, Idaho, 83844-3022, USA*

KLAUS J. STETZENBACH — *Harry Reid Center for Environmental Studies, University of Nevada, Las Vegas, Las Vegas, Nevada 89154-4009, USA*

JIANWU TANG — *Department of Ocean, Earth, and Atmospheric Sciences, Old Dominion University, Norfolk, Virginia 23529-0276, USA*

DAVID VANIMAN — *Group EES-6, Geology, Geochemistry, and Hydrology, MS D462, Los Alamos National Laboratory, Los Alamos, New Mexico 87545, USA*

SCOTT A. WOOD — *Department of Geological Sciences, University of Idaho, Box 44302, Moscow, Idaho, 83844-3022, USA*

ZHONGBO YU — *Department of Geosciences, University of Nevada, Las Vegas, Las Vegas, Nevada 89154-4010, USA*

XIAOPING ZHOU

Harry Reid Center for Environmental Studies and Department of Geosciences, University of Nevada, Las Vegas, Las Vegas, Nevada 89154-4009, USA

Chapter 1

THE ANALYSIS OF PICOGRAM QUANTITIES OF RARE EARTH ELEMENTS IN NATURUAL WATERS

WILLIAM M. SHANNON & SCOTT A. WOOD

Department of Geological Sciences, University of Idaho, Box 443022, Moscow, ID, 83844-3022, USA

1. Introduction

Recent developments in analytical instrumentation, combined with the demonstrated utility of the rare earth elements (REE) as tracers in natural waters, have encouraged a large number of researchers to attempt low-level REE determinations in natural waters. However, there are few, if any, sufficiently detailed guides available that outline the essential requirements for producing reliable REE data for natural waters. In this chapter we attempt to provide such a guide, aimed primarily at those researchers considering entering this research area for the first time.

Ideally, all steps leading to the determination of trace elements such as the REE should be undertaken in dedicated clean-room facilities. However, geochemists sometimes need to determine trace metals to very low levels under less-than-optimum laboratory conditions. We have developed a sample preparation and analytical protocol that permits determination of the REE in a laboratory that is also being used for other projects, which may provide potential sources of contamination. The ability to carry out trace-metal analyses under such conditions is especially useful in an academic laboratory, which can rarely be dedicated to a single project. In this chapter we identify the essential components and conditions needed to determine picogram quantities of the REE in natural waters. We will describe these procedures in the context of our studies of REE in geothermal fluids from New Zealand. In general, REE occur at sufficiently low concentrations in natural waters, or in a sufficiently concentrated matrix, that a separation/preconcentration step is required in their determination. We have found the iron hydroxide co-precipitation technique (section 7.1) to be adequate in most cases. Our discussion therefore assumes the use of this technique, although others, e.g., solvent extraction (Peppard, 1961; Minczewski et al, 1982; Wai, 1992; Shabani et al, 1990; Aggarwal et al, 1996), ion exchange (Minczewski et al, 1982; Fardy, 1992), calcium oxalate co-precipitation (Alfassi, 1992), sorption onto filter supports (Terada, 1992; Shigeru and Katsumi, 1992; Shabani et al, 1992), etc., can also be used.

A number of material, environmental, and quality-control factors make up the components of a methodology for determination of the REE in water samples. These include selection of suitable laboratory surroundings, equipment, consumables, reagents, and standard materials. Equally important are the methodology of sampling, laboratory handling and extraction of the sample, standard preparation, instrument calibration and quality control, analysis of the samples, and processing the raw data into

K.H. Johannesson,(ed), Rare Earth Elements in Groundwater Flow Systems , 1-37.

a final form. Quality control (QC) must be built in to the process so that problems can be identified and traced to their source and so the accuracy and precision of the resulting data can be estimated. We discuss these essential components in more detail in the following sections.

2. Essential Laboratory Space and Equipment

Laboratory space and proper equipment are vital parts of the analytical process. Laboratory traffic patterns, air circulation patterns, and the position of the work space relative to possible contamination sources can make a difference in the level of blank achievable. Any material that will be directly or indirectly in contact with samples is a possible source of contamination. Reusable glassware and plasticware are initially cleaned and then checked for contamination by analysis of method blanks. This is a laborious, but critically important, process. In a multi-user laboratory it is necessary to guard against possible sources of contamination from the work of other users. A single use of cleaned and checked glass or plasticware for other purposes can introduce persistent contamination. Therefore, we have found it necessary to dedicate equipment such as filtration components, centrifuges, automatic pipettes and glassware, and to store separately consumables such as filters, pipette tips, plasticware, reagents, and standards to prevent inadvertent contamination by other users in the laboratory. Below we provide further details on the minimum facilities and equipment required.

2.1. LABORATORY SPACE

The laboratory space should include a sink, a hood, a bench near the sink, a vacuum source near the hood, and storage cabinets. Dust sources should be minimized. Cabinets should have closable doors to reduce dust transport from any REE compounds stored in the lab. The hood and surrounding area should be thoroughly cleaned and reserved for exclusive use each time REE extractions are performed. Working surfaces in our laboratory are slate and are wiped immediately prior to use. If there is corrosion on exposed metal surfaces, then it may be necessary to cover these surfaces with metal-free epoxy based paint. Clean plastic tubs with sealable lids are used to isolate materials to prevent cross contamination.

2.2. LABORATORY EQUIPMENT AND SUPPLIES

2.2.1. *High-purity water source*
A high-purity (at least 17.6-MΩ-cm) water source is an absolute necessity for ultra-trace analysis. In this chapter, we refer to such high-purity water as DIW (de-ionized water). Stored DIW may easily become contaminated by careless manipulation. A DIW supply that continuously re-circulates the DIW through a resin bed is therefore the only alternative in a multi-user lab.

2.2.2. *Analytical balances*
At least one analytical balance should be capable of measurement to 4 decimal places with a capacity of at least 100 grams. Also necessary is a second balance of 2-kg

capacity capable of measurement to 2 decimal places. Accuracy of balances must be checked each day of use with ASTM Class-1 weights spanning the range of expected measurements and the results recorded in a QC log book.

2.2.3. Adjustable-volume air-displacement pipettes

Adjustable-volume repipetters should be dedicated and attached directly to bottles for dispensing reagent liquids. Eppendorf and MLA brand pipettes certified to at least 1% accuracy and capable of delivering volumes from 10 to 100 µL, 100 to 1000 µL and 1 to 5 mL are used to make reagent solutions, standards, add spikes, and dilute samples. Such pipettes should never be inserted directly into any high-purity reagent container. Instead, an appropriate amount of reagent should be dispensed into a clean receptacle from which the desired amount is taken up by the pipette. Pipettes are numbered for identification in log books. Delivery for each pipette must be checked on the analytical balance each day of use at the volume setting used and recorded in the QC log book. Trace-metal certified disposable tips must be used for all 10-100 µL and 100-1000 µL pipettes involved in REE analysis. Macropipettor tips of 5-mL volume graduated in 0.1-mL increments can be used to make reagent solutions and to dilute non-trace cation and anion samples after preparation.

2.2.4. Volumetric glassware

Class A borosilicate volumetric flasks of 500- and 100-mL volumes are used ONLY for preparation of standards. These are preferably closed with Teflon stoppers because they are easily cleaned, make a superior seal, and do not stick in the neck of the volumetric, which should be kept closed during storage. A set of Class A pipettes should be cleaned and set aside for exclusive use. Disposable wipes, (e.g., Kimwipes® Fisher Cat. # 06-666-1A) low-lint, non-abrasive, and preferably metal-free, are used to wipe off the outside of borosilicate pipettes for less critical (e.g., major-element cations and anions) standard preparation.

2.2.5. Reagent, standard and sample collection containers

Teflon FEP (flourinated ethylene polypropylene) bottles of 500-mL (e.g., Fisher 02-923-30C) and 125-mL (e.g., Fisher 02-923-30A) capacity are easily cleaned and should be used to store and dispense fluids, which must have the highest possible purity, for example: acids, high-purity reagents, instrument calibration standards, spike solutions, interference check standards, and internal standards. Low-density polyethylene (LDPE), two-piece wash bottles are used for dispensing DIW and trace-metal grade acids and ammonia. High-density polyethylene (HDPE) bottles of 1000-mL, 500-mL, and 125-mL capacity are used to store stock and calibration standards and for field sampling. These must be cleaned before use (see section 6.1). Two-liter HDPE Erlenmeyer filtration flasks are used to collect the discarded sample fluid during vacuum filtration in the ferric hydroxide co-precipitation procedure. Two-piece polypropylene Buchner funnels are easy to dismantle and clean, and are convenient filter supports in the lab. However, any filtration apparatus can be used as filtration support provided it can be dismantled for cleaning and can stand prolonged storage in an acid bath. Polycarbonate in-line filter holders are used for filtration in the field. HDPE carboys are used to store DIW for rinsing in the field (but in the laboratory DIW is taken only directly from the recirculating source) and to store acidic cleaning solutions for reuse.

2.2.6. *Plasticware*

We use polypropylene centrifuge tubes (e.g., Fisher 05-539-9) with dimensions of 28 x 115 mm and a 50-mL capacity to hold calibration standards in autosampler racks and for sample preparation and short-term storage. These come assembled with caps in sealed packs of 25 and are chosen both for their consistent volume dimensions (top seam corresponds to 50 (± 0.5 mL) and for their very low degree of trace-metal contamination. Polypropylene test tubes in 13 x 100 mm and 16 x 125 mm sizes are used to hold sample solutions in the autosampler racks during instrumental analysis. Polystyrene weighing boats and sample cups are used for weighing solids and for holding small amounts of stock standard solutions, internal standard spikes and reagent acids during sample and standard preparation. These must be kept in a clean location. It is best to use the re-closable centrifuge tubes for temporary storage of ultra-pure reagents involved in the REE extractions. Acetyl Uniwire racks capable of holding 24 50-mL centrifuge tubes are an essential aid in this procedure. These racks are used to hold samples during and after preparation and for storage of the extracts. Plastic polypropylene tubs are used as containers for acid baths, storage of standards and materials to reduce dust contamination, and as secondary containment for storage of toxic or corrosive fluids.

2.2.7. *Vacuum source*

A rotary vacuum pump was connected to a manifold with independently closable valves so that multiple samples can be connected to the same vacuum source. Our low-cost manifold is a circle of reinforced vacuum tubing with four t-connectors branching to the vacuum source and three hoses with in-line valves. The vapor from the manifold is pulled through two polypropylene Erlenmeyer vacuum filtration flasks. The first is a gas bubbler half-filled with water followed by an empty flask (trap) to reduce or prevent corrosive vapors and liquids from reaching the vacuum source. The exhaust side of the vacuum source is vented into the hood but out of the airflow of the REE filtration apparatus.

2.2.8. *Filtration and centrifuges*

Filter materials must introduce minimal blanks during sampling and analytical protocols. Gelman Supor 450TM 0.45-μm polysulfone membranes (P/N 60173) are adequate for filtration of water samples in the field. We have employed pure quartz-microfiber depth-type filters (Whatman P/N 1851-090) in the co-precipitation of the REE in the laboratory using ferric hydroxide. In principle, the Gelman Supor membranes could be used in the collection of the co-precipitate, but in practice these membranes rapidly become clogged, prolonging or halting filtration. Geothermal fluids that have high dissolved silica contents produce voluminous gelatinous precipitates which immediately clog membrane-type filters. Any problem that prolongs the filtration step, necessarily increases exposure of the sample to possible laboratory contamination. The filtration step is one that could benefit most from dedicated clean-room facilities such as a laminar-flow, hepa-filtered hood.

A centrifuge is necessary for the final stage of sample preparation when using the ferric co-precipitation method because contamination (e.g., REE leaching out of the high-purity filters) increases with increasing contact time. Centrifugation speeds the separation of filter and analyte solution. Our centrifuge could accommodate up to eight 50-mL polypropylene centrifuge tubes. This apparatus must be set aside for exclusive use to reduce REE contamination. This is because during centrifugation, the airflow inside the centrifuge can remobilize contamination which can then be driven into the cap and thread areas of the centrifuge tubes. The extracts may subsequently become contaminated when they are decanted into new centrifuge tubes.

2.3. VOLUMETRIC GLASSWARE

When not in use, all volumetric flasks used for standard preparation are stored after adding approximately 5 mL of concentrated trace-metal grade nitric acid. The flasks are stoppered and inverted a few times to coat the inside of the flasks and they are stored upright in polypropylene tubs (to contain accidental spills). The flasks are rinsed four times with DIW just before use for volumetric standard preparation. Class-A borosilicate glass volumetric pipettes are rinsed inside and out, with a 3:1 mixture by volume of 50% hydrochloric and 50% nitric acid followed by DIW prior to and between each use. They are wiped dry with a Kimwipe® for storage and stored with the tips pointed up.

2.4. INITIAL CLEANING OF REUSABLE PLASTICWARE

Reusable plasticware (e.g., Buchner funnels and Erlenmeyer vacuum flasks) are initially cleaned with an anionic surfactant in warm tap water. The plasticware is immediately rinsed with DIW before the tap water can dry onto the surface. The plasticware is then transferred for storage in acid baths or is filled with acid cleaning solution, sealed for storage, and stored upright in polypropylene tubs (to contain accidental spills) until use.

2.5. ACID BATHS

Acid baths are used to soak plastic and glass materials to leach contaminants that are either adsorbed onto surfaces or are present as a consequence of fabrication. In a general-purpose lab the only way to keep laboratory plasticware and glassware clean is to leave it constantly in a dedicated, covered acid bath. A 5-10% nitric 5-10% hydrochloric acid mixture by volume should be used. ACS (American Chemical Society) reagent-grade acids are adequate if items are immediately rinsed with DIW as they are removed from the bath. The Buchner funnels used in the ferric hydroxide co-precipitation procedure are stored in an acid bath. Vacuum filtration flasks are stored filled with acid solution and rinsed with DIW just prior to use. Sample bottles can be filled and allowed to stand in a low-traffic area of the lab.

2.6. PROTECTIVE GLOVES AND CLOTHING

Gloves must be worn for personal protection and to reduce contamination. Gloves must be powder-free and manufactured of a low-contamination material such as nitrile or latex. We prefer latex because of its low metals content, although it is not very resistant to strong acids. Therefore, latex is appropriate only if hands are rinsed often and also immediately after exposure to concentrated acids. Longer term exposure to acids (e.g., immersion in an acid bath) or exposure to acids other than nitric and hydrochloric would require a more resistant glove material such as neoprene, but we have not tested these for use with this method. Splash-proof safety goggles must be used for eye protection. Body powder, skin cream, hairspray, or cosmetics are possible sources of contamination. A clean lab coat used exclusively for REE extractions both protects the user and can help prevent dust on clothing from getting into samples.

3. Quality Control

The goal of the analytical chemist using the analytical procedure described in this chapter is to achieve high accuracy and precision within a particular concentration range for a target (e.g., REE) set of analytes. It is particularly difficult to achieve accuracy and precision at lower detection limits. Methods that are adequate to handle gram quantities, commonly fail for ultra-trace work. Typically, trace quantity losses inherent in most methods become relatively larger as the total analyte quantity becomes smaller. Thus, traces are easily lost during manipulation, and trace contamination and interferences have a correspondingly larger effect. Quality control (QC) protocols help the chemist to quantify accuracy and precision, and to detect contamination and recovery problems. The goal is to reduce contamination and detection limits to as low as reasonably achievable while still attaining good accuracy and precision. To do this, a quality control program should be instituted. Many of these QC procedures are found in U.S. Environmental Protection Agency (EPA) methods. For example, the documents SW846-6020 revision 0, September 1994; EPA-200.8 revision 5.4 May 1994; and 6020 CLP-M version 9.0 describe methods for determination of trace metals in water, wastewater, industrial and organic wastes, and solid wastes. These methods include numerous QC requirements.

3.1. LABORATORY NOTEBOOK

Each person working in a modern laboratory should maintain a laboratory notebook. The notebook should be used to document information and laboratory activities relating to the research project. Lab notebooks should be bound and have sequentially numbered pages. Acid-resistant polypropylene paper is useful in harsh lab environments. Entries should be made only with permanent, black-ink pens. Each page should be dated and initialed. The final entry for each day should also be terminated with date and initials. Inserting of new material and writing over previous entries are practices to be avoided. Changes to numbers, letters, or words should be made using a single line strikethrough that is dated and initialed, and adding the changed entry. Rather than adding or changing entries, it is better to initial and date a new entry and record the new or corrected information and the dates involved.

3.2. CALIBRATION OF ANALYTICAL BALANCES

The laboratory must have a calibrated balance. Standard weights are used to verify balance accuracy each day the balance is used. This can be recorded in your laboratory notebook or you may keep a separate balance logbook. A balance logbook is maintained with columns for the entry date, balance number, check-mass values, observed mass values, and initials of the tester. Typically, a high and a low mass are used, which bracket the expected range of masses to be weighed. The American Society of Testing Materials (ASTM) has published standard limits and procedures for mass and balance calibration. In our laboratory, 1, 10, and 100 gram stainless steel ASTM Class-1 weights are typically used to check our balances.

3.3. CALIBRATION OF PIPETTES

Adjustable pipette settings must be calibrated each time the delivery volume is changed. This is done by weighing successive aliquots of DIW to verify that delivery is accurate. As in the case of analytical balances, a logbook should be kept with columns for the entry date, balance number (the one used to calibrate the pipette), pipette number, desired volume, pipette setting, 3 or more successive aliquot weights, and initials of the tester. A delivery error less than ±1% is sufficient to ensure good accuracy. Most labs do not directly track delivery precision but typically it will be better than the accuracy.

3.4. INSTRUMENT-USE LOGBOOKS

Instrument logbooks are used to track maintenance and use of the instrument. The logbook provides continuity when personnel are replaced and a record of operating conditions, analytical methods, and data files. A modern analytical instrument produces large amounts of data that must be managed. It is a good idea to establish a convention for method and data file names so that data can be easily regenerated if necessary from the raw data files (see below). This is particularly important when data are reprocessed after acquisition. Logging and tracking instrumental tuning parameters helps ensure that goals of instrument performance and stability are being achieved.

3.5. STANDARD LOGBOOKS

These logbooks can be as simple as recording the date a fresh standard is made in the lab notebook. At a minimum, standard bottles should be labelled with the contents, date made, and initials of the person who made the standard. It is also a good idea to write out the recipe of a standard as it is made so that quantities and dilutions can be referenced later and recalculated if necessary. Then, if the solution concentration is found to be suspect, all the information necessary to recalculate the values and avoid having to repeat the analysis will be available. Standards should be prepared fresh, as needed, because the concentration can change on aging. Calibration standards need to be initially verified using a quality-control sample or standard and monitored for stability. For the purpose of establishing standard expiration times, a final aliquot of a standard can be set aside when making a new standard. This aliquot can be analyzed later to monitor long-term stability of that mixture. Long-term stability of a solution kept unused in a sealed bottle may be significantly different from that of a solution in constant use. Care should be taken when preparing mixed standards to ensure that the elements are compatible and stable together.

3.5.1. *Standard expiration dates*
Expiration dates for commercially prepared standards should be the manufacturer's stated expiration date or one year from the date received, whichever is shorter. Dilutions of these standards should be considered to have the same expiration date as the parent or a shorter expiration date determined using the following general

guidelines. Standards that are greater than 100 mg/L in concentration can be stable up to one year. Standards with concentrations over 1 mg/L but less than 100 mg/L should be stable up to 3 months, but commercial, certified standards at 10 mg/L and above are considered valid up to one year. Standards with concentrations less than 1 mg/L are typically assigned expiration dates ranging from a few days to 1 month. Shorter expiration dates are assigned for criteria other than stability (e.g., susceptibility to contamination). We have found that the REE are stable in 4% nitric-1% hydrochloric acid solution for several months; even at ng/L concentrations. Still, it is usually better to make new standards than to try and validate old standards.

3.5.2. *Dealing with multiple-standard recipes*

For a rigorous QC program with many standard and reagent recipes, it becomes tedious to constantly write out component information and recipes, and tracking down the provenance of a standard becomes complex. Thus, any methods that can be used to save time without compromising the rigor of the QC program are useful. The following is useful if a routine protocol requiring complex standard mixtures and reagents will be followed for an extended period of time.

When making a new standard, it is useful to record in a logbook the preparation date, standard identification number, common name, recipe identification number, open date (i.e., first-use date), expiration date, and the user's initials. If recording a commercial standard, then the following should also be included: the date received, source, catalog number, and lot number of the standard. A similar logbook can be used to track reagents such as high-purity acids. The problem of multiple analyte standards can be solved by using a recipe logbook. Multiple analyte standards, whether purchased or made on site should have a recipe recorded with a recipe identification number, common name, start date, expiration date, and final volume. Mixed standards can be made up of many single-analyte standards. Some or all of the standards contributing to a recipe can also be complex. Thus, the identification numbers of all components contributing to the standard should be assigned a list number along with a list of analytes and their respective concentrations in each standard. In addition, the expiration date for the standard and, if applicable, the aliquot of that standard that is present in the final recipe should be listed. The recipe must contain the contributing list numbers and total concentrations of each analyte in the recipe. Acid concentrations and special instructions should be recorded in the notes section. A recipe can be made repeatedly until the recipe expiration date as determined by the component with the shortest expiration date. After logging in or preparing a new standard, label the standard bottle with the standard ID#, common name, receive date, open date, expiration date, and the initials of the responsible parties.

3.6. ESTABLISHING A METHOD OF DATA PROCESSING AND FORMATTING

The raw data from a modern instrument are routinely so complex that the procedures used to process the raw data into final results are now considered by the EPA to be part of the deliverable data package. This means, for example, that archival of raw data necessarily includes the software used to operate the instrument and process the raw data into final results. A legal copy should be made of the software and archived with the raw data generated using that software. Methods, calibration fits, and other software configuration choices that affect the final results should be included with the archived data. When raw instrument data can be reprocessed so as to fit necessary QC criteria, those choices must be preserved as methods or scripts that facilitate producing the exact finished result from the original data. In this way the geochemist also assures an unbiased observer that a consistent set of QC rules have been applied and that deficiencies in an analysis have been acknowledged.

4. Reagents and Materials

For reagents (such as acids) used in analytical protocols, the most important characteristic is usually purity, where knowledge of the precise concentration is of secondary importance. This contrasts with standards for which purity and precisely known concentration are equally important. Excess manipulation in the preparation or transfer of a reagent may introduce contamination. We consider the avoidance of contamination to be more important than being strictly quantitative in transfer and dilution of reagent solutions. However, this does not mean that accuracy need be sacrificed. Procedures can be designed so that quantitative transfer and volumetric dilutions are not necessarily required, thus avoiding additional steps that could lead to contamination. For example, reagent blank solutions are made by adding DIW and appropriate amounts of acids directly into Teflon FEP bottles and diluting with DIW to a mark rather than by using volumetric glassware. In this example, slight inaccuracies in percent acid concentration are less important than achieving the lowest possible blank contamination.

4.1.1. *Deionized water*

As mentioned previously, a high-purity water source is an absolute necessity for ultra-trace analytical work. We use ASTM Type II water (American Society of Testing Methods, D1193) referred to hereafter as DIW. Our DIW is produced by filtration by fiber and activated carbon, followed by reverse osmosis. The product water from the reverse osmosis unit is passed through a high-exchange capacity, mixed-bed ion exchange resin and finally a polishing resin. The last stage continuously re-circulates to maintain water purity and water is dispensed through a 0.2-μm filter. This water typically has at least 17.6-MΩ–cm resistivity (usually > 17.9 MΩ-cm).

4.1.2. *Reagent acids and bases*

Doubly distilled, high-purity Seastar Baseline™ or Fisher Optima™ grade hydrochloric (HCl) and nitric (HNO_3) acids are used for REE standard preparation and sample preservation of REE and cation samples. Trace metal-grade nitric and hydrochloric acids are used for rinsing filtration apparatus. ACS grade acids are used for sample container preparation and in acid baths. Trace metal-grade ammonium hydroxide is used for production of ferric hydroxide precipitate and rinsing of filtration apparatus. Adjustable volume repipetters with closeable tips are attached directly to bottles for dispensing reagent liquids. Reagent bottles that are in use are kept in plastic tubs for secondary containment to reduce the danger of acid or base spills. Repipettors can be fitted with filters for the air that replaces what is dispensed. A plastic beaker is kept with each bottle and tub for excess reagent. On the day of use of a reagent, a small quantity is dispensed and discarded and the dispenser tip rinsed with acid solution and DIW to remove internal and external contamination, which may have settled from the air onto the dispenser tips. Reagent DIW and ammonia solutions in wash bottles should be filled fresh every day to prevent contamination. Acidic solutions in wash bottles for rinsing can be refilled as needed.

4.1.3. *Ferric iron reagent*

The critical aspect for this reagent is that it be of highest purity with respect to REE contamination, not that it be quantitative with respect to iron concentration. This solution should be checked for REE on an ICP-MS before it is used for any samples. We tested and rejected two iron oxide (99.999% pure) sources before we found electrolytic grade iron (99+ pure minus 100 mesh powder) of sufficient purity.

The ferric iron reagent is prepared by dissolving powdered iron in a partially closed, clean Teflon FEP bottle on a hot plate in a hood. To prepare 100 mL, 5.5 to 6.0 grams of powdered iron or a mole-equivalent amount of iron oxide are put in a Teflon FEP bottle and moved to a hood. Iron oxides (which may form from metallic iron or may be in the initial iron reagent) are not easily soluble in nitric acid, so 20 mL of high-purity hydrochloric acid are added and the container is gently warmed on a hotplate in the hood. The Teflon bottle is loosely capped to keep out contaminants, but prevent a build-up of pressure. Next, a few mL at a time, 40 mL of high-purity nitric acid are added (caution, this is a very exothermic reaction!). Loosely cap the bottle while waiting for the iron to dissolve. When all the iron is dissolved, the solution is allowed to cool, transferred to a clean HDPE or Teflon FEP bottle, and diluted to 100 mL (previously marked) with DIW.

4.1.4. *Testing for REE contamination*

The filters, bottles, reagents, and materials that come into contact with the sample fluid during sampling, extraction, or analysis must be checked for REE contamination. We placed filters in 20 mL of high-purity 4% nitric-1% hydrochloric acid solution for several hours and analyzed the resulting solution for REE to determine the blank contribution from different filter materials. It was found that quartz-fiber filters produce a significant REE blank. However, REE contamination from quartz-fiber filters was orders of magnitude lower than REE contamination from borosilicate glass-fiber filters. We added 20 mL of high-purity nitric acid to 10 sample bottles that had been already used for sample collection. The bottles were capped and sealed, allowed to reflux in a warm 50°C bath, and agitated for two weeks. The acid was diluted to 50 mL with DIW and the resultant solution was analyzed for REE. We used these data to demonstrate that REE had not been lost by adsorption to the sample bottle walls. Single-element standard solutions used for internal standards and interference corrections were screened for REE contamination by diluting aliquots of the stock standard solutions to 10 to 100 times the maximum concentration that was used in the analytical procedure.

5. Standards

Standard materials must be of sufficient purity that instrument calibration, internal standardization and interference check solutions are free of interferences. Commercial, mixed, NIST-traceable standards are available, which can be diluted as instrument calibration standards. It is acceptable to make standards from high-purity compounds but still advisable to purchase NIST-traceable standards to run as periodic quality verification during analysis. Single-element standards used for isobaric and spectral interference correction must be of the highest purity. For example, barium standards employed to correct for isobaric interferences in ICP-MS must be as low as reasonably achievable in Eu and Sm contamination (e.g., SPEX Claritas™ standards). Similarly, Nd and Pr standards must be low in Tb and Gd.

5.1. STANDARD PREPARATION PROCEDURE

Multiple-analyte calibration-standard solutions are prepared by combining appropriate volumes of the stock solutions in class-A volumetric flasks (Teflon stoppers are desirable). The purity of water used in making ICP-MS standards is critical to the success of the analysis. A polypropylene or FEP Teflon wash bottle should be reserved for containing the DIW used in making standards. Prior to making one or more standards, empty the wash bottle and rinse repeatedly with DIW before filling with DIW. Set out a temporary waste beaker partially filled with tap water for safe dilution of excess acids and standards. Remove the Teflon stopper and set it aside in a clean weighing boat. Empty the acid in the volumetric flask (see section 2.3) into a waste beaker that is partially full of water so that any acid additions are diluted. Rinse out the flask 4 times with DIW. Add sufficient DIW to fill 20% of the volume of the flask. Then add concentrated HNO_3 and/or concentrated HCl using pipettes or repipettors to result in the required final concentration of nitric (HNO_3) and hydrochloric (HCl) acids. Add appropriate stock standard solutions using a calibrated pipette. The stock standard solutions are added after the acid to avoid hydrolytic precipitation of the analytes upon dilution with DIW.

Solutions and reagents should never be taken directly from the storage bottle but rather should be poured first into a clean secondary container. The aliquot is then withdrawn from the secondary container with a pipette. Always use a new pipette tip and secondary container for each new solution. Pipette tips should be stored in a closed box (e.g., Eppendorf Envirotips and Envirobox) that is opened only when a new tip is needed and tips are never handled with fingers. Read and follow the manufacturer's recommendations for operation of the pipette. Typically, larger volume pipette tips are calibrated to deliver after wetting. Thus, for volumes greater than 50 μL draw some solution into the pipette tip once and then expel or discard it before taking a second aliquot for delivery. Always check the tip after initial wetting and expulsion to be sure that no liquid remains. If the tip does not deliver cleanly then discard it and use another. Manufacturers generally do not recommend pre-wetting the tip for volumes less than 50 μL. Thus, for volumes less than 50 μL, draw the solution up into the pipette tip and deliver directly. Pipettes should never be placed in a horizontal position when they have liquid in their tips. A pipette stand should be employed to keep the pipette near vertical when not in use. Because tilting the pipette affects the calibration volume, a pipette should be held vertically when drawing and delivering solution.

When using disposable pipette tips, deliver the aliquot from the pipette to the neck of the volumetric flask below the level of the ground glass joint without touching the inside of the flask. Otherwise some portion of an aliquot that was added previously may be removed inadvertently. Check the tip after delivery to be sure there is no liquid remaining in the tip. Although it is better to start over, if the tip does not deliver cleanly, a complex standard can be rescued by pulling DIW into the tip and delivering the diluted remainder of the solution into the flask.

When using glass volumetric pipettes, care must be taken to rinse the previously cleaned pipette with DIW between additions and to pre-condition the pipette with the stock standard solution which is then discarded. A fresh aliquot of solution slightly in excess of calibration volume is drawn into the pipette. Let the pipette tip touch the side of the secondary container above the liquid level and let out solution until the volume is exact. Then tilt the pipette slightly so that liquid draws away from the tip and wipe off the outside of the tip with a clean wipe. Deliver the liquid to the volumetric flask only allowing the pipette tip to touch the neck of the flask below the ground-glass joint. Tilt the volumetric slightly so that the pipette can be held vertically for delivery of the last visible liquid. Then wait 20 seconds before touching the tip to the flask a final time.

If adding multiple standards to the same volumetric flask, rinse down the neck of the flask with DIW from the wash bottle between aliquots and swirl the liquid to assist mixing. When the additions are concluded, rinse down the neck of the flask, swirl the liquid to assist mixing, and then make up close to the calibrated volume of the volumetric with DIW directly from the DIW source. Make the final dilution to the mark with DIW from the wash bottle, cap, and mix by inverting several times. Transfer the mixed standard solutions to FEP fluorocarbon (i.e., Teflon) or previously unused HDPE or polypropylene bottles for storage. A bottle may be reused to store only the same standard it originally contained as follows. Rinse well with DIW, rinse once with a small quantity of the standard and discard the rinsate, and transfer the remainder of the standard to the bottle.

5.2. STANDARDS USED FOR REE ANALYSIS

We purchase a 100-mg/L standard containing Sc, Y, and all the REE (e.g., Alfa AESAR Specpure stock #14651) and dilute it to appropriate volumes. We add aliquots of individual 1000-mg/L U, Th, and Zr standards before making the final volumetric dilution. Calibration standards for the ICP-MS are made from two stocks. The 100-mg/L REE standard is one stock and we make a second standard that contains Zr at 100 mg/L and U, and Th at 50 mg/L. A 100-mg/L rhodium standard is made from a SPEX 1000-mg/L stock and stored in a 500-mL Teflon bottle. This standard is used to spike samples in the last stage of extraction to allow the final volume of sample to be calculated from the measured rhodium concentration. Calibration standards are made at 5X dilution intervals (e.g., 200, 40, and 8 µg/L for the REE, Y, Zr, and Sc and half those amounts for Rh, U, and Th) for the top three and then these three standards are each diluted 100X (e.g. 2, 0.4 and 0.08 ug/L) for the lower three standards. All the standards and the blank are made up in 4% nitric plus 1% hydrochloric high-purity acids.

6. Sampling Methods

There are several ways of introducing contamination during sampling. Fortunately, most sample locations have an abundant supply of the water to be sampled and the primary source of contamination is local dust and water carried over from previous locations, both of which are relatively low in soluble REE. With a little care to control dust and rinse sampling apparatus with sample water and DIW, most sources of field contamination can be minimized.

6.1. SAMPLE CONTAINER PREPARATION

We typically employ 1-Liter sample bottles constructed of high-density polyethylene (HDPE) to contain samples destined for REE determination. Teflon bottles may be preferable, especially when dealing with very low concentrations, to avoid sorption losses. However, Teflon bottles are considerably more expensive and we have found HDPE to be adequate for the REE, provided that the samples are sufficiently acidified. We have found that cleaning the bottles with a mixture of >5% nitric + >5% hydrochloric ACS-grade acids works well. After filling the bottles with the acid solution and capping, they were stored in a low-traffic area of the lab for a minimum of 24 hours. After cleaning, the bottles are rinsed several times with DIW and excess water is shaken out. In order to avoid having to measure out acid in the field, the desired volume of high-purity nitric acid is premeasured into the bottles and they are tightly capped and then sealed with Parafilm. Bottles are placed in clean plastic bags for transport to the field. Parafilm sealing is intended primarily to protect the outside of the bottle cap and threads from contamination during transport and storage. A secondary function of the Parafilm is to seal the container if some leakage should occur.

Polycarbonate in-line filter membrane assemblies used for filtration in the field are disassembled and soaked in an acid bath for at least 24 hours, then soaked in DIW and rinsed with DIW just prior to assembly with filters. The silicone rubber gaskets are rinsed briefly with acid and soaked in DIW prior to assembly of the filters. The gaskets should not be soaked in acid as this results in their degradation. After assembly the polycarbonate filters are stored in a clean plastic bag for transport into the field.

6.2. SAMPLE COLLECTION

We routinely collect filtered and unfiltered samples for REE determination. For geothermal fluids, sample collection is accomplished with a polypropylene sampler with a 12-foot handle, and a designated sampling bottle (DSB). The sampler is rinsed with sample down-flow of the sampling point before the sample is collected. The DSB should also be well rinsed with sample prior to using it as a temporary reservoir for filtration of the sample. A hand-cranked MasterFlex L/S-15 pump head with Pt-cured silicone tubing is used to force the sample through an in-line polycarbonate filter holder supporting a Gelman Supor 450TM 0.45-µm polysulfone membrane. We rinse the filter holder and the outside of the tubing with sample before immersion of the tubing into the DSB. Approximately 100 mL of sample are pumped through the tubing and filter before the sample is collected directly from the filter holder into the appropriate sample containers. We rinse the bottle cap off with filtered sample water just before closing and sealing the bottle with Parafilm. Typically we collect the REE samples first, followed by cation and anion samples. We ship the anion bottles to the field filled with DIW which is useful for rinsing the equipment as we set up for sampling. Typically we rinse conductivity, Eh and pH electrodes with DIW before and after sampling to avoid salt build-up and cross contamination. DIW is pumped through the collection apparatus after sample collection is finished. DIW is then used to fill the Pt-cured silicone tubing. We rinse the DSB with DIW and carry it tightly closed and wet or with a little DIW left inside.

A filter pore size of 0.45 μm commonly is considered to be the operational cutoff between dissolved and suspended material. However, such a filter size does not exclude colloidal material. If exclusion of colloidal material is desired, than a filter pore size of less than < 0.1 μm should be used. In this case, forcing 1-Liter water samples through the filter using a hand-cranked pump is tedious and time-consuming, so a motor-driven pump may be desirable. For samples containing a lot of suspended matter, it may be necessary to change the filter membrane a number of times during the filtration of a single liter of sample.

6.3. FIELD MEASUREMENTS

In order to be able to interpret the results of REE determinations on natural waters and to be in a position for thermodynamic modeling, a number of physicochemical parameters must be determined in addition to the REE concentrations. The following measurements are made in the field at each site: temperature, conductivity, and pH. We found a three-person team to be most time-efficient. One person made temperature and conductivity measurements and collected the sample fluid. A second person operates the pump to filter the anion, cation and filtered REE samples. The third person prepared the pH apparatus for measurement and recorded all the other measurements in the field notebook. Determinations of pH and conductivity are made at, or as close as possible to, the emergence temperature of the fluid. We make these measurements by suspending the sensors directly in the water if possible or in the freshly rinsed and filled designated sample bottle otherwise. We found that the 12-foot handle of the sampler is very useful to extend the reach of the temperature and conductivity probes. Care should be taken that the pH electrode does not interfere with conductivity measurements by measuring separate aliquots (small amounts of the KCl filling solution typically leak out of the electrode liquid junction during pH measurement). By the same token, pH (or Eh) measurements should not be made on the same sample aliquot to be used for subsequent cation or anion determinations. The pH buffers used to calibrate the electrode should be maintained at as close as possible to the emergence temperature of the fluids. This is accomplished by suspending the pH buffers in centrifuge tubes in a bath filled with the sample fluid and kept hot by refilling as often as necessary. The pH electrode is also put in a centrifuge tube filled with the sample solution in the same bath. The tube is filled with fresh solution just prior to measurement of the pH. The pH typically is measured by recording the mV reading of the sample fluid and the mV readings of the thermally equilibrated pH 4, 7, and 10 buffers. The pH is calculated later by linear regression of the mV readings against polynomial functions of pH versus temperature for each buffer. For a field pH electrode a wide temperature range, fast response, impact ruggedness, and resistance to dissolved metal or sulfide poisoning are all desirable, if not essential, qualities. We use a rugged-bulb, double-junction, variable-temperature combination glass electrode (AccuTpH+, Fisher P/N 13-620-184).

Alkalinity measurements are made by Gran titration within 72 hours (typically the same night) using a plastic burette-based field titration kit. Certified sodium hydroxide solutions were purchased for standardization of acid solutions used for alkalinity titrations.

7. Sample Preparation in the Laboratory

As already noted, many natural waters have REE concentrations sufficiently low that some form of pre-concentration is necessary. Also, in many cases, the REE may need to be separated from potentially interfering elements. We have found the ferric hydroxide co-precipitation method to be the most convenient, generally applicable method, and our version of this method is described below. In a few cases we have encountered some difficulties with the ferric hydroxide co-precipitation, and so we also describe an alternative method involving solvent extraction.

7.1. REE PRECONENTRATION BY FERRIC HYDROXIDE CO-PRECIPITATION

The REE, Y, Th and U in most of the samples we have worked with have been preconcentrated for analysis by the classic method of co-precipitation with iron hydroxides. This method has been investigated more recently for efficiency in recovery of ultra-trace quantities of metals (Bonner and Kahn, 1951; Minczewski, 1982; Alfassi, 1992) and has been used in the recent geochemical literature (Laul et al., 1988; Lepel et al, 1989; Fee et al., 1992; Johannesson and Lyons, 1994; Klinkhammer et al., 1994; van Middlesworth and Wood, 1998; Bau, 1999; Shannon et al, 1999; Shannon et al, 2001). A 50-mL aliquot of each 1-liter REE sample is saved for possible later analysis, against the possibility of using a more sensitive instrument in the future, and to have a back-up in case something goes wrong during sample preparation. The remaining sample is weighed to determine the ultimate concentration factor. Samples are screened for concentration and those with sufficient REE contents are determined by ICP-MS without a pre-concentration step. For samples requiring preconcentration, an aliquot containing 30- to 60-mg of ferric iron is added to the sample bottles and they are agitated to mix the sample. This is followed by addition of 50 mL of 22% trace-metal-grade ammonium hydroxide to neutralize the 2% acid and raise the pH to between 8 and 9 (note: the exact amount of ammonium hydroxide to be added depends on the volume of the REE sample and the exact amount of the acid added to preserve it. As noted below, a final pH of 8-9 is critical to obtaining quantitative recovery of the REE without fractionation). The sample bottle is capped and agitated. The iron precipitates as a reddish-brown to yellow, gelatinous flocculate. The bottle is allowed to stand for at least an hour while the ferric hydroxide precipitate settles to the bottom of the sample bottle.

The iron hydroxide precipitate is collected by filtration. However, membrane filters of the type used to filter the REE samples in the field are too rapidly clogged due to the co-precipitation of dissolved silica, which is typically present in high concentrations in geothermal fluids (however, this is not as much of a problem for most low-temperature surface waters). Borosilicate glass-fiber filters as a group are too contaminated with REE to be useful. We found that pure quartz-fiber depth-type filters give satisfactory results provided special precautions are followed to reduce method blanks. The quartz-fiber filters are immersed, standing edgewise, in a 1-liter beaker filled with a hot 3:1 mixture of 50% hydrochloric and 50% nitric acid, made from ACS reagent-grade acids, and stored there prior to use. The elapsed time from collection on the filter to re-dissolution and separation from the filter material should be a maximum of 2 hours. The filtration apparatus comprises a plastic, 90-mm Buchner-type filter which can be disassembled for cleaning, a 2-liter Nalgene filtration flask and a connecting seal. The filter assembly is soaked in a dilute HCl-HNO_3 acid bath between each filtration. The apparatus is connected via a plastic valve to a vacuum manifold. The assembled apparatus and filter are rinsed with trace-metal grade hydrochloric followed by DIW. The apparatus is rinsed a second time with 2% trace-metal grade ammonia followed by DIW. The sample is decanted onto the filter and the collected precipitate is rinsed with 2% trace-metal grade ammonia.

Upon completion of filtration, the filter with the damp Fe-hydroxide precipitate is put into a 50-mL polypropylene centrifuge tube. The transfer of the damp filter is a step with high potential for contamination. The "working hand", covered with a clean latex glove, is dipped in warm 50% HCl-HNO_3 solution and rinsed with DIW. The other hand (also gloved) and a pair of tweezers are dipped in a dilute HCl-HNO_3 solution bath and rinsed with DIW after the working glove is clean. Both hands are shaken dry. Transfer is accomplished by using the tweezers to lift up one edge of the filter. The working glove is then used to roll the filter into a cylinder and transfer the filter to a labeled centrifuge tube. If precipitate is observed beneath the filter or around the edges, the filter can be used to police the precipitate onto the filter. The blank will be lower if the filter is not torn in the process. The precipitate is then taken into solution with 20-40 mL of Seastar Baseline 4% nitric-1% hydrochloric acid solution. We typically add a 1000-ng rhodium spike to the sample at this point to correct for dilution errors caused by variable water retention in the precipitate. Gentle rocking can be used to speed dissolution of the precipitate but avoid disaggregating the filter material as this will result in difficult centrifuge separation and a higher blank. Upon complete dissolution of the Fe-hydroxides (about half an hour), the liquid is centrifuged and decanted into a clean polypropylene centrifuge tube and capped until analysis. van Middlesworth and Wood (1998) showed that, when pH was adjusted to between 8 and 9, recovery of the REE was better than 90%. We have made further improvements to their method over time. Method detection limits and spike recoveries were demonstrated with 300 mg of added Si to simulate the natural samples of geothermal waters (see discussion in section 9.1, and Table 1). Our recoveries, and those obtained by van Middlesworth and Wood (1998), are consistent with those reported by Buchanan and Hannaker (1984), who also demonstrated recoveries of REE up to 99% when pH was adjusted to 8.6 and higher.

7.2. ALTERNATE REE EXTRACTION METHOD

Although the iron hydroxide co-precipitation method is relatively straightforward and effective in the majority of cases, it was not successful in eliminating isobaric interference from Ba in those samples containing relatively high concentrations of this element. Moreover, as noted above, precipitation of large amounts of silica from higher-temperature geothermal waters is also a problem. To reduce interference by silica and barium, we tested an extraction method based on the partitioning of the REE between the acidified aqueous sample and a mixture of mono- and bis-2-ethylhexyl phosphate esters (Aggarwal et al, 1996; Shabani et al, 1990, Minczewski et al, 1982). These have the form of $(RO)_n$-PO-$(OH)_{(3-n)}$ where n equals 1 or 2, corresponding to the mono- or bis-2-ethylhexyl phosphate (MEHP and DEHP). The mixed esters are not available in the USA. However, we were able to obtain a supply of mixed ethylhexyl phosphate esters by directly ordering from Merck in Germany. The Merck material is a mixture comprising 45% mono- and 55% bis- ethylhexyl phosphate esters. The ester mixture was dissolved in heptane and shaken with the pH-adjusted (2% nitric acid; pH$\leq$1)

Table 1 Isotopes, Counting Times, Detection Limits, and Total Analytical Error

Spike / DF / Date annotations per group: December 1998 MDL's section — Spike 4 ng/kg, DF 0.04x, Date 3/26/01. March 2001 IDL's section — Spike 8 ng/kg, DF 0.02x, Date 3/26/01. May 2001 MDL's section — Spike 0.2 ng/kg, DF 0.02x, Date 5/7/01.

Isotope	count time	Type	REC %	December 1998 MDL's, LMB's run March 2001 — Avg LMB ng/kg	Avg MDL ng/kg	MDL ng/kg	CV %	TAEM %	MDL pmole/kg	March 2001 IDL's and estimated MDL's — Avg IDL ng/kg	IDL ng/kg	CV %	TAEM %	est MDL ng/kg	est MDL pmole/kg	May 2001 MDL's — Avg MDL ng/kg	MDL ng/kg	ratio w/w	CV %	TAEM %	MDL pmole/kg
^{139}La	2.70		117	3.00	7.77	9.64	34	115	69.4	9.3	0.921	3	17	0.0184	0.133	0.522	0.265	0.8	16	166	1.91
^{140}Ce	1.80		141	5.96	7.86	1.98	9	98	14.1	9.3	1.318	5	17	0.0264	0.188	0.733	0.338	0.6	15	272	2.41
^{141}Pr	3.60		98	0.53	3.40	0.45	5	16	3.19	7.5	0.739	3	7	0.0148	0.105	0.277	0.067	3.0	8	40	0.47
^{146}Nd	1.80		101	2.28	4.81	1.16	7	22	8.04	9.6	1.362	4	21	0.0272	0.189	0.674	0.768	0.3	36	267	5.33
^{147}Sm	3.60		-	0.68	4.30	0.58	6	10	3.88	12	3.425	9	55	0.0685	0.456	0.347	0.154	1.3	14	77	1.02
^{152}Sm	1.80		109	0.67	4.33	0.75	7	11	5.00	11	2.342	7	34	0.0468	0.312	0.363	0.263	0.8	23	92	1.75
^{151}Eu	3.60		98	0.15	3.37	0.56	7	17	3.66	9.7	1.236	4	21	0.0247	0.163	0.289	0.061	3.3	7	46	0.40
^{160}Gd	1.80		96	0.78	4.05	0.81	6	6	5.13	9.4	2.709	9	20	0.0542	0.345	0.376	0.130	1.5	11	90	0.83
^{159}Tb	3.60		98	0.13	3.56	0.50	7	13	3.17	8.1	0.601	2	3	0.0120	0.076	0.252	0.059	3.4	7	28	0.37
^{163}Dy	1.80		92	0.83	3.86	0.68	6	7	4.19	9.3	1.680	6	18	0.0336	0.207	0.394	0.107	1.9	9	98	0.66
^{165}Ho	2.70		100	0.14	3.84	0.45	7	8	2.72	8.2	0.723	3	4	0.0145	0.088	0.272	0.067	3.0	8	37	0.40
^{166}Er	1.80		95	0.38	3.58	0.60	8	13	3.59	9.4	1.445	5	18	0.0289	0.173	0.299	0.122	1.6	13	53	0.73
^{169}Tm	3.60		102	0.08	3.40	0.55	7	16	3.23	8.1	0.484	2	2	0.0097	0.057	0.263	0.069	2.9	8	33	0.41
^{174}Yb	1.80		93	0.41	3.61	0.80	7	12	4.61	9.2	1.677	6	17	0.0335	0.194	0.382	0.225	0.9	19	98	1.30
^{175}Lu	3.60		103	0.08	3.36	0.44	6	17	2.52	8.1	0.318	1	2	0.0064	0.036	0.262	0.070	2.9	8	33	0.40
^{89}Y	2.70		55	3.73	5.26	1.03	5	32	11.5	8.3	0.626	2	5	0.0125	0.141	0.879	0.548	0.4	20	350	6.16
^{232}Th	2.70		103	0.71	2.06	0.26	5	6	1.11	6.0	0.803	4	50	0.0161	0.069	0.236	0.118	1.7	16	141	0.51
^{238}U	2.70		100	0.41	2.44	0.40	6	23	1.69	4.5	0.554	4	14	0.0111	0.047	0.300	0.127	1.6	13	204	0.53

Internal Standards and Ba were counted at 0.90 seconds per mass
All table values (except the first %REC) apply to an HP model 4500 ICP-MS
Isotopic counting times are totals of counts on three central peaks with 5 replicates per mass counted

aqueous sample in a separatory funnel. The REE are extracted into the organic phase The aqueous phase is discarded and the organic phase is transferred into a smaller separatory funnel. The REE are back-extracted into the aqueous phase by addition of *n*-octanol and 6 N hydrochloric acid. A rhodium spike is added to this extract to obtain a dilution factor upon analysis. The dissolved *n*-octanol in the extract is removed by shaking with three aliquots of *n*-heptane. The final cleaned up volume is approximately 15 mL. An additional evaporation step to eliminate the acid was not deemed necessary. The extract is transferred to a centrifuge tube for storage until analysis.

Using the method outlined above, we encountered difficulties in obtaining a method blank less than 10^6 times chondrite. The mixed esters we obtained from two different manufacturing sources are both contaminated with HREE (Er, Tm, Yb, Lu) as well as Y and U. Cleanup of the reagents is difficult and resulted in detection limits no better than the iron hydroxide coprecipitation. Finally, we found the method unsatisfactory for continental chloride brines from the Salton Sea geothermal area owing to high (1000's mg/kg) concentrations of transition metal cations such as Fe, Mn, As, Pb, and Zn. These saturate the exchange capacity of the phosphate esters and prevent or greatly reduce the recovery of La through Ho. A method where the DEHP esters are added onto a C18 filter disk has potential for lower contamination because it allows the disk and DEHP esters to be acid cleaned prior to extraction (Shabani et al., 1992). However, the saturation of exchange capacity by loading of transition metals remains a problem.

8. Measurement of Method Performance

Method performance is affected by the sum of variation from REE contamination and losses due to material components and manipulation by the analyst, and those caused by analytical interferences. There are several sources of interference that can occur. The following definitions and tools can help the analyst determine sources of interference and whether they are obtaining the best possible analysis. The performance measurement tools can be used to quantify the precision of an analysis, track down problems and assure confidence in method detection limits and total error of measurement.

8.1. INTERFERENCES

Interferences include: (1) Isobaric mass interference caused by molecular species, oxides, doubly-charged ions, or isotopes that have mass/charge ratios coinciding with that of the analytical mass; (2) Space-charge interference where the high ionic abundance of ions with higher mass/charge ratios interferes with the passage of ions with lower mass/charge ratios through the electronic lenses of the mass spectrometer; (3) Physical interferences caused by sample nebulization and transport processes; (4) Chemical interferences caused by unusual matrix compositions; (5) Memory interferences caused by residual accumulation of previous samples in the sample introduction system. When isobaric mass corrections are applied, there is a need to verify their accuracy by analyzing interference check solutions (e.g., Ba and Nd solutions).

8.1.1. *Physical Interferences*

These are effects associated with the sample nebulization and transport processes. Changes in viscosity and surface tension can cause significant inaccuracies, especially in samples that contain high dissolved solids or acid content, hydrophobic organics (oil), or surfactants. These can change the droplet size distribution and may also alter the wetting characteristics of the spray chamber, which may change the total amount of material reaching the plasma. Also, high dissolved solids content can result in salt build-up, which affects aerosol characteristics and flow rate and causes instrumental drift. The above problems can be controlled by using mass-flow controllers, using a high-solids nebulizer, using appropriate internal standard elements, increasing rinse times, using an anionic surfactant rinse solution, diluting the sample, or some combination of the preceding.

8.1.2. *Chemical and isobaric interferences*

Normally, isobaric interferences (i.e., formation of molecular ions, oxides, and doubly-charged ions) and solute vaporization effects can be minimized by careful selection of operating conditions (incident power, observation position, and so forth), but principally by reduction of water load on the plasma (e.g., desolvation) and dilution of the sample. Ionization interferences and solute vaporization effects are typically avoided because of the low dissolved solids (e.g., < 0.2 %) load of ICP-MS samples. Chemical interferences are highly dependent on matrix type and the specific analyte and are not discussed except to note that they can exist.

8.1.3. *Memory interferences*

These result when analytes in a previous sample contribute to the signals measured in a new sample. Memory effects can result from sample contamination of the uptake tubing to the nebulizer and from the accumulation of sample material in the nebulizer, spray chamber, and plasma torch and plasma sampling cones. The site where these effects occur is dependent on the individual element and can be minimized by flushing the system with a rinse blank between samples. A rinse blank can be the same as or different from the calibration blank and may, for example, include a surfactant or a different acid mixture. The rinse flush time should be established for a sample containing the analytes at the upper limit of their linear dynamic range (LDR) or 10 times the levels normally encountered. Following a normal sample analysis period, the length of time is recorded for successive measurements made on a blank solution until the analyte concentration has been reduced to within a factor of 2 of the method detection limit (MDL see 8.2.5). Until such a rinse time is established, a rinse time of at least 60 seconds between all samples and standards is suggested. If a memory interference is suspected for a sample following analysis of a sample with a high concentration, the sample should be re-analyzed after a long rinse period. A short rinse with a preliminary rinse solution comprised of 10 to 20% nitric acid may help decrease required rinse times. A good rinse solution should contain an anionic surfactant to condition the nebulizer and spray chamber and the acid concentration should be slightly higher than that in the samples. Prolonged use of high acid concentrations may adversely affect the peristaltic pump tubing by decreasing resiliency or increasing susceptibility to contamination. High acid concentrations in samples should be avoided and may require matrix matching because they are functionally equivalent to solutions with high dissolved solids. A large difference between the rinse and analysis pump rates may introduce noise stemming from hysteresis effects in the Tygon tubing walls. Therefore, it may be necessary to increase the read delay time before analysis to reduce the hysteresis effects.

8.2. DETECTION OF MATRIX INTERFERENCES

The tools described below can be used to identify interferences that are related to the bulk sample composition, which is commonly called the sample matrix. The composition of the bulk sample can produce interferences during sample collection, processing, and analysis. In geothermal fluids, matrix compositions that might affect results are those with high total dissolved solids (e.g., borate, sulfate, chloride, bicarbonate, sulfide, iron, manganese, or silica). Other things that might affect results from geothermal wells are anti-scaling reagents and organic tracers. The tools described below are typically used in the environmental analysis industry to quantify matrix interferences. The tools described in this section differ from the performance measurement tools (see section 8.3) because they incorporate the error of the entire method and the influence of the sample composition into the resulting measurements. Each of these tools measures different aspects of matrix interference. The matrix dilution and analyte addition tests help detect matrix interference at the instrument, are easy to do, and are immediately useful. The matrix spike focuses on matrix interference with spike recovery during processing of the sample prior to analysis. The replicate tests are less matrix-dependent and more a measure of the skill of the analyst but matrix interferences can still introduce variability that is not observed in replicate blanks and spiked blanks.

We recommend the collection of sample replicates for quantifying variability. The matrix dilution and analyte addition tests are good general-purpose tools for detecting matrix interference at the instrument. The processing of matrix spike samples is appropriate when it is suspected that sample composition is interfering with recovery of the analytes of interest.

8.2.1. *Matrix Sample (M)*

The matrix sample is a sample selected to be representative of the general expected composition for a group of samples. A matrix sample is also one that has been chosen for calculating spike recovery (see 8.2.3) in the presence of possible "matrix" interference. This matrix sample is used in the matrix dilution and matrix duplicate methods described below.

8.2.2. *Matrix duplicates (MDUP)*

The matrix duplicate is a second aliquot of the matrix sample, which is processed separately with identical procedures through the entire analytical procedure, treated exactly as a sample, including exposure to all glassware, equipment, solvents, reagents, and internal standards that are used with other samples. Analysis of the M and MDUP are used to verify precision of the method when matrix interference may affect the result. The relative percent deviation, calculated as shown below based on the analysis of M and MDUP, represents the smallest difference that can be distinguished between two samples by the analytical protocol in question.

(1)   RPD = 100 x [(MDUP x mddf) - (M x df)] / {[(MDUP x mddf) + (M x df)] / 2}

where: RPD = relative percent deviation
       M = matrix sample concentration
       df = preparation dilution factor for M
       MDUP = duplicate sample concentration
       mddf = preparation dilution factor for MDUP

8.2.3. Matrix spike (MS) and matrix spike duplicate (MSD)

This is a test for matrix interference upon the entire method by testing the recovery of a matrix spike. Known quantities of the method analytes are added in the laboratory to an aliquot of matrix sample. The matrix spike (MS) sample is processed through the method, treated exactly as a sample, including exposure to all glassware, equipment, reagents, solvents, and internal standards that are used with other samples. The MS and a duplicate (MSD) are used to determine whether the sample matrix contributes bias to the sample results. The un-spiked concentrations (e.g., M or MDUP) must be determined from a separate aliquot and the measured MS and MSD values corrected for the background concentrations to calculate spike recovery. The percent spike recovery (R) of the MS and MSD can be calculated in the same manner as for the RPD of the M and MDUP.

(2)   R = 100 x {[MS - (M x msdf / df)] / s}

where: R = percent spike recovery
       M = sample matrix background concentration
       df = preparation dilution factor for M
       MS = fortified sample matrix concentration
       msdf = preparation dilution factor for MS or MSD
       s = concentration equivalent of analyte added to fortify the sample matrix.

8.2.4. Analyte addition

This is a test for matrix interference at the point of analysis. An analyte standard is added to a portion of a prepared sample, or its dilution. The analyte addition should produce a minimum level of 10 and a maximum of 100 times the MDL. If the resulting analyte addition is < 20% of the sample analyte concentration, then dilute the sample so that the above addition limits can be met. If recovery of the analyte is not within the limits of 75% to 125% recovery, a matrix effect should be suspected. Dilution of the sample is the easiest way to fix the problem if matrix effects are observed. Sometimes changing to a more suitable internal standard is sufficient. Also, high acid concentration causes matrix effects, which can be decreased by dilution of the sample.

8.2.5. Matrix dilution (MDIL)

This is another test for matrix interference at the point of analysis. The matrix sample is diluted five-fold and then treated with identical procedures to that for a sample, including exposure to all glassware, equipment, solvents, reagents, and internal standards that are used with other samples. If the analyte concentration is sufficiently high (minimally, a factor of 50 above the MDL in the original solution but < 90% of the LDR limit), an analysis of a 1:5 dilution should agree (after correction for the five-fold dilution) to within ±10% of the original determination. If not, a matrix interference effect should be suspected for the analyte. The matrix dilution recovery should be calculated as follows:

(3) $R = 100 \times \{(MDIL - M) / [(MDIL + M) / 2]\}$

where: R = percent recovery
         M = matrix sample concentration
         MDIL = concentration five-fold dilution of matrix sample adjusted for
dilution

8.3. PERFORMANCE MEASUREMENT TOOLS

8.3.1. *Instrument detection limits (IDL) and method detection limit (MDL)*

The MDL is defined as "…the minimum concentration of an analyte that can be identified, measured, and reported with 99% confidence that the analyte concentration is greater than zero…" (Code of Federal Regulations, volume 40, Part 136, Appendix B — Definition and Procedure for the Determination of the Method Detection Limit — Revision 1.11). Prior to determination of the MDL, it is expeditious to estimate the instrument detection limit (IDL) by analyzing 7 to 10 replicates of the calibration blank. Calculate the estimated IDL as follows:

(4) $IDL = t\sigma$

where t is the Students' t value for a 99% confidence level (e.g., t = 2.896 for nine replicates, 2.998 for eight replicates, and 3.143 for seven replicates) and σ is the estimate of the standard deviation with n-1 degrees of freedom.

The actual IDL can now be measured by starting from the estimated IDL concentrations. To do this, prepare a standard with the concentration of each analyte at 3-5 times but no more than 10 times the estimated IDL. Divide this IDL standard into 7 to 10 aliquots, and analyze them using the method instrumental conditions. Perform the above calculation and report the concentration values in the appropriate units.

To establish the method detection limits of a procedure one must determine the MDL for each analyte. Sample solutions or representative blanks are spiked at a concentration that, when carried through the procedure, will read two to three times the IDL at the instrument. For example, to determine MDL values for aqueous samples, take seven to ten sample bottles and clean them. Fill them with DIW and add all reagent acids, fortify them by spiking with a standard, process through the entire analytical protocol and analyze the MDL samples. Calculate the MDL just as the IDL with an appropriate Students't and the sigma of the MDL samples.

If the relative standard deviation (RSD) from the analysis of the seven aliquots is < 10%, the concentration used to determine the analyte MDL may have been inappropriately high for the determination. If so, this could result in the calculation of an unrealistically low MDL. Determination of the MDL in spiked DIW represents a best-case situation and does not reflect possible matrix effects of real-world samples. Because silica was a known interferent in determination of REE in geothermal fluids, we made up MDL samples that had been spiked with appropriate amounts of a plasma-grade silicon standard (see section 9.1). However, repeated successful routine analysis of spiked blanks or even spiked samples will give confidence in the MDL value determined in reagent water.

The MDL's should be determined initially and then re-determined periodically. For example, when a new person uses the method, or whenever, in the judgement of the lab supervisor, a change in the laboratory equipment or environment, instrument hardware or operating conditions suggests that they should be re-determined. Analysis of an IDL standard solution on a more frequent basis can confirm the continued performance of the analytical instrument. Loss of IDL performance can alert the analyst that instrument maintenance is necessary. We also routinely prepare and analyze spiked MDL samples to monitor method performance.

8.3.2. *Primary quantitative limit (PQL)*

Once the MDL has been determined it is useful to establish a PQL. The PQL is the concentration limit above which the total error of analysis is relatively low (e.g., less than 10%). Typically the PQL is set at three to four times the MDL.

8.3.3. *Linear dynamic range (LDR)*

The LDR is concentration range over which the instrument response to an analyte remains linear as verified (at the upper end) by analysis of a linear range standard (LRS). Follow the instrument manufacturer's recommendation for determination of LDR. If too wide an LDR is chosen then the accuracy at the low end will be compromised. The maximum instrumental value that can be reported outside of bracketing calibration standards can be determined by analysis of an appropriate linear range standard (LRS) at the time of analysis. The LRS should read back to within 5% of the certified value for an analyte to be reported at instrument readings higher than the highest calibration standard.

8.3.4. *Field or trip blanks*

While sampling in the field, it is good practice periodically to designate a set of the cleaned sample bottles (section 6.1) to be trip blanks. It is important to treat trip blanks exactly as a normal sample. Thus, our trip blanks contained acid because it was part of our sample collection method. These are opened in the field and then resealed. After transport back to the lab they are filled with DIW instead of sample and from that point on they are treated precisely as a normal sample. Another useful type of field blank is one in which DIW is carried in one cleaned bottle from lab to field, and then subjected to the same processing in the field as the samples, including filtration from the designated sample bottle into a second cleaned, acid filled bottle. This blank is also subsequently treated in the same manner as samples in the laboratory. Field blanks are used to help determine if method analytes or other interferences are present in the cleaned sample bottles or introduced from the field environment. A well chosen field blank, supports the validity of your final results.

8.3.5. *Laboratory method blank (LMB)*

The laboratory method blank is an aliquot of DIW that is processed through the method, treated exactly as a sample, including exposure to all bottles, glassware, equipment, solvents, reagents, and internal standards that are used with other samples. The LMB is used to determine if method analytes or other interferences are present in the laboratory environment, reagents, or apparatus. We routinely prepare and run LMBs to monitor our performance. Commonly, the LMB concentrations are above the IDL. This is true in the case of the light REE even when we are using high-purity quartz-fiber filters. The LMB concentration, however, should routinely cluster at or below the MDL. LMB concentrations can be tracked over time and used to demonstrate that the MDL's are being achieved. If LMB's are significantly higher than the MDL then the MDL needs to be re-established.

8.3.6. *Laboratory control sample (LCS)*

The LCS is an aliquot of DIW to which known quantities of the method analytes are added in the laboratory. The LCS is processed through the method, treated exactly as a sample, including exposure to all bottles, glassware, equipment, solvents, reagents, and internal standards that are used with other samples. The LCS is used to verify accuracy of the method and over time, the bias and precision of the method. The standard deviation (sigma) should be used to establish an ongoing precision statement (see TAEM below) for the level of concentrations included in the LCS.

8.3.7. *Total analytical error measurement (TAEM)*

The total analytical error of a method can be estimated by analyzing a group of replicate laboratory control samples (LCS or LC samples) that have analyte concentrations that are at least 100 times the MDL. These samples should be processed through the entire method procedure so that all sources of error in the method are represented in the TAEM. The TAEM is calculated so as to include both the variation and bias of the measurements from the expected value as shown below.

(5) $\text{TAEM} = 100 \times \{[\sigma^2 + (\overline{X} - S)^2]^{1/2}\}/S$

Where σ is the standard deviation of the measurements, $\overline{X}$ is the mean of the measurements for each method analyte, and S is the expected value of the spike for that method analyte. The TAEM can be calculated without bias by omitting the ($\overline{X}$ - S) term from the calculation. The TAEM simplifies to the coefficient of variation (CV) when every S is replaced by $\overline{X}$ in the equation.

9. Instrumental Analysis

9.1. ANALYSIS OF REE, Y, Th, AND U BY ICP-MS

We currently employ a Hewlett Packard model 4500 ICP-MS, housed at Washington State University, for the determination of the REE. Typically, oxide formation as measured by $^{156}[CeO]/^{140}Ce$ is less than 0.5%, equivalent to $^{248}[ThO]/^{232}Th$ of less than 1%. Interference corrections are made for BaH on La, BaO on Nd, Sm, and Eu, BaCl on Yb and Lu, PrO on Gd, and NdO on Tb. Correction is made by periodically running both a 20-mg/L Ba standard and a mixed 200-μg/L Pr-Nd standard. The interference corrections for Pr and Nd are straightforward but the corrections for Ba are slightly more complex (see below). We use Ru, In, and Re as internal standards for the REE and Rh (spiked for a 20-mL volume to equal 50-μg/L) to correct for sample extract volume. On occasion, Co and Bi are measured as internal standards for Sc, U, and Th. However, Bi can be abundant enough to be a problem and in such cases we used Re. We measure ^{147}Sm (free of $Ba^{16}O$ interference) for comparison with more sensitive ^{152}Sm to evaluate the BaO correction for ^{152}Sm and ^{151}Eu. We correct ^{160}Gd for ^{160}Dy using ^{163}Dy and measure ^{178}Hf to correct Yb for isobaric ^{174}Hf interference. Calibration verification is monitored and updated by periodic analysis of two or more mid-level calibration standards and the calibration blank. The internal standards are used to compensate for instrument drift.

Barium abundances in some natural samples can produce Ba concentrations as high as 20 ppm in the final extract solution. The proportions of BaO generated in the plasma may increase or decrease over the course of the analysis independently of internal standard behavior. Periodic measurement of a pure 10 to 20 ppm Ba standard are used to generate correction factors to apply to ^{139}La, ^{146}Nd, ^{147}Sm, ^{175}Lu, and particularly ^{152}Sm and ^{151}Eu measurements. The ionization of Ba to BaO is about twice that of Ce to CeO under our running conditions. If oxide generation can be kept below 0.5% as measured by the CeO/Ce ratio, then correction for BaO interference is typically adequate for REE concentrations greater than 10^{-4} chondrite. Examples of these corrections are shown in Fig. 1 for a solution of Ba and a mixed solution of Pr plus Nd. Note that the purity of the Pr-Nd mixture is not as good as that of the Ba standard. A sample containing about 20 mg/L of Ba in the final extract is shown with and without isobaric interference corrections in Fig. 2. Note how ^{139}La, ^{146}Nd, ^{147}Sm, ^{174}Yb, and ^{175}Lu are less adversely affected than ^{152}Sm and ^{151}Eu by the Ba interference. For most samples, only the latter two are significantly affected. Typically, for a good instrument run, the percent oxide generation does not change significantly over the course of the analysis, no interpolation is necessary, and the correction from the first Ba standard is sufficient for the whole run. Some groups of samples cause drift and the percentage of oxide generation changes over the course of the analysis. For each adversely affected analysis, a weighting is chosen to interpolate between a Ba standard run at the beginning and end of instrumental analysis. A single weighting factor is chosen on a sample by sample basis so that ^{147}Sm and ^{152}Sm are approximately equal. This weighting factor applies successfully to the other isobaric interferences because they are all dependent linear combinations of the isotopic abundances coupled to isobaric interferences from a single element, Ba. This makes possible a consistent correction to apply particularly to ^{151}Eu, which otherwise could not be estimated, but also to ^{139}La, ^{146}Nd, ^{174}Yb, and ^{175}Lu.

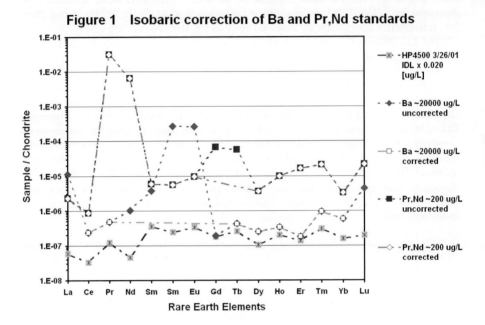

Figure 1. Isobaric corrections for Ba, Pr and Nd. The figure shows interferences from corrected and uncorrected Ba and mixed Pr + Nd standards. Symbol key: solid diamond = uncorrected Ba; empty (white) square = corrected Ba; solid square = uncorrected Pr + Nd; empty (white) diamond = corrected Pr + Nd. Note for corrected values: The empty (white) symbols are absent because values are below the MDL. Estimated MDL (star-in-square, dashed line) plotted for reference.

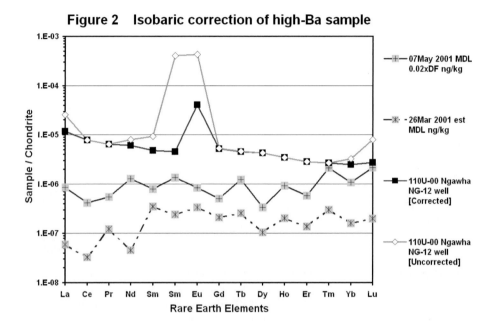

Figure 2. Plot of Ngawha NG-12 geothermal well unfiltered samples. Corrected (solid black square) and uncorrected (empty white diamond) for isobaric interference from barium. BaO interferes with ^{146}Nd, ^{147}Sm, ^{152}Sm, and ^{151}Eu; BaH interferes with ^{139}La; BaCl interferes with ^{174}Yb and ^{175}Lu. Nd and Pr are also corrected but corrections are too small to be visible. Actual (plus-in-square, solid line) and estimated (star-in-square, dashed line) MDL's are plotted for reference.

All data reduction is accomplished off-line using a standardized computer spreadsheet. Isotopic masses of REE, internal standards, and masses measured to correct isobaric interferences are listed in Table 1 along with instrument detection limits (IDL) method detection limits (MDL) and total analytical error measurements (TAEM) estimated from standard replicates and calculated from standards and MDL samples. Replicates of the lowest calibration standard (80 ng/L), which are 20 to 250 times the IDL, result in a TAEM that is typically less than 5%.

Linear, quadratic, and other calibration curves tend to give poor results over some part of the working analytical range. We have found that a log-linear calibration curve (formulas 6, 7) gives the best fit over this range of calibration and has the advantage that a log-linear plot is easier to visually evaluate for fit (Fig. 3).

(6) $\text{Log}(Y) = [\text{slope}(\text{Log}[(X - \text{bias})/S]) + \text{intercept}]$

(7) $Y = 10^{[\text{slope}(\text{Log}[(X - \text{bias})/S]) + \text{intercept}]}$

where: Y = the known and predicted concentrations in calibration units (e.g., μg/L)
         X = observed intensity in counts per second
         bias = baseline offset intensity value in counts per second
         S = interpolated intensity value of bracketing internal standards in counts per second
         intercept = regression intercept of X for known Y
         slope = regression slope of X for known Y

Instrumental drift will occur for geothermal samples because the combination of high iron and high silica in the extracts armors the sample cones. The analytical error due to instrumental drift can be averaged over the instrumental run by including the periodic calibration standards and blank as continuing calibration. This distributes the analytical drift error over the entire instrument run. The two internal standards bracketing the mass of each REE are represented as a weighted sum to be used as the internal standard for that REE. We adjust the percentage weights of the bracketing internal standards to equally minimize the %RSD of the mid and low standards. Because the baseline counts vary slightly from run to run, we first subtract the lowest counts-per second sample from the counts per second of every analysis for that analyte before applying the internal standardization.

9.2. DISCUSSION OF IDL's AND MDL's

Method detection limits and spike recoveries in simulated geothermal fluids were demonstrated by addition of 60 mg of ferric iron and 300 mg Si to 7 replicates with each REE spiked at 4 ng/L and a final volume of 40 mL equivalent to a 25-fold concentration factor. This results in expected concentrations at the instrument of 100 ng/L. These were prepared and analyzed in December of 1998 along with 6 laboratory method blank (LMB) samples and all of them were later reanalyzed in late March of 2001 (Table 1, Fig. 4). The 4-ng/L spike is about 1 to 9 times the resulting MDL for both years. The average spike recoveries for 1998 and 2001 are in good agreement in that the ratio of 2001 and 1998 spike recoveries ranges from 92% to 109% with the exception of La 117%, Ce 141%, and Y 55%. As proficiency with the method improved, the time of contact between the quartz-fiber filters and sample decreased. The agreement of average spike recoveries in two measurements separated by two years and using different instruments suggests that the stability of the REE extracts is very good. No Rh was added to these samples to correct for final extract volume. Not correcting for the final dilution volume increases the variation and bias of the results, which in turn may increase the MDL's and TAEM's. The resulting MDL's are low enough that contamination from the quartz fiber filter media is evident in the recoveries of La, Ce, Nd, Y, and U. Excluding La, Ce, Nd, Y, and U; the average spike recoveries have a low bias due to lack of dilution correction but recoveries are better than 80% and averaged 94%. The 2001 TAEM for these MDL samples are less than 20% excepting La 115%, Ce 98%, Nd 22%, Y 32% and U 23%, which suffer from contamination from the quartz filters.

The relatively high MDL's of the 4-ng/L spike, when the corresponding IDL's are a factor of 2 to 14 times lower, suggests that the MDL's are higher than they could be if the method were improved. For comparison, the 4 ng/L spike would read 100 µg/L at the instrument and the average recovery of the 80 µg/L low calibration standard is typically within 3-5% of the correct value. The 1998 MDL's decrease as much as 5-fold if the dilution factors are corrected to make the average of the odd atomic number REE (Pr, Eu, Tb, Ho, Tm, and Lu) (100% abundance and higher counting times) come out to the expected average. This suggested that use of a Rh dilution spike to calculate final dilution volumes could result in lower MDL's. At the same time, the 4-ng/L MDL's were run in March of 2001; we also ran an 8-µg/L REE standard to determine an IDL. This concentration is similar to running MDL's spiked at 0.16 ng/L. The CV's were all less than 9%. The respective TAEM's were as high as 55% and as low as 2% when the IDL's were 2 times and 25 times lower than the 8 µg/L REE standard concentration.

A new set of MDL's was prepared and run in May of 2001 (Table 1, Fig. 4). These MDL samples (without added silica) were spiked 20-fold lower, with only 0.2-ng/L of REE, prior to extraction and post-spiked with 1000 ng of Rh (see section 7.1) for dilution correction. In addition, the final dilution volume was cut in half to 20 mL and only 30 mg of ferric iron were added to keep the final iron content lower. These MDL's represented the extraction procedure used for all samples that we collected in 2000 and beyond. The dilution corrected volumes ranged from 21.2 to 23.3 mL with the respective mean and standard deviation of 22 mL and 0.69 mL. The May 2001 MDL's are sufficiently low that the 0.2-ng/L spike represents less than 4 times the MDL. Typically, the TAEM decreases from 100% at the MDL to less than 10% for concentrations that are over 10 times the MDL. TAEM values greater than the expected amount are caused by calibration bias or contamination. The CV for 0.2-ng/L REE spikes are typically less than 20% except for Nd 36% and Sm 23%. The high CV's and TAEM's for Nd were traced to Nd contamination from a laboratory experiment. The TAEM's that are less than 100% are acceptable given that the concentrations are less than 4 times the MDL. The larger TAEM values for La, Ce, Nd (in addition to the above contamination), Y, Th, and U are probably caused by contamination from the quartz-fiber filters.

It is useful to compare MDL's from different batches and conditions to see if differences can be explained. The May 2001 MDL's were 2 to 9 (excepting La, which

Shannon and Wood

Figure 3   Log-Linear REE Calibrations

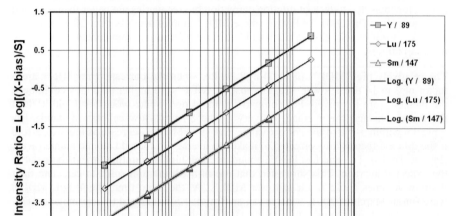

Figure 3. Plot of Log-Linear REE calibrations. Curves with higher intensity ratios have better detection limits. Thus, Y has the lowest and Sm has the highest MDL. Calibration curves for the remaining REE fall between Lu and Sm.

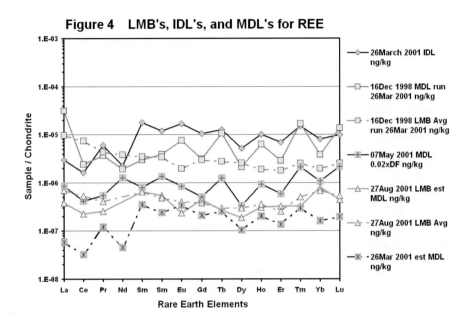

Figure 4. REE IDL's, estimated MDL's, and actual MDL's.

was 36 times lower) times lower than the December 1998 MDL's. However, if the dilution factors of the 1998 MDL sample results run in 2001 are adjusted to optimize the odd REE as discussed above, and if the 2-fold higher concentration factor is taken into account, then (with the exception of La) the MDL's from the 1998 samples are very similar to those re-determined in May, 2001. We can conclude with steps taken to improve the method. First, and IDL and MDL were determined on the less sensitive Perkin-Elmer Elan ICP-MS. Later the MDL was re-measured on the same solutions on the more sensitive HP-4500 ICP-MS. An 8 µg/L derived IDL determined at the same time suggested that the MDL could be lower. Steps were taken to reduce volume, reagent additions, and spike amount. Finally the new MDL does actually improve over the old MDL. Excepting for La, Ce, Nd, Lu, Y, and U the new MDL is only about 7 times higher than that estimated from the 8 µg/L IDL standard. This suggests that the current MDL is as low as can be achieved with the current method and instrumental parameters. Further improvement might come from decreasing the final extract volume and finding a filter material with acceptable performance and a lower blank.

10. Evaluation of REE Data

The REE data obtained via analysis can be evaluated for contamination by a single REE or small group of REE by first normalizing the REE data to average chondrite (e.g., Boynton, 1984) and then plotting the results. This smoothes out the Oddo-Harkins effect (i.e., the higher abundance of elements with even atomic number relative to elements of odd atomic number). We choose to standardize with equal concentrations of each REE at each level of calibration because it is comparatively simple to make equal concentration standards. The Oddo-Harkins effect assists in detection of laboratory contamination. Chondrite-normalized REE patterns are comparatively smooth if the sample comes from natural sources, but show the Oddo-Harkins effect if there is carryover from a single or mixed REE source in which the REE occur at non-natural relative abundances. Due to the Oddo-Harkins effect, most of the odd atomic number REEs have only one or two natural isotopes, lower abundance, and no isobaric overlap with the even atomic number REE. Table 1 shows counting times for the REE isotopes used during sample analysis. The odd atomic number REEs are counted twice as long as the even atomic number REE. This makes for chondrite-normalized IDL's that are relatively smooth under ideal operating conditions but which are still slightly higher (worse) for the odd atomic number REE (Fig. 4). However, the longer counting times on the odd atomic number REE alerts us to instrumental contamination problems. Using our instrumental running conditions, we find that if the odd atomic number REE results begin to spike up on natural samples, then the sample cones have probably become contaminated by the periodic calibration check standards. Very thorough cleaning is then necessary to remove the sintered REE contaminated iron silicate coating which forms on the sample cones.

Other projects in our lab use specific single rare earth sources that typically exhibit purity of 99.99+%. Therefore, contamination from one of these projects shows up as a spike of that specific REE but normally does not significantly affect the adjacent REE values. We have found that dust and aerosol sources are the primary vectors for REE contamination. Even slow evaporation of REE solutions will result in REE contamination spikes if the evaporation is not done in a separate room or hood. A sample with laboratory Nd contamination is shown in Fig. 5. In this case we would not report a value for Nd. Geothermal fluid samples were collected from the Iodine Pool at

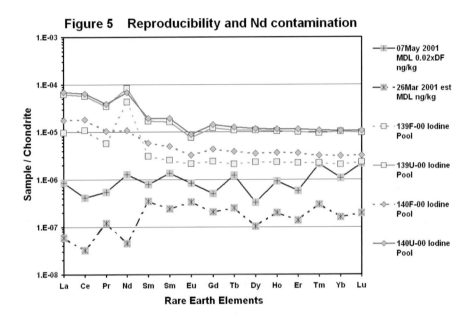

Figure 5. Plot showing laboratory contamination from Nd and reproducibility of replicate samples. Year 2000 samples from Iodine Pool, Waimangu, New Zealand. Samples taken in same pool about 6 meters apart. The unfiltered samples are solid lines and corresponding filtered samples are dashed lines. Note that Nd contamination seems larger in one of the filtered samples but is actually of similar magnitude to those in the unfiltered samples. Actual (plus-in-square, solid line) and estimated (star-in-square, dashed line) MDL's are plotted for reference.

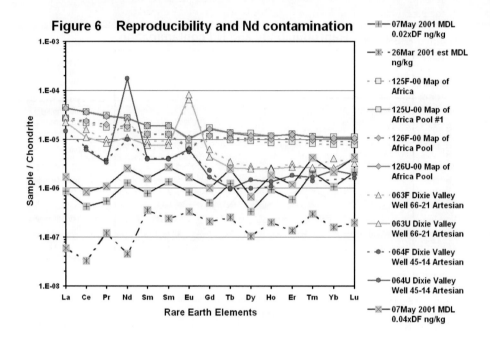

Figure 6. The square and diamond symbols are Year 2000 duplicate samples from the Map of Africa Pool, Orakeikorako, New Zealand. The samples were taken at the same location and time. The unfiltered samples are solid lines and corresponding filtered samples are dashed lines. The Map of Africa Pool samples (squares and diamonds) are more identical than those of the Iodine Pool samples in Fig. 5. The Dixie Valley, Nevada wells (triangles and circles) show Nd contamination. Note how Dixie Valley samples have positive heavy REE slopes near the MDL. The MDL's are plotted for reference as follows: 0.04xMDL = x-in-square, solid line; 0.02xMDL = plus-in-square, solid line; 0.02x estimated MDL = star-in-square, dashed line.

Waimangu in New Zealand. These samples were taken at approximately the same time from locations about 6 meters apart. They were all affected slightly during extraction by laboratory Nd contamination. The HREE of the filtered samples are just above the 0.02x dilution factor (DF) MDL (equivalent to a 50-fold concentration of the sample) and probably reflect the actual sample REE contents. Two other geothermal samples were collected from the Map of Africa Pool at Orakeikorako in New Zealand. These were collected at the same location and at the same time. Note the slightly closer match of the REE data for both the filtered (F) and unfiltered (U) samples (Fig. 6).

Dilution factors can be important in interpretation of REE data. Two geothermal well samples from Dixie Valley were extracted using the same method and 0.04x DF as the 1998 MDL's (Fig. 6). These also have laboratory Nd contamination from Nd-phosphate that was calcined in a furnace, without use of a hood, in the same room where these water samples were extracted. Because the concentration factor is only 25-fold these are better represented by the slightly higher 0.04xDF MDL. Note how the HREE slope increases from Tb through Lu and parallel the 0.04xDF MDL (Fig. 6). This is common for results that are near the MDL and the HREE slope could just as likely be negative. All one can say about the HREE so close to the MDL is that they are probably not higher than the MDL and they could be significantly lower.

11. Summary and Conclusions

We have shown how the REE can be determined down to 10-100 picograms per liter in natural waters in a general-purpose academic laboratory using ferric hydroxide co-precipitation followed by ICP-MS. Successful analysis of the REE in such an environment requires many quality control measures to ensure that detection limit, precision, and blank goals are being attained.

Acknowledgments

This study was funded by the U.S. Department of Energy Geothermal Reservoir Technology Program (Contract DE-FG07-98ID13575). The Geoanalytical Lab Facility at Washington State University provided access to their ICP-MS. The following individuals aided in the development of the analytical protocols described here: Charles Knaack, Kate Fulcher, Xiaoping Jin, Brittany Brown, Shara Leavitt, Dave Cook, and Chad Ross. Leslie Baker, Tom Williams, Brendan Twamley, and Kirsten Nicholson provided editorial assistance and moral support.

References

Aggarwal, J. K., Shabani, M. B., Palmer, M. R., and Ragnarsdottir, K. V., (1996) Determination of the rare earth elements in aqueous samples at sub-ppt levels by inductively coupled plasma mass spectrometry and flow injection ICPMS. Analytical Chemistry, 68, 4418-4423.

Alfassi, Z.B., (1992) Preconcentration by Coprecipitation of Trace Elements (in) Preconcentration Techniques for Trace Elements, Z.B. and Alfassi, C.M. Wai (ed), CRC Press, Inc., Boca Raton, p. 3-99.

Bau, M., (1999) Schavaging of dissolved yttrium and rare earths by precipitating iron oxyhydroxide: Experimental evidence for Ce oxidation, Y-Ho fractionation, and lanthanide tetrad effect. Geochimica et Cosmochimica Acta, 63, 67-77.

Bonner, N.A. and Kahn, M., (1951) Behavior of Carrier-Free Tracers, (in) Radioactivity Applied to Chemistry, Wahl, A.C. and Bonner, N.A. (ed), John Wiley and Sons, Inc., New York, p. 102-178.

Boynton, W.V., 1984. Cosmochemistry of the rare earth elements: Meteorite studies. In P. Henderson (ed.), Rare Earth Element Geochemistry: 63-114. Amsterdam, Elsevier.

Buchanan, A.S. and Hannaker, P. (1984) Inductively coupled plasma spectrometic determination of minor elements in concentrated brines following precipitation. Analytical Chemistry, 56, 1379-1382.

Fardy, J.J., (1992) Preconcentration of Trace Elements by Ion Exchangers (in) Preconcentration Techniques for Trace Elements, Z.B. and Alfassi, C.M. Wai (ed), CRC Press, Inc., Boca Raton, p. 181-210.

Fee, J.A., Gaudette, H.E., Lyons, W.B., and Long, D.T., (1992) Rare-earth element distribution in Lake Tyrrell groundwaters, Victoria, Australia. Chemical Geology, 96, 67-93.

Johannesson, K.H. and Lyons, W.B., (1994) The rare earth element geochemistry of Mono Lake water and the importance of carbonate complexing. Limnology and Oceanography, 39, 1141-1154.

Klinkhammer, G.P., Elderfield, H., Edmond, J.M., and Mitra, A., (1994) Geochemical implications of rare earth element patterns in hydrothermal fluids from mid-ocean ridges. Geochimica et Cosmochimica Acta, 58, 5105-5113.

Laul, J.C., Smith, M.R., Lepel, E.A., and Maiti, T.C., (1988) Natural Radionuclides of Uranium and Thorium Series and Rare Earth Elements in Hot Brines and Cores from the Salton Sea Geothermal Field (Analog Study). Technical Report PNL-SA-14775.

Lepel, E.A., Laul, J.C., and Smith, M.R., (1989) Rare Earth Element Patterns in Brine Groundwaters (Analogue Study). (in) Radioactive Waste Management and the Nuclear Fuel Cycle. Harwood Academic Publishers, p. 367-377.

Minczewski, J., Chwastowska, J., and Dybczynski, R., (1982) Separation and Preconcentration Methods in Inorganic Trace Analysis, Ellos Horwood Series in Analytical Chemistry, Translation Editor M.R. Masson, John Wiley and Sons, Halsted Press, ISBN 0-470-27169-8 543 pp.

Peppard, D.F., (1961) Separation of Rare Earths by Liquid-Liquid Extraction. (in) The Rare Earths, Editors F.H. Spedding and A.H. Daane, John Wiley and Sons, Inc., pp 38-54.

Shabani, M. B., Akagi, T., Shimizu, H., and Masuda, A., (1990) Determination of Trace Lanthanides and Yttrium in Seawater by Inductively Coupled Plasma Mass Spectrometry after Preconcentration with Solvent Extraction and Back-Extraction. Analytical Chemistry, 62, 2709-2714.

Shabani, M. B., Akagi, T., Shimizu, H., and Masuda, A., (1992)Preconcentration of Trace Rare-Earth Elements in Seawater by Complexation with Bis(2-ethylhexyl) Hydrogen Phosphate and 2-Ethylhexyl Phosphate Absorbed on a C_{18} Cartridge and Determination by Inductively Coupled Plasma Mass Spectrometry. Analytical Chemistry, 64, 737-743.

Shannon, W.M., Wood, S.A., Brown, K., Arehart, G. (1999) Preliminary measurements of concentrations of lanthanide elements in geothermal fluids from the Taupo Volcanic Zone, New Zealand. *Proceedings, Twenty-Fourth Workshop on Geothermal Reservoir Engineering Stanford University, Stanford, California, January 25-27*, 1999 SGP-TR-162, 227-235.

Shannon, W.M., Wood, S.A., Brown, K., and Arehart, G. (2001) REE Contents and Speciation in Geothermal Fluids from New Zealand. *Proceedings, Tenth Internation Symposium on Water-Rock Interaction, Villasimius, Italy, June 10-15, 2001.*

Shigeru, T., and Katsumi G., (1992) Preconcentration of Trace Elements by Filter Papers (in) Preconcentration Techniques for Trace Elements, Z.B. and Alfassi, C.M. Wai (ed), CRC Press, Inc., Boca Raton, p. 427-444.

Taylor, S. R., and McLennan, S. M., (1985) The Continental Crust: Its Composition and Evolution. Blackwell Scientific Publ., Oxford, England, 312 p.

Terada, K., (1992) Preconcentration by Sorption (in) Preconcentration Techniques for Trace Elements, Z.B. and Alfassi, C.M. Wai (ed), CRC Press, Inc., Boca Raton, p. 211-241.

van Middlesworth, P. E. and Wood, S. A. (1998) The aqueous geochemistry of the rare earth elements and yttrium. Part 7. REE, Th and U contents in thermal springs associated with the Idaho Batholith. *Applied Geochemistry* **13**, 861-884.

Wai, C.M., (1992) Preconcentration of Trace Elements by Solvent Extraction. (in) Preconcentration Techniques for Trace Elements, Z.B. and Alfassi, C.M. Wai (ed), CRC Press, Inc., Boca Raton, p. 101-132.

Wood, S.A., Palmer, D.A., Wesolowski, D.J., and Bénézeth, P., (in preparation) The Aqueous Geochemistry Of The Rare Earth Elements And Yttrium. Part XI. The Solubility Of $Nd(OH)_3$ And Hydrolysis Of Nd^{3+} From 30 To 290°C At Saturated Water Vapor Pressure With In-Situ pH Measurement.

Zief, M. Mitchell, J.W., (1976) Contamination Control in Trace Element Analysis. Volume 47 (in) Chemical Analysis. A Series of Monographs on Analytical Chemistry and its Applications. P.J. Elving, I.M. Kolthoff (ed). Wiley-Interscience, John Wiley and Sons, Inc., 262 p.

Chapter 2

DETERMINATION OF 56 TRACE ELEMENTS IN THREE AQUIFER- TYPE ROCKS BY ICP-MS AND APPROXIMATION OF THE RELATIVE SOLUBILITIES FOR THESE ELEMENTS IN A CARBONATE SYSTEM BY WATER-ROCK CONCENTRATION RATIOS

CAIXIA GUO[1], KLAUS J. STETZENBACH[1], & VERNON F.HODGE[2]

[1]*Harry Reid Center for Environmental Studies, University of Nevada-Las Vegas, Box 454009, Las Vegas, NV 89154-4009, USA*
[2]*Department of Chemistry, University of Nevada-Las Vegas, Box 454003, Las Vegas, NV 89154-4003, USA*

Abstract

Rock digestion methods, including lithium metaborate fusion and three microwave digestion procedures, are evaluated for the analysis of the rare earth elements (REEs) and 42 other trace elements in samples of Paleozoic carbonate, Paleozoic quartzite and shale from southern Nevada, and the U.S. Geological Survey standard for diabase, W-2. The trace elements were determined in the dissolved rock samples by low resolution inductively coupled plasma-mass spectrometry (ICP-MS). The lithium metaborate fusion method gave excellent recoveries of the REEs and most other trace elements for W-2. However, the method yielded relatively poor recoveries for some trace elements such as Cr, Zn, and Pb. Two of the three microwave digestion procedures performed better overall than the fusion method. Potential interferences from the suspected major components (Si, Cl, Fe, Sr, and Ba) and interferences of the light rare earth element oxides on the heavy REEs are discussed. Since the rock type may determine the chemical composition of the associated groundwater, the relative solubilities for the elements under study were approximated for a carbonate system by water-rock concentration ratios.

1. Introduction

The long history of testing nuclear weapons at the Nevada Test Site (NTS), located in the southwestern portion of the state of Nevada, USA, and the proposed high-level nuclear waste repository at Yucca Mountain, at the southwestern edge of the NTS, have stimulated much of the hydrologic studies in the surrounding regions (Eakin, 1966; Winograd and Friedman, 1972; Winograd and Thordarson, 1975; Thomas et al., 1989; Johannesson et al., 1996b and 1997a; Stetzenbach et al., 1999 and 2001). The measurement of trace element concentrations in groundwater from springs in southern Nevada and Death Valley, California, was initiated in our laboratory in 1992, in an attempt to determine if trace metal concentration fingerprints, with as many elements as possible, could be used to study groundwater mixing and flow. This approach was chosen because several researchers had suggested that the trace elements present in

K.H. Johannesson,(ed), Rare Earth Elements in Groundwater Flow Systems , 39-65.
© *2005 Springer. Printed in the Netherlands.*

groundwater were most likely derived from the rocks through which the water flowed; thus, water traversing aquifers dominated by different rock types would exhibit different trace element fingerprints (Garrels and Mackenzie, 1967; White et al., 1980; Frape et al., 1984; Hem, 1985; Smedley, 1991). The analytical instrument required for such studies had to have very low elemental detection limits and measure a large number of elements. To that end, an inductively coupled plasma-mass spectrometer (ICP-MS) was purchased in 1991. Our research began by sampling several springs at locations surrounding the NTS and analyzing the filtered (0.45☞m)-acidified samples for the rare earth elements (REEs, 14 including lanthanum) and 42 other trace elements by ICP-MS. The results of our investigations (Stetzenbach et al., 1994, 1999, 2001; Johannesson et al., 1994, 1996a and b, 1997a and b, 2000; Kreamer et al., 1996; Hodge et al., 1996, 1998) have generally reinforced existing hypotheses about the origin and flow of groundwaters in the region, hypotheses built on data obtained from the measurements of the major cations (K, Na, Mg, and Ca), stable isotopes of water, and physical hydrology measurements (for example: Eakin, 1966; Winograd and Friedman, 1972; Winograd and Thordarson, 1975; Dudley and Larson, 1976; Winograd and Pearson, 1976).

The next logical step was to look at the trace-element-concentration fingerprint of aquifer rocks. Since they had to be completely dissolved before ICP-MS analysis, various methods of sample dissolution were examined. These included lithium metaborate fusion, open-beaker digestion, and microwave digestion; however, each method had its limitations (Jarvis and Jarvis, 1992; Totland et al., 1992). Of these methods, high temperature lithium metaborate fusion has been touted as the best method for dissolving solid geologic materials because highly refractory components such as zircon are completely decomposed (Feldman, 1983; Totland et al., 1992; Jarvis and Jarvis 1992). However, a major disadvantage of this method is that it produces analytical samples with high amounts of dissolved solids which can foul the interface cones of the ICP-MS. Due to the high sensitivity of the ICP-MS and the relatively high concentrations of most elements in the rocks, sample dilution can mitigate this problem. Open-beaker digestion of rock samples with strong acids such as $HCl+HNO_3$, even with HF, often leaves insoluble material, which may lead to erroneous results (Gromet and Silver, 1983). Microwave oven digestions, however, can be conducted under more rigorous conditions of temperature and pressure, and have therefore become widely used (Nadkarni, 1984; Gilman and Engelhart, 1989; CEM, 1991a and b; Totland et al., 1992, 1995). Therefore, we decided to evaluate the ICP-MS results from the lithium metaborate fusion method and three somewhat different microwave oven dissolution methods for sample preparation prior to ICP-MS analysis. To test the absolute recoveries of the elements, the geostandard, U.S. Geological Survey (USGS) diabase W-2, was subjected to the four dissolution procedures (lithium metaborate fusion and three microwave digestions). Additionally, three different rock samples were collected from the study area, Paleozoic limestone, quartzite, and shale, to serve as proxies for aquifer material.

2. Experiemental

2.1. INSTRUMENTS

Trace element concentrations in the digests of rock samples were determined with a Perkin-Elmer ICP-MS model ELAN 5000 (Norwalk, CT), which was equipped with an ultrasonic nebulizer (CETAC Technologies, Inc., Omaha, NB, Model U-5000 AT) and an active-film multiplier detector. In general, the elements were quantified using the most abundant isotopes, providing that there were no spectral interferences (see Table 1). For elements for which more than one isotope was monitored, the average concentration is reported. For example, Nd has seven isotopes, of which [143]Nd and [146]Nd are free from isobaric interferences.

Table 1. Choice of isotopes for ICP-MS analysis of rock samples

Element	Isotope	Abundance (%)	Comments
Li	7	92.58	most abundant free isotope
Be	9	100	monoisotopic
Al	27	100	monoisotopic
Ti	49	5.51	less prone to interference
V	51	99.76	most abundant free isotope
Cr	52	83.76	most abundant free isotope
Mn	55	100	monoisotopic
Co	59	100	monoisotopic
Ni	60	26.23	most abundant free isotope
Cu	63	69.09	most abundant free isotope
Zn	66	27.81	most abundant free isotope
Ga	71	39.6	most abundant free isotope
Ge	74	36.54	most abundant free isotope
Rb	85	72.15	most abundant free isotope
Sr	86	9.86	less prone to interference
Y	89	100	monoisotopic
Zr	90	51.46	most abundant free isotope
Nb	93	100	monoisotopic
Mo	97	9.46	less prone to interference
Ru	99	12.72	most abundant free isotope
Rh	103	100	monoisotopic
Pd	105	22.23	most abundant free isotope
Ag	107	51.82	most abundant free isotope
Cd	111	12.75	most abundant free isotope
In	115	95.72	most abundant free isotope
Sn	120	32.85	most abundant free isotope
Sb	121	57.25	most abundant free isotope
Te	125	6.99	most abundant free isotope
Cs	133	100	monoisotopic
Ba	137	11.32	most abundant free isotope

Table 1. (Continued)

Table 1. (Continued)

Element	Isotope	Abundance (%)	Comments
La	139	99.911	most abundant free isotope
Ce	140	88.48	most abundant free isotope
Pr	141	100	monoisotopic
Nd	143	12.17	most abundant free isotope
Nd	146	17.62	most abundant free isotope
Sm	149	13.83	most abundant free isotope
Eu	153	52.18	most abundant free isotope
Gd	157	15.68	most abundant free isotope
Tb	159	100	monoisotopic
Dy	163	24.97	most abundant free isotope
Ho	165	100	monoisotopic
Er	166	33.41	most abundant free isotope
Er	167	22.94	most abundant free isotope
Tm	169	100	monoisotopic
Yb	174	31.84	most abundant free isotope
Lu	175	97.41	most abundant free isotope
Hf	178	27.14	most abundant free isotope
Ta	181	99.988	most abundant free isotope
W	182	26.41	most abundant free isotope
Re	187	62.93	most abundant
Ir	193	62.7	most abundant free isotope
Pt	195	33.8	most abundant free isotope
Au	197	100	monoisotopic
Tl	205	70.5	most abundant free isotope
Pb	208	52.3	most abundant free isotope
Bi	209	100	monoisotopic
Th	232	100	monoisotopic
U	238	99.276	most abundant free isotope

An MDS-2100 microwave oven (CEM Corporation, Matthews, NC) with a capacity of 12 samples was used for rock digestions. A Hermolyne Model 48000 muffle furnace (Hermolyne Corporation, Dubuque, IA) was employed in the lithium metaborate fusions.

2.2 REAGENTS

Mixed-element standards, used to calibrate the ICP-MS, were purchased from Perkin-Elmer Corporation (Norwalk, CT), whereas National Institute for Standards and Technology (NIST)-traceable quality assurance calibration check samples were purchased from High Purity Standards (Charleston, SC) or NIST (Gaithersburg, MD, i.e., NIST "Trace Elements in Water" No. 1643C).

Ultrapure nitric acid (16 **M**) and hydrochloric acid (12 **M**) were double subboiling distilled in quartz, while hydrofluoric acid (29 **M**) was double subboiling distilled in Teflon7 (purchased from Seastar Chemicals, Seattle, WA). Boric acid, AR grade, was obtained from Mallinckrodt Specialty Chemicals (Paris, KY). LiBO$_2$, AR grade, was obtained from Baker (J. T. Baker, Inc., Phillipsburg, NJ). Distilled water, used to prepare all solutions, was obtained by passing deionized water through a Barnstead (Dubuque, IA) Nanopure water system and then distilling the Nanopure water in an all-glass still.

2.3. SAMPLE COLLECTION AND PREPARATION

Three rock samples collected from southern Nevada and one rock standard reference sample (USGS W-2 diabase) were examined in this study. The Nevada samples included: 1) a Paleozoic carbonate from the Aysees Member of the Ordovician Antelope Valley Limestone (collection location 36°26.40' N 115°14.83' W), which is considered to be characteristic of the regional Paleozoic Carbonate aquifer of southern and eastern Nevada; 2) a Paleozoic era quartzite from the Upper Devonian to Mississippian Eleana Formation (collection location 37°03.77' N 116°11.36' W); and 3) a shale from the Bright Angel Shale formation located east of Las Vegas, Nevada (collection location 36° 13.21' N 115°03.14' W) (Figure 1).   The three rock samples were first crushed to 70-mesh using a Bico rock-crushing machine (Vibratory Pulverizer, Burbank, CA) housed in the Geoscience Department at the University of Nevada, Las Vegas.

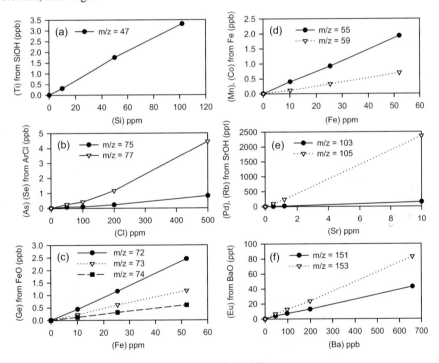

Figure 1. Potential interferences from Si, Cl, Fe, Sr, and Ba

2.4. SAMPLE DIGESTION

An aliquot of each of the Nevada rock samples, USGS diabase standard W-2, as well as a method blank was digested by four procedures, alkali fusion and three different microwave oven routines.   Each sample (except W-2) was prepared in duplicate.

2.4.1. *Alkali fusion (Method 1)*
One-tenth gram of crushed rock sample was mixed with 0.4 g of $LiBO_2$ in clean graphite crucibles.   The graphite crucibles were placed in a muffle furnace at 1050°C for 20 min.   The crucibles were removed from the furnace and the molten sample beads immediately poured into 50.0 g of 12.5% v/v ultrapure HNO_3, contained in 100-mL Teflon7 beakers,   then mixed on a hot plate using a Teflon7 coated magnetic stir bar until the beads dissolved completely (usually ~30 min.).   The solutions were then filtered through Nuclepore7 0.4µm poly-carbonate filters into 60-mL polyethylene bottles in order to remove any suspended carbon particles before analysis by ICP-MS. Immediately before analysis, the samples were diluted with distilled water (18.3 MΩ-cm or better).   The final dilution was generally 50,000 for the rare earth elements and 10,000 for all other trace elements except Ag, Cd, In, Bi, Ru, Rh, Pd, Te, Ir, Pt, and Au which were run at a dilution of 1,000 (Guo, 1996).   All dilutions were done by weight using an analytical balance (Sartorius Research R 200D, readability 0.1mg).

2.4.2. *Microwave oven digestion*
Three different microwave oven digestion procedures were chosen from the CEM Corporation manual (CEM, 1991a, b).   These three procedures were developed and recommended for digestion of limestones, zeolite minerals, and shales; hereafter, we will refer to them as "Methods 2, 3, and 4", respectively.

Method 2 - Limestone.   Method 2 calls for 0.5 g of 70-mesh of each crushed rock sample to be placed into Teflon7-lined microwave digestion vessels.   Next, 2.5 mL of 16 **M** ultrapure HNO_3 and 2.5 mL of 29 **M** ultrapure HF were added to the digestion vessels, whereupon the mixtures were allowed to stand for 15 minutes or until any reaction subsided before vessels were sealed with Teflon7-lined caps with pressure relief valves that were connected by tubes to an overflow container.   The largest and most reactive sample was placed in a control vessel with a modified cap assembly to monitor temperature and pressure levels.   Then the samples were heated to 148°C at 52.8% power (for 8 sample vessels) and pressurized to 8.96 x 10^5 Pa (130 psi) for 15 minutes for a total run time of 25 min. (CEM, 1991a).   Once the heating program was completed, the vessels were cooled for a minimum of 5 min., placed in a laboratory hood, and manually vented before they were opened.   At this point, 30 mL of 4% boric acid solution were then added to each sample to complex the remaining HF.   Each vessel was then resealed, placed back into the microwave oven, heated to a temperature of 60°C at 100% power, and pressurized to 3.45 x 10^5 Pa (50 psi) (pressurized time = 5 min.) for a total of 5 min. This step is required in order to eliminate the HF attack on the quartz torch of the ICP-MS (Nadkarni, 1984).   The samples were then cooled for a minimum of 5 min. and removed from the oven.   The solutions were inspected visually to verify that they were free of particulate matter, and then quantitatively transferred with three 1% HNO_3 rinses into 125 mL polyethylene bottles.   The samples were then weighed and diluted as necessary prior to analysis by

ICP-MS.

Method 3 - Zeolite.   One-quarter gram of crushed rock was placed into a Teflon7-lined vessel along with 6 mL of 16 **M** ultrapure HNO_3, 3 mL of 12 **M** ultrapure HCl, and 2 mL of 29 **M** ultrapure HF.   Hereafter, Method 3 was virtually identical to Method 2, except for slight differences in the microwave heating program.   For this method, the samples were heated to 137°C at 56% power and pressurized to 5.52 x 10^5 Pa (80 psi) for 15 min for a total run time of 25 min.   After heating and cooling, 50 mL of a 4% boric acid solution was added to each sample before the vessels were resealed. The second heating program of Method 3 was the same as for Method 2.   The samples were transferred to 125 mL polyethylene bottles, weighed and diluted as necessary before analysis.

Method 4 - Shale.   In Method 4, 0.2 g of crushed rock was placed in a vessel along with 7 mL of 29 **M** ultrapure HF, 3 mL of 16 **M** ultrapure HNO_3, and 1 mL of 12 **M** ultrapure HCl. Each sample was then heated to 156°C at 52% power and pressurized to 5.52 x 10^5 Pa (80 psi) for 10 min. for a total run time of 15 min.   The samples were again cooled and vented prior to the addition of 50 mL of a 4% boric acid solution. At this point Method 4 followed the sequence of heating as discussed for Method 2. The samples were placed in 125 mL polyethylene bottles and treated as in Method 2.

2.5. SAMPLE ANALYSIS

The sensitivity of the ICP-MS was optimized for each analytical run, using a tuning solution containing 10 ppb ($\mu g\ L^{-1}$) of each of the following elements: Mg, Cu, Cd, Pb, Sc, Rh, Tl, Ce, Tb, Ba, and Ge.   The following criteria were used for tuning: (1) Mg, 5 x 10^5 counts per second (cps); Rh, 3 x 10^6 cps; and Pb, 5 x 10^5 cps.   To demonstrate the instrument stability, 70 replicates of the tuning solution should result in relative standard deviation (RSD) of less than 10%.   If the RSD=s were greater than 10%, corrective action should be taken.   Moreover, the CeO^+/Ce^+ ratio was required to be less than 3%, the BaO^+/Ba^+ less than 2%, and the Ba^{++}/Ba^+ ratio not over 3%. Fulfilling these operating conditions provides the best balance between maximum sensitivity and minimum oxide formation [Perkin Elmer 1992].   Mass calibration of the ICP-MS was performed periodically using the same tuning solution.   The criteria were: Mg, 23.93-24.03 atomic mass units (amu); Rh, 102.86-102.96 amu; and Pb, 207.93-208.03 amu.

The ICP-MS was calibrated with four or five concentrations of four mixed-element standard solutions, designated Set 1, Set 2, Set 3, and Set 4 (Table 2).   Each set was prepared in 1% v/v ultrapure HNO_3.   The concentrations used were in ppb ($\mu g\ L^{-1}$): Set 1, 0.050, 0.460, 1.000, and 2.300; Set 2, 0.050, 0.410, 1.000, 5.000, and 10.000; Set 3, 0.050, 0.500, 1.000, 7.000, and 14.000; Set 4, 0.050, 0.500, 1.000, 5.000, and 14.000. The calibration curve was forced through zero (the calibration blank).   The calibrations for Sets 1-4 elements were verified by the analysis of solutions prepared from NIST traceable solutions from High-Purity Standards (Charleston, SC) or NIST SRM 1643C.
   All sample concentration measurements were required to be below the highest concentration on the calibration curve.

Table 2. Method detection limits (MDLs) for ICP-MS

Element		Solution (ppt)	Rock equivalent (ppm)	Element		Solution (ppt)	Rock equivalent (ppm)
					Li	19	0.19
					Be	50	0.50
					Al	30	0.30
	Y	1.4	0.07		V	15	0.15
	La	2.3	0.12		Cr	41	0.41
	Ce	3.2	0.16		Mn	13	0.13
	Pr	3.0	0.15		Co	9	0.09
	Nd	9.0	0.45		Ni	39	0.39
	Sm	9.0	0.45		Cu	90	0.90
	Eu	3.2	0.16		Zn	52	0.52
Set 1	Gd	9.5	0.47	Set 2	Ga	14	0.14
	Tb	1.6	0.08		Rb	16	0.16
	Dy	4.8	0.24		Sr	27	0.27
	Ho	1.6	0.08		Ag	40	0.024
	Er	4.8	0.24		Cd	34	0.034
	Tm	0.6	0.03		In	13	0.013
	Yb	3.4	0.17		Cs	8	0.08
	Lu	0.6	0.03		Ba	32	0.32
	Th	4.6	0.23		Tl	15	0.15
					Pb	21	0.21
					Bi	11	0.011
					U	13	0.13
	Ru	12	0.012				
	Rh	4.6	0.005		Ti	28	0.28
	Pd	16	0.016		Ge	14	0.14
	Sn	11	0.11		Zr	7.2	0.07
Set 3	Sb	12	0.012	Set 4	Nb	2.5	0.03
	Te	30	0.03		Mo	15	0.15
	Hf	5.0	0.05		Ta	8.1	0.08
	Ir	3.2	0.003		W	12	0.12
	Pt	19	0.019		Re	5.7	0.006
	Au	10	0.01				

* MDL — (t × s × concentration of known)/intensity of known;
  t = student's value for 98% confidence level (t = 3.14 for seven replicate)
  s = standard deviation estimated with n-1 degrees of freedom
** Rock equivalent in Set 1 — MDL × 50,000 (Dilution factor)
** Rock equivalent in Set 2,3,4 — MDL × 1,000 or 10,000 (Dilution factor)

Two internal standards of Pt and Tb were used: Pt was used with Set 1 (the REEs), and Tb was used for Sets 2, 3, and 4. For each set of analyses, the internal standard was added to the calibration solutions and the blank, as well as all of the samples. The concentrations of the internal standards used were 12.7 ppb for Tb and 76.0 ppb for Pt. In this work, if the absolute intensities of internal standard decreased to less than 60% or increased to more than 125% of the original response in the reagent blank (as allowed by U.S. EPA Method 200.8), the run was terminated and rerun   (Long and Martin, 1991)   However, the average change in internal standard intensity in this study was below 20%.

A mid-range standard, as a continuing calibration check, was run every three to five samples.   QA/QC required the results to be within ±10% of the value determined during the initial calibration.   If the deviation was greater than 10%, the instrument was recalibrated and samples were rerun.

3.   Results and Discussion

3.1 FACTORS AFFECTING THE MEASUREMENT RESULTS

Detection limits for the ICP-MS and possible isobaric interferences are discussed in this section.

3.1.1. *Detection limits*
In order to determine the instrumental detection limits (IDLs) of the elements, known amounts of each element were added to distilled/deionized water to make solutions with concentrations from two to five times the estimated detection limits.   Seven aliquots of each of these solutions were then analyzed from which the standard deviation (S) for each element was calculated.   The one-sided t - value for 7 - 1 = 6 degrees of freedom, at the 98% confidence level is given as 3.14 (Long and Martin, 1991) The IDLs (S x 3.14) are listed in Table 2.   It was necessary to adjust the IDLs for these rock analyses to account for the dilutions made during sample preparation.   Thus, for a given element, the rock equivalent IDL is simply the IDL times the dilution factor.   As mentioned above, the factor for Set 1, primarily the REEs, was 50,000, and for Sets 2, 3, and 4, the dilution factors ranged from 1,000 to 10,000.   Consequently, the detection limits of the rock equivalent in Set 1 range from 30 ppb for Tm and Lu to 470 ppb for Gd.   The rock equivalent IDLs for Set 2 range from 11 ppb for Bi to 900 ppb for Cu, whereas the rock equivalent IDLs for Sets 3 and 4 range from 3 ppb for Ir to 280 ppb for Ti.

3.1.2. *Potential interelement interferences from Si, Cl, Sr, Fe and Ba (false positive)*
Quartzite, shale, and Paleozoic limestone are enriched in Si, Fe, Cl, Sr, and Ba (Hurlbut and Klein, 1977), and relatively high concentrations of these elements may produce false positive signals. These isobaric interferences could be caused by elemental isobars or the formation of chemical species such as oxides.   In order to experimentally test this potential, standard solutions of Si, Fe, Cl, Sr, and Ba at concentrations typical to analytical solutions of the rock digest samples were prepared in 1% v/v ultrapure HNO_3 (Guo, 1996). These solutions were then analyzed by ICP-MS for all 56 elements in the same manner described above for the rock samples. The results are reported in Figure 1.

Figure 1a shows the false positive concentrations of Ti at mass 47 (^{47}Ti) produced when solutions of Si were introduced into the ICP-MS. Due to the relatively high levels of Si in most rocks, the significant levels of $SiOH^+$ (mass 47) formed in the ICP-MS could impair the accurate measurement of Ti. Fortunately, the concentrations of Ti in the rock sample solutions were high enough that when the rock sample solutions were diluted

50,000 to 600,000-fold in order to measure Ti concentrations, the Si concentrations were diluted to less than ~1 ppm, therefore, Si interference was not a problem for Ti quantitation. However, in addition to $SiOH^+$, phosphorus monoxide (PO^+), which may reflect phosphate (which was not measured) in the sedimentary rock samples examined in this study, also has a mass of 47.   Therefore, we chose to quantify Ti at mass 49.

Far more serious interferences result from $ArCl^+$ on both As at mass 75 and Se at mass 77 (Figure 1b). Argon, the dominant species present in the plasma, combines with Cl to form $^{40}Ar^{35}Cl^+$ at mass 75 and $^{40}Ar^{37}Cl^+$ at mass 77. The presence of Cl in the sample solutions is due to sample preparation where HCl was used in two of the microwave digestions (methods 3 and 4) and, to a lesser degree, in the rock itself. Because the concentrations of As and Se in the USGS W-2 were found to be a factor of 2 to 4 times higher than previously reported, we have not reported values for either As or Se in this paper.

Figures 1c and d show the false concentrations of Ge at masses 72, 73, and 74, Mn at mass 55, and Co at mass 59 from solutions containing only iron.   When iron (i.e., ^{54}Fe, ^{56}Fe, ^{57}Fe) in the rock samples combines with oxygen and/or hydrogen in the plasma stream, interferences can occur at masses 55, 59, 72, 73, and 74.   However, at the 10,000 fold dilutions used for the measurement of Ge, Mn and Co, the Fe concentrations in these analytical samples are probably less than 5 ppm and will not cause false positive measurements.   Moreover, in the case of Ge, mass 74 was selected for quantitation because it experiences the smallest interference from Fe.

Platinum group metals are very rare; they occur at exceptionally low concentrations in rocks and are even lower in ground waters (Stetzenbach et al., 1994; Wood and Vlassopoulos, 1990; Cook et al, 1992; Wood et al., 1992).   The question arose in this study as to whether signals recorded by ICP-MS for Pd isotopes were from Pd or the result of isobaric interferences.   Palladium has six isotopes (^{102}Pd, ^{104}Pd, ^{105}Pd, ^{106}Pd, ^{108}Pd, and ^{110}Pd), of which ^{105}Pd is the only one that has no elemental isobaric interferences.   Although quantitation of ^{105}Pd could be impacted from the singly charged oxide of Y (i.e., $^{89}Y^{16}O^+$), the Y concentrations are relatively low in our analytical samples (less than 40 ppb).   Therefore, interference from Y on ^{105}Pd was ignored.   Another potential isobaric interference on mass 105 is $SrOH^+$ from the most abundant Sr isotope, ^{88}Sr (relative abundance 82.56%). A 1/1000 dilution would also eliminate this interference (Figure 1e).   Another platinum group element, Rh, is monoisotopic at mass 103, and therefore its quantitation by ICP-MS could be influenced by interferences from the oxide of ^{87}Sr and the hydroxide of ^{86}Sr (relative abundances 7.02% and 9.86%, respectively).   Again, even though the Sr concentrations in these rock samples range from 18.5 ppm up to 325 ppm, the concentrations of Sr in the analytical solution are less than 0.3 ppm.

The interferences of Ba isotopes on Eu at mass 151 and 153 are well-known [Jarvis et al., 1989; Stetzenbach et al., 1994].   Barium has seven naturally occurring isotopes: ^{130}Ba, ^{132}Ba, ^{134}Ba, ^{135}Ba, ^{136}Ba, ^{137}Ba, and ^{138}Ba. Both ^{135}Ba and ^{137}Ba combine with oxygen in the plasma stream to form BaO^+ with masses of 151 and 153, respectively, which can interfere with ^{151}Eu and ^{153}Eu quantification (Figure 1f).   Measurement of Eu was free of significant interferences from Ba owing to a dilution factor of 50,000 for the sample solutions, even though the concentration of Ba in these rocks ranges from

48.6 ppm to 1098 ppm. For example, in the shale, the concentration of Ba for Method 2 is 1100 ppm and 22 ppb in the analytical solution, which would yield a false positive for Eu of about 5 ppt.   The actual concentration of Eu in the analytical solutions was approximately 11 ppb or about 2200 times higher than the expected Ba interference. The doubly charged Ba ion is also known to cause interferences with [69]Ga but not [71]Ga (Stetzenbach et al., 1994). Therefore, we selected mass 71 for the quantitation of Ga in this study.

3.1.3. Oxide formation as a function of argon flow rate

Oxides of REEs such as Sm, Eu, Dy, Ho, Er, Tm, and Yb are not sources of significant interferences in the analysis of common rocks and minerals (Dulski, 1994), however, oxides from Ba, Ce, Pr, Nd, Gd, and Tb can complicate the measurements of the REEs. The argon gas flow rate to the nebulizer in the ICP-MS has a significant effect on the formation of oxides (Longerich et al. 1987).   To determine the yield of these oxide ions, single element solutions containing 200 ppb of Ba and 20 ppb of the selected REEs were prepared.   These solutions were run using three different argon flow rates to the nebulizer.   The data in Table 3 clearly show that the oxide yields increase with increasing   nebulizer   gas flow rate.   For example, the formation of $^{141}Pr^{16}O^+$, which

Table 3. % Oxides formation as a function of nebulizer argon flow rate

Element	Mass	Oxides forming	% Oxide formation		
			Nebulizer flow rate (L/min)		
			0.417	0.517	0.600
Eu	151	BaO	0	0.1	0.1
Eu	153	BaO	0	0.1	0.1
Gd	157	PrO	0.6	1.3	6.1
Gd	156	CeO	0.8	1.1	28
Gd	158	CeO	1.1	1.0	24
Tb	159	NdO	1.0	1.7	3.2
Dy	161	NdO	0	2.2	5.2
Dy	162	NdO	1.3	1.3	2.5
Yb	171	GdO	0	0.6	2.1
Yb	172	GdO	0	0.5	2.5
Yb	173	GdO	1.2	0.5	3.6
Yb	174	GdO	0.6	0.7	2.4
Lu	175	TbO	0.5	0.5	2.4

Concentration of Ba is 200 ppb
Concentrations of Ce, Nd, Gd, Tb, and Pr are 20 ppb

would impact ^{157}Gd, varies from 0.6% at 0.417 L argon/min to 6.1% oxide at 0.600 L argon/min.   Terbium-159 oxide formation, which would create false positive measurement for ^{175}Lu, increases from 0.5% at 0.417 L argon/min to 2.4% of the oxide at 0.600 L argon/min.   Thus, based on these results, a nebulizer gas flow rate of 0.517 L argon/min was selected for our analysis of the rock digests, because higher sensitivity was produced and the formation of oxides was below 1% (see Table 3).   The exceptions include $^{145}Nd^{16}O^+$ and $^{146}Nd^{16}O^+$, which are slightly higher than 1% and can affect ^{161}Dy and ^{162}Dy measurements.   Therefore, ^{163}Dy, which is free of oxide interference was selected for quantitation in this study.

3.2 COMPARISON OF USGS W-2 REFERENCE VALUES BY LITHIUM METABORATE FUSION AND MICROWAVE OVEN DIGESTION TECHNIQUES

Accuracy is a measure of how close the analytical data are to the Atrue@ composition of the sample and, in practice, is evaluated by the analysis of standard reference materials (SRMs). A sample of the USGS standard reference material W-2, which is a diabase (igneous rock), was used for each digestion technique. Table 4 gives the values and shows the mean and standard deviation of the 56 elements in W-2 by the fusion and the three microwave oven digestion techniques and the concentrations reported by USGS. The reported values with standard deviation are recommended values, and others without standard deviation are information values. The results for Y, La, Ce, Pr, Nd, Sm, Eu, Gd, Tb, Dy, Ho, Er, Tm, Yb, Lu, and Th by the fusion technique (Method 1) and microwave Methods 2 (Limestone) and 3 (Zeolite) are in excellent agreement with those reported by USGS. However, the values for Method 4 (Shale) are considerably lower, which may be due to the shorter run time.

The internal consistency of the REEs data set can be visualized by normalizing the REE concentrations to shale abundances (Sholkovitz, 1988; Hanson, 1980; Jarvis, 1988). Figure 2 is a shale-normalized plot of the REE values for W-2 obtained by fusion and the microwave techniques. The fusion and Methods 2 and 3 give normalized results that are nearly identical to each other and to the USGS results, whereas the values obtained using Method 4 are consistently low.

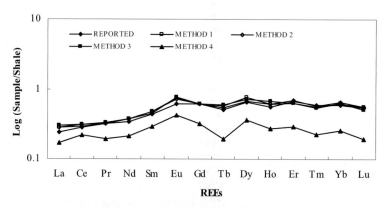

Figure 2. A shale-normalized plot of the REE values for W-2

Table 4. Concentrations of elements in W-2 by four sample
preparation techniques (ppm)

Element	USGS	Method 1 (M ± S.D)	Method 2 (M ± S.D)	Method 3 (M ± S.D)	Method 4 (M ± S.D)	MDL
Li 7	9.6±0.54	nm	6.9±0.3	7.4±0.4	3.9±0.4	0.19
Be 9	1.3	1.34 ±0.05	0.9 ±0.3	1.3 ±0.3	2.5 ±0.1	0.50
Al 27	81800 ±84	83900 ±800	84000 ±600	84000 ±1000	55700 ±700	0.30
Ti 49	6360 ±60	6720 ±100	7100 ±100	6900 ±112	6600 ±200	0.28
V 51	260 ±12	285 ±5	285 ±2	277 ±2	243 ±3	0.15
Cr 52	92 ±4.4	70 ±1	96 ±3	91 ±1	105 ±1	0.41
Mn 55	1290 ±31	1370 ±20	1350 ±10	1320 ±3	1060 ±20	0.13
Co 59	43 ±2.1	31.3 ±0.3	44.6 ±0.4	44.5 ±0.4	53.2 ±0.1	0.09
Ni 60	70 ±2.5	50.6 ±0.3	70.9 ±0.3	72 ±1	85.3 ±0.9	0.39
Cu 63	110 ±4.9	75 ±2	101 ±1	100 ±1	114 ±2	0.90
Zn 66	80 ±2	62 ±1	77 ±2	81 ±2	84 ±1	0.52
Ga 71	17 ±0.89	15.0 ±0.3	18.1 ±0.3	18.1 ±0.4	17.6 ±0.3	0.14
Ge 74		1.3 ±0.1	1.5 ±0.1	1.1 ±0.1	2.1 ±0.2	0.14
Rb 85	21 ±1.1	17.3 ±0.3	19.5 ±0.2	19.8 ±0.3	5.5 ±0.1	0.16
Sr 86	190 ±3	175 ±3	200 ±5	191 ±2	117 ±2	0.27
Y 89	23 ±1.6	24.1 ±0.4	22.2 ±0.1	22.5 ±0.2	7.0 ±0.2	0.07
Zr 90	100 ±4	85.5 ±0.9	92 ±3	94 ±2	89 ±2	0.07
Nb 93	7.9	6.9 ±0.1	6.9 ±0.2	7.3 ±0.1	8.2 ±0.1	0.03
Mo 97		0.54 ±0.05	0.31 ±0.03	0.64 ±0.03	0.65 ±0.05	0.15
Ru 99		<DL	<DL	0.015±0.001	0.018±0.007	0.012
Rh 103		<DL	0.007±0.001	0.007±0.001	<DL	0.005
Pd 105		<DL	<DL	<DL	<DL	0.016
Ag 107		<DL	<DL	<DL	<DL	0.024
Cd 111		<DL	0.07±0.01	0.10±0.01	0.06±0.03	0.034
In 115		<DL	0.051±0.001	0.068±0.002	0.05±0.01	0.013
Sn 120		3.04 ±0.09	1.65 ±0.03	2.02 ±0.04	2.1 ±0.1	0.11
Sb 123	0.79	0.8 ±0.1	0.8 ±0.1	0.75 ±0.01	0.86 ±0.07	0.012
Te 125		<DL	<DL	<DL	<DL	0.03
Cs 133	0.99	0.76 ±0.02	0.88 ±0.03	0.92 ±0.02	<DL	0.08
Ba 137	170 ±11	155 ±1	172 ±2	171 ±2	124 ±2	0.32
La 139	10 ±0.59	11.7 ±0.3	11.6 ±0.2	12.1 ±0.4	6.9 ±0.4	0.12
Ce 140	23 ±1.5	25.5 ±0.6	24.4 ±0.8	25.5 ±0.4	18.3 ±0.4	0.16
Pr 141		3.31 ±0.08	3.24 ±0.05	3.26 ±0.05	2.00 ±0.15	0.15
Nd143,146	13 ±1	14.0 ±0.3	14.2 ±0.3	14.2 ±0.5	8.1 ±0.3	0.45
Sm 149	3.3 ±0.13	3.4 ±0.2	3.61 ±0.07	3.6 ±0.3	2.2 ±0.1	0.45
Eu 153	1.0 ±0.06	1.2 ±0.2	1.2 ±0.1	1.2 ±0.1	0.68 ±0.04	0.16
Gd 157		4.0 ±0.4	3.9 ±0.2	4.0 ±0.5	2.1 ±0.3	0.47
Tb 159	0.63	0.71±0.06	0.67 ±0.04	0.7 ±0.1	0.24 ±0.06	0.08
Dy 163	3.6 ±0.8	4.2 ±0.1	3.7 ±0.3	3.9 ±0.4	2 ±1	0.24
Ho 165	0.76	0.83±0.08	0.82 ±0.02	0.77 ±0.04	0.37 ±0.05	0.08
Er 166,167	2.5	2.4 ±0.1	2.6 ±0.2	2.4 ±0.1	1.1 ±0.2	0.24
Tm 169	0.38	0.36±0.03	0.37 ±0.01	0.34±0.05	0.14 ±0.03	0.03
Yb 174	2.1 ±0.2	2.2 ±0.1	2.3 ±0.2	2.1 ±0.3	0.91 ±0.08	0.17
Lu 175	0.33	0.35 ±0.02	0.34 ±0.02	0.31 ±0.06	0.12 ±0.02	0.03
Hf 178	2.6 ±0.18	2.65 ±0.08	2.61 ±0.07	2.69 ±0.09	2.65 ±0.04	0.05
Ta 181	0.5	0.54±0.05	0.53 ±0.05	0.49 ±0.01	0.47 ±0.09	0.08
W 182		0.37 ±0.05	0.32 ±0.04	0.25 ±0.02	0.41 ±0.05	0.12
Re 187		<DL	0.01 ±0.001	0.028 ±0.001	<DL	0.006
Ir 193		<DL	<DL	<DL	<DL	0.003
Pt 195		<DL	<DL	0.036 ±0.005	0.051 ±0.003	0.019
Au 197		<DL	<DL	<DL	<DL	0.01
Tl 205		0.22 ±0.05	0.199 ±0.002	0.49 ±0.02	2.1 ±0.4	0.15
Pb 208	9.3	4.8 ±0.1	7.8 ±0.2	7.4 ±0.1	6.2 ±0.2	0.21
Bi 209		<DL	0.024 ±0.001	0.045 ±0.007	0.09 ±0.01	0.011
Th 232	2.4 ±0.1	2.1 ±0.2	2.4 ±0.1	2.3 ±0.2	0.2 ±0.2	0.23
U 238	0.53	0.48 ±0.02	0.53 ±0.01	0.50 ±0.03	0.47 ±0.03	0.13

W-2 USGS DIABASE;

nm — not measured;

MDL — method detection limit.

The fusion technique is not necessarily applicable for the determination of all of the other trace elements measured in this study (Table 4). Although the values of some elements such as Be, Al, Ti, V, Mn, Sr, Y, Sb, Ba, Hf, Ta, and U differ by less than 10% from the USGS value, this is not the case of other elements such as Cr, Co, Ni, Cu, Zn, Cs, and Pb, which are lower by more than 20% from the USGS values. The high temperatures of the metaborate fusion have been reported to prevent the accurate determination of these elements (Totland et al., 1992). Results for Ge, Mo, Sn, Pr, Gd, and W were determined in W-2 by fusion, but USGS values are not available for comparison. Upper limits are reported for 12 other elements not reported by the USGS. As shown in Table 4, the results of trace element quantitations by Methods 2 and 3 are in excellent agreement with the reference values reported by USGS. Of the other 23 USGS reported trace elements, 20 have the measured values by Methods 2 and 3 within 10% of the USGS value. Lithium, Be, and Pb deviate by more than 20% from the USGS values. Lead values for Methods 2 and 3 of 7.8 ppm and 7.4 ppm, respectively, were lower than the value of 9.3 ppm reported by USGS. These values are the same as the 7.7 ppm reported by Jenner (Jenner et al., 1990) and are much closer to the USGS value than the 4.8 ppm obtained by Method 1 (LiBO$_2$ fusion). Lead readily forms volatile species, which are subsequently lost from the sample (Jarvis, 1990). Finally, despite the fact that Method 4 gives unsatisfactory results for many of the trace elements, those that are within about 10% of the USGS values include Ti, V, Cu, Zn, Ga, Nb, Hf, and Ta.

3.3 RESULTS FOR THE RARE EARTH ELEMENTS AND 42 OTHER TRACE ELEMENTS IN ROCKS FROM SOUTHERN NEVADA BY FUSION AND MICROWAVE OVEN DIGESTION TECHNIQUES

Two sedimentary rocks (limestone and shale) and one metamorphosed sedimentary rock (quartzite) collected from southern Nevada were also analyzed for all 56 trace elements using the fusion technique (Method 1) and three microwave oven digestion techniques Methods 2, 3, and 4. The results for the REEs using Methods 1, 2, and 3 are in excellent agreement. Figure 3 (a), (b) and (c) shows shale-normalized REE plots. The overlap of symbols for Methods 1, 2, and 3 illustrates the excellent inter-method reproducibility. The overlap of normalization for the quartzite is not as good as that for limestone and shale. Again, results for Method 4 are substantially lower than those determined by Methods 1, 2, and 3. For the limestone, the concentrations of REEs were not detectable by this method. Examining Tables 5-7, it is apparent that Y concentrations are higher by Method 1 than by Methods 2 and 3. This was not observed when comparing the results for the USGS standard rock, W-2.

The data for the non-rare earth elements in Tables 5-7 by Methods 1, 2, and 3 are in good or excellent agreement, whereas data from Method 4 are not. Indeed, the results for most elements by Methods 1, 2, and 3 are within 5% to 10%, except for Be, Tl, Zr, Nb and Sn (Tables 5-7). The values for Be, for example, are low by Method 2 in quartzite and shale and high by about a factor of two for Method 4. We suggest that, in general, the values obtained by Methods 1 and 3 are better because these two methods   provided accurate results for Be in W-2 (Table 4). Thallium concentrations

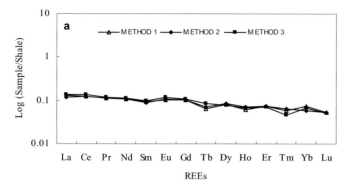

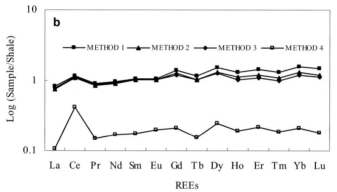

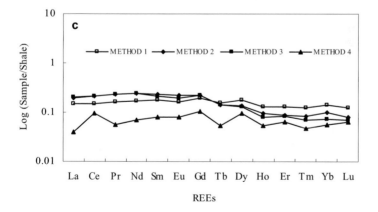

Figure 3. (a) Shale-normalized REE plots of Paleozoic limestone;
(b) Shale-normalized REE plots of Bright Angel shale;
(c) Shale-normalized REE plots of Quartzite

Guo et al.

Table 5. Concentrations of elements in Paleozoic limestone by four sample preparation techniques (ppm)

Element	Method 1 (M ± S.D)	Method 2 (M ± S.D)	Method 3 (M ± S.D)	Method 4 (M ± S.D)	MDL
Li 7	nm	2.3±0.4	2.5±0.3	1.7±0.5	0.19
Be 9	0.6±0.1	0.6±0.2	0.8±0.2	2.2±0.3	0.50
Al 27	5050±40	5040±50	4890±30	400±90	0.30
Ti 49	241±6	243±6	231±5	245±4	0.28
V 51	5.9±0.1	5.7±0.3	7.4±0.4	8.4±0.2	0.15
Cr 52	7.6±0.4	6.5±0.1	5.6±0.2	3.8±0.2	0.41
Mn 55	55.3±0.8	74±2	66±3	27.7±0.5	0.13
Co 59	1.03±0.01	2.99± 0.09	3.08±0.06	1.20±0.02	0.09
Ni 60	8.5± 0.2	7.9±0.4	6.3±0.1	3.5±0.2	0.39
Cu 63	2.3±0.8	2.53±0.09	2.56±0.09	0.25±0.14	0.90
Zn 66	6.0±0.3	5.3±0.3	4.7±0.4	7.1±0.2	0.52
Ga 71	0.84±0.03	1.16±0.02	0.98±0.04	0.40±0.02	0.14
Ge 74	0.14±0.03	0.17±0.03	0.24±0.02	1.2±0.1	0.14
Rb 85	8.4±0.4	11.5±0.2	11.8±0.2	10.0±0.2	0.16
Sr 86	302±6	325±5	330±5	26.7±0.5	0.27
Y 89	3.2±0.2	2.78±0.07	2.76±0.06	<DL	0.07
Zr 90	41±1	37.7±0.2	49.6±0.3	35.3±0.1	0.07
Nb 93	0.87±0.07	1.58±0.06	0.94±0.02	1.89±0.04	0.03
Mo 97	0.25±0.07	0.15±0.06	0.17±0.03	0.40±0.05	0.15
Ru 99	<DL	<DL	<DL	<DL	0.012
Rh 103	<DL	<DL	<DL	<DL	0.005
Pd 105	<DL	<DL	<DL	<DL	0.016
Ag 107	<DL	<DL	<DL	<DL	0.024
Cd 111	<DL	<DL	<DL	<DL	0.034
In 115	<DL	<DL	<DL	<DL	0.013
Sn 120	0.50±0.03	0.12±0.01	0.17±0.01	0.24±0.01	0.11
Sb 123	<DL	0.047±0.002	0.062±0.001	0.113±0.005	0.012
Te 125	<DL	<DL	0.039 ±0.004	0.05 ±0.01	0.03
Cs 133	0.08±0.01	0.13±0.01	0.12±0.01	0.15±0.01	0.08
Ba 137	45.2±0.8	49±1	48.6±0.7	4.4±0.1	0.32
La 139	5.46± 0.09	4.92± 0.04	5.50± 0.04	<DL	0.12
Ce 140	10.4± 0.1	10.4± 0.1	11.1± 0.3	<DL	0.16
Pr 141	1.12± 0.07	1.15± 0.06	1.19± 0.05	<DL	0.15
Nd 143,146	4.2± 0.2	4.2± 0.3	4.4± 0.2	<DL	0.45
Sm 149	0.71± 0.08	0.69± 0.07	0.75± 0.08	<DL	0.45
Eu 153	0.17±0.02	0.17± 0.02	0.20±0.02	<DL	0.16
Gd 157	0.65±0.06	0.65± 0.07	0.69± 0.03	<DL	0.47
Tb 159	0.08± 0.02	0.09± 0.01	0.10± 0.01	<DL	0.08
Dy 163	0.46± 0.05	0.46± 0.04	0.43± 0.04	<DL	0.24
Ho 165	0.08± 0.01	0.10± 0.01	0.09± 0.02	<DL	0.08
Er 166,167	0.28± 0.04	0.28± 0.04	0.26± 0.05	<DL	0.24
Tm 169	0.038± 0.009	0.040± 0.003	0.030± 0.006	<DL	0.03
Yb 174	0.25± 0.03	0.21± 0.03	0.23± 0.02	<DL	0.17
Lu 175	0.03± 0.01	0.03±0.01	0.03± 0.01	<DL	0.03
Hf 178	0.76±0.03	0.49±0.02	0.43±0.01	0.05±0.01	0.05
Ta 181	<DL	<DL	<DL	<DL	0.08
W 182	0.34±0.02	0.32±0.030	0.26±0.02	0.39±0.05	0.12
Re 187	<DL	0.012±0.001	0.006 ±0.001	<DL	0.006
Ir 193	<DL	<DL	<DL	<DL	0.003
Pt 195	<DL	<DL	<DL	<DL	0.019
Au 197	<DL	<DL	<DL	<DL	0.01
Tl 205	0.20±0.04	0.21±0.02	0.3±0.3	1.4±0.2	0.15
Pb 208	0.24±0.03	2.03±0.07	2.04±0.08	0.50±0.04	0.21
Bi 209	<DL	<DL	<DL	<DL	0.011
Th 232	0.76± 0.04	0.77± 0.08	0.81± 0.07	<DL	0.23
U 238	0.60±0.03	0.62±0.03	0.61±0.03	0.37±0.01	0.13

nm — not measured;

MDL — method detection limit

Table 6. Concentrations of elements in quartzite by four sample
preparation techniques (ppm)

Element	Method 1 (M ± S.D)	Method 2 (M ± S.D)	Method 3 (M ± S.D)	Method 4 (M ± S.D)	MDL
Li 7	nm	6.1±0.5	6.9±0.5	7.6±0.6	0.19
Be 9	1.00±0.08	0.63±0.06	1.1±0.2	2.1±0.4	0.50
Al 27	8200±100	8310±70	8400±100	8370±80	0.30
Ti 49	657±9	580±10	550±20	570±10	0.28
V 51	12.1±0.1	13.3±0.2	17.9±0.4	19.1±0.4	0.15
Cr 52	21.3±0.3	23.5±0.3	25.2±0.7	27.7±0.4	0.41
Mn 55	536±4	690±8	735±8	780±10	0.13
Co 59	2.21±0.05	2.71±0.06	2.94±0.07	2.90±0.09	0.09
Ni 60	14.2±0.2	19.2±0.4	21.6±0.5	20.3±0.4	0.39
Cu 63	6.9±0.1	6.4±0.1	6.6±0.2	5.1±0.2	0.90
Zn 66	40±1	47±1	50±2	49±1	0.52
Ga 71	1.58±0.04	2.16±0.09	2.40±0.05	2.20±0.04	0.14
Ge 74	0.9±0.1	1.3±0.2	1.6±0.4	3.2±0.1	0.14
Rb 85	2.26±0.02	3.20±0.02	3.21±0.10	2.78±0.08	0.16
Sr 86	14.6±0.2	18.5±0.3	22.2±0.3	18.3±0.6	0.27
Y 89	4.90±0.17	2.77±0.09	2.72±0.08	1.96±0.06	0.07
Zr 90	91±3	68±4	88±8	42.2±0.5	0.07
Nb 93	1.8±0.1	1.3±0.1	2.0±0.1	1.8±0.1	0.03
Mo 97	1.9±0.1	1.9±0.1	2.1±0.1	2.1±0.1	0.15
Ru 99	<DL	<DL	0.05±0.01	0.07±0.01	0.012
Rh 103	<DL	<DL	<DL	<DL	0.005
Pd 105	<DL	<DL	0.029±0.003	<DL	0.016
Ag 107	<DL	0.057±0.001	0.054±0.002	0.082±0.005	0.024
Cd 111	<DL	0.35±0.01	0.37±0.01	0.36±0.01	0.034
In 115	<DL	<DL	<DL	<DL	0.013
Sn 120	0.73±0.05	0.52±0.04	0.83±0.05	0.87±0.04	0.11
Sb 123	0.29±0.04	0.30±0.05	0.37±0.04	0.23±0.03	0.012
Te 125	<DL	<DL	<DL	<DL	0.03
Cs 133	0.17±0.01	0.19±0.01	0.18±0.02	0.21±0.02	0.08
Ba 137	33.9±0.2	34.8±0.2	36.5±0.7	39±1	0.32
La 139	6.1±0.1	8.1±0.4	8.3±0.2	1.6±0.1	0.12
Ce 140	12.5±0.2	17.4±0.8	17.6±0.4	7.9±0.1	0.16
Pr 141	1.63±0.08	2.33±0.09	2.35±0.07	0.57±0.05	0.15
Nd 143,146	6.4±0.4	9.3±0.6	9.2±0.4	2.6±0.2	0.45
Sm 149	1.3±0.2	1.7±0.2	1.6±0.2	0.6±0.1	0.45
Eu 153	0.27±0.03	0.35±0.02	0.31±0.05	0.14±0.03	0.16
Gd 157	1.2±0.1	1.3±0.1	1.4±0.2	0.7±0.1	0.47
Tb 159	0.19±0.03	0.18±0.02	0.18±0.02	0.07±0.01	0.08
Dy 163	1.0±0.1	0.8±0.1	0.7±0.1	0.5±0.1	0.24
Ho 165	0.17±0.01	0.13±0.01	0.11±0.02	0.07±0.03	0.08
Er 166,167	0.48±0.06	0.35±0.06	0.32±0.09	0.24±0.04	0.24
Tm 169	0.08±0.01	0.05±0.01	0.04±0.02	0.03±0.02	0.03
Yb 174	0.49±0.03	0.34±0.01	0.26±0.09	0.21±0.04	0.17
Lu 175	0.08±0.01	0.05±0.01	0.04±0.01	0.04±0.01	0.03
Hf 178	0.76±0.03	0.94±0.03	0.85±0.02	1.27±0.03	0.05
Ta 181	0.23±0.02	0.29±0.09	0.28±0.09	0.30±0.04	0.08
W 182	7.9±0.1	11.5±0.3	10.5±0.4	10.9±0.4	0.12
Re 187	<DL	0.008 ±0.001	0.029 ±0.004	<DL	0.006
Ir 193	<DL	<DL	0.005±0.001	0.006±0.002	0.003
Pt 195	<DL	<DL	0.068 ±0.005	0.024±0.004	0.019
Au 197	<DL	<DL	1.1±0.1	1.4±0.3	0.01
Tl 205	0.23±0.06	0.17±0.03	0.32±0.04	1.20±0.19	0.15
Pb 208	0.94±0.07	2.35±0.09	2.24±0.03	2.11±0.06	0.21
Bi 209	<DL	0.019 ±0.001	<DL	<DL	0.011
Th 232	1.33±0.06	1.49±0.04	1.46±0.09	<DL	0.23
U 238	0.61±0.02	0.54±0.03	0.45±0.03	0.50±0.02	0.13

nm — not measured;
MDL — method detection limit

Table 7. Concentrations of elements in shale by four sample
preparation techniques (ppm)

Element	Method 1 (M ± S.D)	Method 2 (M ± S.D)	Method 3 (M ± S.D)	Method 4 (M ± S.D)	MDL
Li 7	nm	16.9±0.8	15.7±0.5	15.7±0.5	0.19
Be 9	1.7±0.1	1.4±0.3	2.0±0.3	3.0±0.3	0.50
Al 27	36300±200	39000±200	38600±400	37800±500	0.30
Ti 49	2140±40	2140±30	1850±20	2160±50	0.28
V 51	15.9±0.4	18.9±0.3	19.9±0.3	27.5±0.7	0.15
Cr 52	15.1±0.4	17.1±0.3	16.0±0.3	21.2±0.7	0.41
Mn 55	1139±4	1300±40	1340±8	1320±30	0.13
Co 59	2.8±0.1	3.6±0.1	3.6±0.1	5.1±0.1	0.09
Ni 60	7.1±0.3	6.0±0.1	5.9±0.3	7.9±0.4	0.39
Cu 63	5.8±0.3	13.8±0.3	2.7±0.2	0.9±0.3	0.90
Zn 66	12.5±0.5	13.8±0.3	14.4±0.3	18±1	0.52
Ga 71	6.4±0.1	8.7±0.1	8.20±0.1	9.3±0.2	0.14
Ge 74	1.1±0.1	1.5±0.1	1.4±0.4	4.6±0.2	0.14
Rb 85	89±4	121±2	123±1	126±4	0.16
Sr 86	120±5	150±5	138±3	149±4	0.27
Y 89	53.3±0.9	36.5±0.4	36.5±0.4	3.4±0.1	0.07
Zr 90	364±4	361±5	339±5	450±1	0.07
Nb 93	7.9±0.2	7.2±0.3	7.5±0.3	10.8±0.5	0.03
Mo 97	0.67±0.08	0.68±0.05	0.7±0.07	0.9±0.1	0.15
Ru 99	<DL	<DL	0.03±0.002	<DL	0.012
Rh 103	<DL	<DL	<DL	<DL	0.005
Pd 105	1.7±0.1	0.36±0.01	0.61±0.01	0.58±0.05	0.016
Ag 107	<DL	<DL	<DL	<DL	0.024
Cd 111	<DL	0.05±0.01	0.05±0.01	<DL	0.034
In 115	<DL	0.025±0.001	0.033±0.001	0.027±0.001	0.013
Sn 120	1.04±0.09	0.74±0.05	1.1±0.1	1.6±0.2	0.11
Sb 123	0.29±0.04	0.28±0.01	0.31±0.01	0.3±0.02	0.012
Te 125	<DL	0.03±0.01	0.07±0.01	0.09±0.04	0.03
Cs 133	2.28±0.05	2.82±0.08	2.88±0.03	2.96±0.04	0.08
Ba 137	1041±6	1100±10	1100±20	1010±10	0.32
La 139	33.2±0.6	30.5±0.4	31.7±0.2	4.4±0.1	0.12
Ce 140	96±1	90±2	91±1	34.0±0.4	0.16
Pr 141	9.1±0.2	8.6±0.2	8.8±0.3	1.5±0.1	0.15
Nd 143,146	36±1	35±1	35±1	6.4±0.3	0.45
Sm 149	7.9±0.4	7.6±0.4	7.9±0.8	1.3±0.2	0.45
Eu 153	1.73±0.05	1.67±0.05	1.66±0.08	0.31±0.08	0.16
Gd 157	8.9±0.2	8.1±0.3	7.6±0.5	1.4±0.2	0.47
Tb 159	1.41±0.05	1.24±0.04	1.25±0.08	0.19±0.02	0.08
Dy 163	8.3±0.2	7.2±0.3	7.0±0.3	1.4±0.1	0.24
Ho 165	1.75±0.08	1.49±0.07	1.35±0.09	0.26±0.02	0.08
Er 166,167	5.4±0.2	4.5±0.2	4.1±0.2	0.81±0.07	0.24
Tm 169	0.82±0.05	0.69±0.08	0.62±0.07	0.12±0.03	0.03
Yb 174	5.5±0.1	4.6±0.3	4.0±0.1	0.79±0.06	0.17
Lu 175	0.89±0.05	0.73±0.03	0.69±0.04	0.11±0.02	0.03
Hf 178	10.6±0.4	9.2±0.4	7.3±0.5	10.4±0.8	0.05
Ta 181	1.05±0.04	1.73±0.07	1.36±0.09	1.8±0.3	0.08
W 182	2.4±0.1	2.9±0.1	2.9±0.2	2.6±0.2	0.12
Re 187	<DL	0.008±0.0003	0.033 ±0.003	0.09±0.01	0.006
Ir 193	<DL	<DL	<DL	0.25±0.06	0.003
Pt 195	<DL	<DL	<DL	<DL	0.019
Au 197	0.18±0.04	<DL	4±1	3±1	0.01
Tl 205	0.31±0.05	0.60±0.04	0.92±0.01	1.8±0.4	0.15
Pb 208	4.8±0.1	15.3±0.3	14.6±0.2	13.7±0.2	0.21
Bi 209	<DL	0.16±0.02	0.3±0.01	0.2±0.01	0.011
Th 232	14.2±0.2	15.3±0.3	15.1±0.3	<DL	0.23
U 238	3.4±0.1	3.7±0.1	3.3±0.1	3.3±0.1	0.13

nm — not measured;
MDL — method detection limit

vary somewhat in the three rock samples. The Tl values, for example, obtained by Methods 1 and 2 are in reasonable agreement in the limestone and quartzite but vary substantially (>3 σ) in the shale sample. Thallium values determined by Method 4 are always significantly higher.

3.4 COMPARISON OF THE TRACE ELEMENT CONCENTRATIONS OF THE THREE ROCK TYPES FROM SOUTHERN NEVADA

In general, the shale sample exhibits the highest concentrations of most of the trace elements, whereas the limestone sample has the lowest concentrations (Table 8). The presence of higher concentrations of many trace elements in shale compared to carbonate rock is well known (Drever, 1997). Titanium concentrations, for example, are 2,140 ppm in the shale, 580 ppm in the quartzite, and 243 ppm in the limestone. Concentrations of REEs are likewise highest in the shale. The shale normalized values for limestone trend downward from Ce to Lu, as do those for the quartzite sample, while there appears to be enrichment in the HREEs in the shale sample. A few elements such as Cr, Ni, Cu, Zn, Mo, and W, are somewhat higher in the quartzite than either the shale or limestone samples. Over all, Al, Sr, Ba, Ti, Mn, and Zr have the highest concentrations in all three rocks types.

Table 8. Comparison of trace-element concentrations among rock samples

Element	Limestone ppm	Quartzite ppm	Shale ppm	Ratio Shale/ Limestone	Ratio Shale/ Quartzite	Ratio Quartzite/ Limestone
Li	2.3	6.1	16.9	7.3	2.8	2.7
Be	0.6	0	*1.7	2.8	1.7	1.7
Al	5040	8310	39000	7.7	4.7	1.6
Ti	243	580	2140	8.8	3.7	2.4
V	5.7	13.3	18.9	3.3	1.4	2.3
Cr	6.5	23.5	17.1	2.6	0.7	3.6
Mn	74	690	1300	17.6	1.9	9.3
Co	2.99	2.71	3.6	1.2	1.3	0.9
Ni	7.9	19.2	6.0	0.8	0.3	2.4
Cu	2.53	6.4	2.7	1.1	0.4	2.5
Zn	5.3	47	13.8	2.6	0.3	8.9
Ga	1.16	2.16	8.7	7.5	4.0	1.9
Ge	0.17	1.3	1.5	8.8	1.2	7.6
Rb	11.5	3.20	121	10.5	37.8	0.3
Sr	325	18.5	150	0.5	8.1	0.06
Y	2.78	2.77	36.5	13.1	13.2	1.0
Zr	37.7	*91	361	9.6	4.0	2.4
Nb	1.58	*1.8	7.2	4.6	4.0	1.1
Mo	0.15	1.9	0.68	4.5	0.4	12.7
Ru	<0.012	<0.012	<0.012			
Rh	<0.005	<0.005	<0.005			

58 Guo et al.

Table 8. (Continued)

Element	Limestone ppm	Quartzite ppm	Shale ppm	Ratio Shale/ Limestone	Ratio Shale/ Quartzite	Ratio Quartzite/ Limestone
Pd	<0.016	<0.016	0.36			
Ag	<0.024	<0.024	<0.024			
Cd	<0.034	<0.034	<0.034			
In	<0.013	<0.013	0.025			
Sn	0.12	0.52	*1.04	8.7	2.0	4.3
Sb	0.047	0.30	0.28	8.7	0.9	6.4
Te	<0.03	<0.03	0.03			
Cs	0.13	0.19	2.82	21.7	14.8	1.5
Ba	49	34.8	1100	22.4	31.6	0.7
La	4.92	8.1	30.5	6.2	3.8	1.6
Ce	10.4	17.4	90	8.7	5.2	1.7
Pr	1.15	2.33	8.6	7.5	3.7	2.0
Nd	4.2	9.3	35	8.3	3.8	2.2
Sm	0.69	1.7	7.6	11.0	4.5	2.5
Eu	0.17	0.35	1.67	9.8	4.8	2.1
Gd	0.65	1.3	8.1	12.5	6.2	2.0
Tb	0.09	0.18	1.24	13.8	6.9	2.0
Dy	0.46	0.8	7.2	15.7	9.0	1.7
Ho	0.10	0.13	1.49	14.9	11.5	1.3
Er	0.28	0.35	4.5	16.1	12.9	1.3
Tm	0.040	0.05	0.69	17.3	13.8	1.3
Yb	0.21	0.34	4.6	21.9	13.5	1.6
Lu	0.03	0.05	0.73	24.3	14.6	1.7
Hf	0.49	0.94	9.2	18.8	9.8	1.9
Ta	<0.08	0.29	1.73		6.0	
W	0.32	11.5	2.9	9.1	0.3	35.9
Re	0.012	0.008	0.008	0.7	1.0	0.7
Ir	<0.003	<0.003	<0.003			
Pt	<0.019	<0.019	<0.019			
Au	<0.01	<0.01	<0.01			
Tl	0.21	*0.23	0.60	2.9	2.6	1.1
Pb	2.03	2.35	15.3	7.5	6.5	1.2
Bi	<0.011	0.019	0.16			
Th	0.77	1.49	15.3	19.9	10.3	1.9
U	0.62	0.54	3.7	6.0	6.9	0.9

Data used by method 2 of microwave oven digestion
*Data used by fusion method
<Below detection limit

3.5   CACULATION   OF   THE   RELATIVE   SOLUBILITIES   OF   THESE
       ELEMENTS IN A CARBONATE SYSTEM BASED ON WATER-ROCK
       CONCENTRATION RATIOS

Many investigators have argued convincingly that the chemical composition of
groundwaters are strongly related to the chemical composition of the rocks through
which they flow as a result of solid-liquid reactions such as weathering reactions,
mineral precipitation/dissolution reactions, and ion exchange reactions (e.g., Garrels
and Mackenzie, 1967; Frape et al., 1984; Welch et al., 1988; Banner et al., 1989;
Thomas et al., 1989; Smedley, 1991; Gosselin et al., 1992; Johannesson et al., 1997b).
The average concentrations of trace elements in the carbonate spring waters (Hodge et
al., 1996) and the limestone rock sample from this study are presented in Table 9. The
mean values of the "carbonate" groundwaters were derived from the analysis of 23
springs sampled five times over a three-year period from 1992-1994.   Eighteen of the
springs are located in the Ash Meadows National Wildlife Refuge in southern Nevada
and the remaining springs are located in Death Valley National Park in eastern
California.   The pH of the spring waters ranged from 7-8. When the concentrations of
the trace elements in these carbonate waters (in ppb) are divided by the concentrations
in the limestone (in ppb) (Table 9 and Figure 4), one obtains a measure of the apparent
solubilities of these elements resulting from the complex water-rock interactions over
perhaps thousands of years.   On this basis (ppb water/ppb rock), the most soluble
elements are Mo, Li and Cs,   which are found to be present in the spring waters at
roughly one twentieth the concentrations found in the rock.   Of the top ten most
soluble elements: three are members of the alkali metal family, Li, Cs, and Rb; two
are members of the alkali earth elements, Ba and Sr; four are elements which potentially
form oxyanions in solution, Mo, Sb, Ge, and W; and U which is often postulated to
exist in carbonate waters as the uranyl carbonate complex anion (Starik and Kolyadin,
1957).   In contrast, the least soluble elements, Al, Th, La and REEs, and Zr, are
present in the water at concentrations six to seven orders of magnitude less than in the
rock.   It   is   apparent   from   Figure   4   that   the   rock   normalized

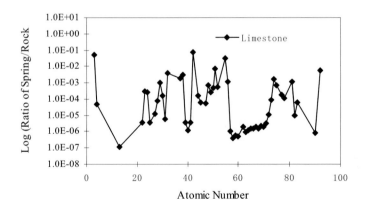

Figure 4. Comparison of trace element concentrations in spring waters
and associated rock

Table 9. Comparison of trace elements in spring waters and rock

Rock type		Carbonate		
Element	Atomic #	Rock (ppm)	*Springs (ppb)	Spring/Rock
Li 7	3	2.3E+00	1.2E+02	5.1E-02
Be 9	4	6.0E-01	<2.7E-02	(4.5E-05)
Al 27	13	5.0E+03	5.9E-01	1.2E-07
Ti 49	22	2.4E+02	9.1E-01	3.7E-06
V 51	23	5.7E+00	1.7E+00	2.9E-04
Cr 52	24	6.5E+00	1.7E+00	2.6E-04
Mn 55	25	7.4E+01	2.6E-01	3.6E-06
Co 59	27	3.0E+00	3.7E-02	1.2E-05
Ni 60	28	7.9E+00	5.9E-01	7.4E-05
Cu 63	29	2.5E+00	2.5E+00	9.9E-04
Zn 66	30	5.3E+00	8.5E-01	1.6E-04
Ga 71	31	1.2E+00	7.0E-03	6.0E-06
Ge 74	32	1.7E-01	6.5E-01	3.8E-03
Rb 85	37	1.2E+01	2.2E+01	1.9E-03
Sr 86	38	3.3E+02	1.1E+03	3.2E-03
Y 89	39	2.8E+00	9.8E-03	3.5E-06
Zr 90	40	3.8E+01	4.7E-02	1.3E-06
Nb 93	41	1.6E+00	<5.6E-03	(3.5E-06)
Mo 97	42	1.5E-01	1.2E+01	7.7E-02
Ru 99	44	<1.2E-02	<2.0E-03	(1.7E-04)
Rh 103	45	<5.0E-03	3.0E-04	(6.0E-05)
Ag 107	47	<6.0E-02	<3.2E-03	(5.3E-05)
Cd 111	48	<3.4E-02	2.5E-02	(7.3E-04)
In 115	49	<1.3E-02	<3.4E-03	(2.6E-04)
Sn 120	50	3.3E-01	1.5E-01	4.7E-04
Sb 123	51	4.7E-02	3.5E-01	7.5E-03
Te 125	52	<3.0E-02	1.7E-02	(5.5E-04)
Cs 133	55	1.3E-01	4.0E+00	3.1E-02
Ba 137	56	4.9E+01	5.4E+01	1.1E-03
La 139	57	4.9E+00	5.1E-03	1.0E-06
Ce 140	58	1.0E+01	4.2E-03	4.0E-07
Pr 141	59	1.2E+00	6.5E-04	5.6E-07
Nd 143	60	4.2E+00	2.2E-03	5.2E-07
Sm 149	62	6.9E-01	1.4E-03	2.0E-06
Eu 153	63	1.7E-01	1.7E-04	9.8E-07
Gd 157	64	6.5E-01	7.4E-04	1.1E-06
Tb 159	65	9.0E-02	1.3E-04	1.4E-06
Dy 163	66	4.6E-01	6.7E-04	1.4E-06
Ho 165	67	1.0E-01	2.0E-04	2.0E-06
Er 166	68	2.8E-01	4.5E-04	1.6E-06
Tm 169	69	4.0E-02	8.5E-05	2.1E-06
Yb 174	70	2.1E-01	4.2E-04	2.0E-06
Lu 175	71	3.2E-02	1.0E-04	3.3E-06
Hf 178	72	4.9E-01	<5.4E-03	(1.1E-05)
Ta 181	73	<8.0E-02	7.4E-03	(9.3E-05)
W 182	74	3.2E-01	5.3E-01	1.7E-03
Re 187	75	1.2E-02	8.2E-03	6.8E-04
Ir 193	77	<3.0E-03	<5.8E-04	(1.9E-04)
Pt 195	78	<1.9E-02	<2.0E-03	(1.1E-4)
Tl 205	81	2.1E-01	2.5E-01	1.2E-03
Pb 208	82	2.0E+00	<1.9E-02	(9.4E-06)
Bi 209	83	<1.1E-01	<6.3E-03	(5.7E-05)
Th 232	90	7.7E-01	6.5E-04	8.4E-07
U 238	92	6.2E-01	3.3E+00	5.4E-03

Data were from Ash Meadow and Death Valley area
<Below detection limit
Parentheses — approximation
Rock data were used by limestone program of microwave oven digestion

REE values show increasing solubility with atomic number or smaller ionic radius (the exception being Sm).   The relative values of the REEs range from a low of 0.4 x10^{-6} for Ce, up to 3x10^{-6} for Lu.   This trend is further demonstrated by plotting the apparent solubilities against the radii of the tripositive ions, Figure 5.   Thus, over time, during the evolution of this Aaverage@ carbonate water, the HREEs have accumulated to a greater extent in solution than the LREEs.

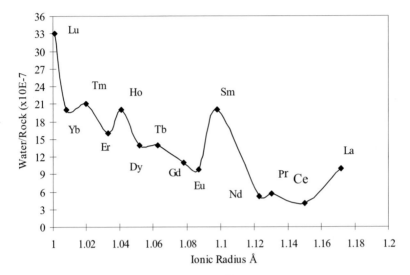

Figure 5. Solubility of REEs, as Water/Rock (x10^{-7}) in carbonate water (pH 7-8) versus ionic radius for the (+3) ions

3. Conclusions

This study focused on comparing the performance of rock digestion techniques including a commonly used fusion method (lithium metaborate flux) and three microwave oven methods, coupled with ICP-MS analysis to measure 56 different trace elements in three different aquifer type rocks from southern Nevada and in the USGS W-2 standard rock.   In addition, a comparison of the concentrations of 54 trace elements between ground waters and the associated rock was also carried out.   As a result of these studies, the following conclusions have been drawn:

(1) The lithium metaborate fusion technique with the ICP-MS can be used successfully to quantify the REEs and other trace elements such as Be, V, Sr, U, Sb, Hf, Nb, and Ta in sedimentary rocks (i.e., limestone, quartzite, and shale).   Unfortunately, the high temperatures of the fusion method prevent the accurate determination of other trace elements like Pb, Cr, Co, Ni, Cu, Zn and Gd.
(2) Of the microwave techniques, Methods 2 and 3 produced excellent results for the USGS reference standard, and consistent results for the three rocks from southern Nevada, whereas Method 4 generally gave unsatisfactory results. Overall, the microwave oven Methods 2 and 3 performed better than the fusion method

(3) For the three sedimentary rocks analyzed from southern Nevada, the shale sample generally exhibits the highest trace element concentrations, followed by the quartzite and limestone.

(4) The concentrations of the trace elements in carbonate waters collected from springs in Ash Meadows and Death Valley were compared to the limestone rock=s chemical composition. The results show that the carbonate waters are approximately two to seven orders of magnitude lower the concentrations of the 54 trace elements. The ten most soluble elements are Mo, Li, Cs, Sb, U, Ge, Sr, Rb, W and Ba. The least soluble elements are Al, Th, La, REEs and Zr. There is a steady increasing trend in solubility of the REEs, excluding Sm, with atomic number or decreasing ionic radii.

Acknowledgments

This work was supported by the U.S. Department of Energy (cooperative Agreement DE-FC08-93NV11399). We are especially grateful to Mr. J. Tang for assistance with figure, table, and text formatting.

References

Banner, J.L. Wasserburg, G.J., Dobson, P.F., Carpenter, A.B. and Moore, C.H. 1989. Isotopic and Trace element constrains on the origin and evolution of saline ground waters from central Missouri. *Geochim. Cosmochim, Acta.* **53**, 383-398.

CEM Corporation 1991a. Microwave sample preparation system, Matthews, NC, U.S.A.

CEM Corporation, 1991b. Operation Manual. Matthews, NC, USA.

Cook, N.J., Wood, S.A. and Zhang, Y. 1992. Transport and fixation of Au, Pt and Pd around the Lac Sheen Cu-Ni-PGE occurrence in Quebec, Canada. *J. Geochem. Explor.* **46**, 187.

CRC Press, 1985-1986. Handbook of Chemistry and Physics, 66th Edition, ed. R.C. Weast.

Drever, J.I. 1997. The geochemistry of natural water. Simon and Schuster, Upper Saddle River, NJ.

Dulski, P. 1994. Interferences of oxide, hydroxide and chloride analyte species in the determination of rare earth elements in geological samples by inductively coupled plasma- mass spectrometry. *Fresenlus J. Anal. Chem.* **350,** 194-203.

Dudley, Jr., W.W.and Larson, J.D. 1976. Effect of irrigation pumping on desert pupfish habitats in Ash Meadows, Nye County, NV. U.S. Geol. Surv. Prof. Pap. 927, 52 pp.

Eakin, T.E. 1966. A regional interbasin groundwater flow system in the White River area, southeastern Nevada. Water Res. 2, 251-271.

Farnham, I.M., Stetzenbach, K.J., Singh, A.K., and Johannesson, K.H. 2000. Deciphering groundwater flow system in Oasis Valley, Nevada, using trace element chemistry, multivariate statistics, and geographical information system. Mathematical Geology, vol. 32, No. 8, 943-968.

Feldman, C. 1983. Behavior of trace refractory minerals in the lithium metaborate fusion-acid dissolution procedure. *Anal. Chem.* **55,** 2451-2453.

Frape, S.K., Fritz, P., and Nutt, R.H. 1984. Water-rock interaction and chemistry of groundwaters of the Canadian Shield. *Geochim. Cosmochim. Acta.* **48**, 1617-1627.

Garrels, R.M. and MacKenzie, F.T. 1967. Origin of the chemical composition of some springs and lakes, equilibrium concepts in natural water systems. Am. Chem. Soc. Adv. Chem. Ser. V. 67, pp. 222-242.

Gilman, L.B. and Engelhart, W.G. 1989. Recent advances in microwave sample preparation. Spectroscopy V. 4, No. 8, 14-21.

Gosselin, D.C., Smith, M.R., Lepel, E.A., and Laul, E.A. 1992. Rare earth elements in chloride-rich ground water, Palo Duro Basin Texas USA. *Geochim. Cosmochim. Acta.* **56**, 1495-1505.

Gromet, L.P. and Silver, L.T. 1983. Rare earth element distributions among minerals in a granodiorite and their petrogenetic implications, *Geochim. Cosmochim.Acta.* **47**, 925-939.

Guo, C.X. 1996. Determination of fifty-six elements in three distinct types of geological materials by inductively coupled plasma-mass spectrometry. M.S. thesis, University of Nevada Las Vegas.

Hanson, G.N. 1980. Rare earth elements in petrogenetic studies of igneous systems, *Ann. Rev. Earth Planet. Sci.* **8**, 371.

Hem, J.D. 1985. Study and interpretation of chemical characteristics of natural waters. U.S. Geol. Survey. Water Supply Pap. 2254, 3rd ed.

Hodge, V.F., Johannesson, K.H., and Stetzenbach, K.J. 1996. Rhenium, molybdenum, and uranium in groundwater from the southern Great Basin, USA: Evidence for conservative behavior, *Geochim. Cosmochim. Acta.* **60**, 3197-3214.

Hodge, V.F., Stetzenbach, K.J., and Johannesson, K.H. 1998. Similarities in the chemical composition of carbonate groundwaters and seawater. *Environ. Sci. Technol.* **32**, 2481-2486.

Hurlbut, Jr., C.S., and Klein, C. 19777 Manual of Mineralogy, 19th Edition, John Wiley and Sons, New York.

Jarvis, I. and Jarvis, K.E. 1992. Plasma spectrometry in the earth sciences: techniques, applications and future trends. *Chem. Geol.* **95**, 1-33.

Jarvis, K.E., Gray A.L., and Mcmurdy, E. 1989. Avoidance of spectral interference on europium in inductively coupled plasma mass spectrometry by sensitive measurement of the doubly charged ion. *J. Anal. At. Spectrom.* **4**, 743-747.

Jarvis, K.E. 1988. Inductively coupled plasma mass spectrometry: A new technique for the rapid or ultra-trace level determination of the rare-earth elements in geological materials. *Chem. Geol.* **68**, 31-39.

Jarvis, K.E. 1990. A critical evaluation of two sample preparation techniques for low-level determination of some geologically incompatible elements by inductively coupled plasma-mass spectrometry. *Chem. Geol.* **83**, 89-103.

Jenner, G.A., Longerich, H.P., Jackson, S.E., and Fryer, B.J. 1990. ICP-MS - A powerful tool for high precision trace-element analysis in Earth sciences: Evidence from analysis of selected USGS reference samples. *Chem. Geol.* **83**, 133-148.

Johannesson, K.H., Lyons, W.B., Fee, J.H., Gaudette, H.E., and McArthur, J.M. 1994. Geochemical processes affecting the acidic ground waters of Lake Gilmore, Yilgarn Block, Western Australia: A preliminary study using neodymium, samarium, and dysprosium. *J. Hydrol.* **154**, 271-289.

Johnnesson, K.H., Stetzenbach, K.J., Hodge, V.F. and Lyons, W.B. 1996a. Rare earth element complexion behavior in circum-neutral pH groundwater: Assessing the role of carbonate and phosphate ions. *Earth Planet. Sci. Lett.* **139**, 305-319.

Johannesson, K., Stetzenbach, K.J., and Hodge, V.F. 1996b. Speciation of the rare earth element neodymium in ground water of the Nevada Test Site and Yucca Mountain and implications on actinide solubility. *J. Hydrol.* **178**, 181-204.

Johannesson, K.H., Stetzenbach, K.J., Hodge, V.F., Kreamer, K., and Zhou, X. 1997a. Delineation of groundwater flow systems in the southern Great Basin using aqueous rare earth element distributions. *Ground Water.* **35**, 807-819.

Johannesson, K.H., Stetzenbach, K.J. and Hodge, V.F. 1997b. Rare earth-elements as geochemical tracers of regional groundwater mixing. *Geochim. Cosmochim. Acta.* **61**, 3605-3618.

Johannesson, K.H., Zhou, X., Guo, X., Stetzenbach, K.J., and Hodge, V.F. 2000. Origin of rare earth element signatures in groundwaters of circumneutral pH from Southern Nevada and Eastern California, USA. *Chem. Geol.* **164**, 239-257.

Kreamer, D.K., Hodge, V.F., Rabinowitz, I., Johannesson, K.H., and Stetzenbach, K.J. 1996. Trace elements geochemistry in water from selected springs in Death Valley National Park, California. *Ground Water* **34**, 95-103.

Long, S.E., and Martin, T.D. 1991. Determination of trace elements in waters and wasters by inductively coupled plasma-mass spectrometry. U.S. Environmental Protection Agency.

Longerich, H.P., Fryer, B. J., Strong, D.F. and Kantipuly, C.J. 1987. Effects of operation conditions on the determination of the rare earth elements by inductively coupled plasma mass spectrometry (ICP-MS). *Spectrochimica Acta.* **42B**, 75-92.

Nadkarni, R.A. 1984. Applications of microwave oven sample dissolution in analysis. *Anal. Chem.* **56**, 2233-2237.

Perkin Elmer 1992. Reference manual ELAN 5000 inductively coupled plasma-mass spectrometer. Norwalk, Connecticut, U.S.A.

Sholkovitz, E.R. 1988. Rare earth elements in the sediments of the North Atlantic Ocean, Amazon Delta, and East China Sea: Reinterpretation of terrigenous input patterns to the ocean. *Am. J. Sci. v.* **288**, 236-281.

Smedley, P.L. 1991. The geochemistry of rare earth elements in ground water from the Carnmenellis area, Southwest England. *Geochim. Cosmochim. Acta.* **55**, 2767-2779.

Starik I. E., and Kolyadin, L. B. 1957. The occurrence of uranium in ocean water. *Geochemistry* **3**, 245-256.

Stetzenbach, K.J., Amano, M., Kreamer, D.K., and Hodge, V.F. 1994. Testing the limits of ICP-MS: Determination of trace elements in ground water at the part-per-trillion level. *Ground Water* **32**, 976-985.

Stetzenbach, K.J., Farnham, I.M., Hodge, V.F., and Johannesson, K.H. 1999. Using multivariate statistical analysis of groundwater flow in a regional aquifer. *Hydrol. Processes.* **13**, 2655-2673.

Stetzenbach, K,J., Hodge, V.F., Guo, X., Farnham, I.M., and Johannesson, K.H. 2001. Geochemical and statistical evidence of deep carbonate groundwater within overlying volcanic rock aquifers/ aquitards of Southern Nevada, USA. *J. Hydrol.* **243**, 254-271.

Thomas, J.M., Welch, A.H., and Preissler, A.M. 1989. Geochemical evolution of ground water in Smith Creek Valley-A hydrologically closed basin in central Nevada, USA. *Appl. Geochem. V.* **4**, 493-510.

Totland, M.M., Jarvis, I., and Jarvis, K.E. 1992. An assessment of dissolution techniques for the analysis of geological samples by plasma spectrometry. *Chem. Geol.* **95,** 35-64.

Totland, M.M., Jarvis, I., and Jarvis, K.E. 1995. Microwave digestion and alkali, fusion procedures for the determination of the platinum-group elements and gold in geological materials by Welch, A.H., Lico, M.S.and Hughes, J.L. 1988 Arsenic in ground water of the western United States. *Ground Water.* **26,** 333-347.

White, A.F., Classen, H.C., and Benson, L.V. 1980. The effect of dissolution of volcanic blass on the water chemistry in a tuffaceous aquifer, Rainier Mesa, Nevada. U.S. Geol. Surv. Water Supply Pap. 2535, 34 pp.

Winograd, I.J., and Friedman, I. 1972. Deuterium as a tracer of groundwater flow, southern Great Basin, Nevada and California. *Geol. Soc. Am. Bull.* **83,** 3691-3708.

Winograd, I.J. and Thordarson, W. 1975. Hydrogeologic and hydrochemical framework, south-central Great basin, Nevada, California, with special reference to the Nevada Test Site, U.S. Geol. Surv. Prof. Pap. 712-C, 125 pp.

Winograd, I.J. and Pearson, R.J. 1976. Major carbon-14 anomaly in a regional carbonate aquifer: possible evidence for megascale channeling, south central Great Basin. *Water Resours.* **Res.** 12, 1125-1143.

Wood, S.A. and Vlassopoulos, D. 1990. The dispersion of Pt, Pd and Au in surficial media about two PGE-Cu-Ni prospects in Quebec. *Can. Mineral.* **28,** 649.

Wood, S.A., Mountain, B.A., and Pan, P. 1992. The aqueous geochemistry of platinum, palladium, and gold: Recent experimental constraints and a re-evaluation of theoretical predictions. *Can. Mineral.* **30,** 955.

Chapter 3

RARE EARTH, MAJOR, AND TRACE ELEMENT GEOCHEMISTRY OF SURFACE AND GEOTHERMAL WATERS FROM THE TAUPO VOLCANIC ZONE, NORTH ISLAND NEW ZEALAND

ROBYN E. HANNIGAN

Arkansas State University, Department of Chemistry, PO Box 419 State University, Arkansas, USA

Abstract

Geothermal fluids, surface water, and sediments were collected from the Taupo Volcanic Zone (TVZ) on the North Island of New Zealand. This region is characterized by hydrothermal activity with groundwaters feeding Lake Taupo, Waikato River, and the Wairakei and Wai-O-tapu hydrothermal fields of Rotorua.   pH values ranged from near neutral (6.5-7.2) to alkaline (7.3-8.5). Surface water temperatures were between 15-16°C, while geothermal fluid temperatures ranged between 22 and 90°C. The major element chemistry of these waters describes the sample populations as geothermal and rhyolitic-sourced with some samples representing a mixture of these end-members. Using the rare earth element (REE) composition of the geothermal and surface waters, the competitive influences of mixing and water-rock interactions on fractionation were explored. Comparison of the REE composition of geothermal fluids, surface water, and sediments/ precipitates, indicate that temperature has a strong effect on REE fractionation in the sampled fluids. Eu- and Ce-anomalies (Eu/Eu*, Ce/Ce*) are visible in several samples and are related both to the inheritance of Eu and Ce abundances from geologic host materials and fractionation of these elements during water-rock interactions. These two anomalies are inversely related in the waters and sediments with sediments showing a positive relationship between Eu/Eu* and Ce/Ce* and waters showing an equally significant negative correlation between these two variables.   The unique relationships between the Eu- and Ce- anomalies are indicative of the chemistry of the fluids and the nature of their interaction with the aquifer material such that precipitation of material during cooling led to the development of negative Eu-anomalies under reducing conditions. The REE, trace element and major element data demonstrate that competitive processes, including water-rock interaction at depth with aquifer wall-rock, differential leaching of REE from mineral phases and precipitation of secondary minerals from geothermal fluids result in highly evolved fluids discharging into the surface waters of the TVZ.

1. Introduction

Rare earth elements (REE) are uniquely suited to the study of a variety of geological processes. The distinctive chemistry of the REE across the series from La to Lu is applied to sediment provenance (Taylor and McLennan, 1985; Wombacher and Muenker,

K.H. Johannesson,(ed), Rare Earth Elements in Groundwater Flow Systems , 67-88.
© 2005 *Springer. Printed in the Netherlands.*

2000; Ding et al., 2001; Nyakairu and Koeberl, 2001), chemical weathering (Morey and Setterholm, 1997; Viers et al., 1997; Land et al., 1999; Hannigan and Sholkovitz, 2001) and, groundwater chemistry (Smedley, 1991; Johannesson et al., 1996; Johannesson et al., 1997a,b; Johannesson et al., 1999; Dia et al., 2000). The relative abundance of the REE in sediments, geothermal fluids, and surface waters is used here to define the competitive processes controlling REE fractionation in a suite of samples collected within the Taupo Volcanic Zone (TVZ), New Zealand (Figure 1).

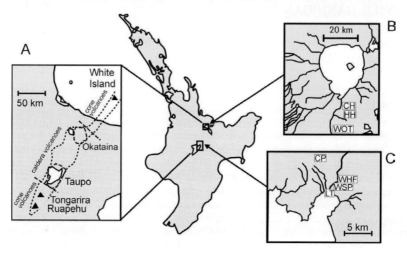

Figure 1. Map showing sample locations. A. Taupo Volcanic Zone. Samples were collected from two volcanic centers, Taupo and Rotorua-Okataina. B. Map showing sample locations in the Rotorua-Okataina center (WOT – Wai-O-tapu, HH – Hot Hole, CH – Cold Hole; for sample details see Table 1). C. Sample locations from the Taupo center (LT – Lake Taupo, WSP – Waikato at Spa Park, WHF – Waikato above Huka Falls; for sample details see Table 1).

River REE studies demonstrate that chemical weathering of the continents leads to fractionation between the dissolved REE composition of the water and the watershed bedrock (Elderfield et al., 1990; Goldstein and Jacobsen, 1987; 1988a, b; Sholkovitz, 1992, 1995). REE geochemistry of the Waikato River within the TVZ shows significant fractionation relative to other world rivers (Figure 2). The overall pattern of the Waikato is similar to the HREE enriched pattern of average global river water (Goldstein and Jacobsen, 1987, 1988b) and to that of the Mississippi. The Eu-anomaly (Eu/Eu* - defined as $Eu_N/(Sm_N*Gd_N)^{0.5}$, where N is the chondrite-normalized value) as well as the lower overall REE concentrations, however, underscores the importance of water-rock interactions as a significant control on REE composition.

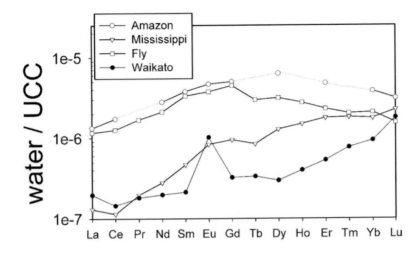

Figure 2. Representative REE geochemistry of some world rivers. Data are normalized to upper continental crust to highlight the fractionation of the REE. Upper continental crust values are from Taylor and McLennan (1995). The Waikato River shows significant fractionation relative to other world rivers. The overall pattern of the Waikato is similar to the HREE enriched pattern of average global river water (Goldstein and Jacobsen, 1987, 1988b) and to that of the Mississippi with the exception of the positive Eu/Eu* of the Waikato, which is the result of preferential removal of Eu from plagioclase during water-rock interactions at depth.

To date there has not been a study of the relative influence of the rock type on the REE geochemistry in this region. Previous geochemical studies focus on the assessment of inputs of alkali metals (e.g., Schouten, 1983; Timperley, 1983; Timperley and Huser, 1994; Eser and Rosen, 1999) to the region's fresh water and geochemical structure of geothermal zones (Giggenbach et al., 1994; Simmons et al., 1994). Graham (1992) showed that the $^{87}Sr/^{86}Sr$ ratios of primary alkali chloride waters are higher than the aquifer rocks, necessitating significant water-rock interaction in the central and western portions of the Rotorua geothermal field (Figure 1B). Li and Rb data of the Waikato River indicate that, in addition to direct injection of geothermal fluids, there is significant water-rock interaction contributing to the chemistry of TVZ surface waters (Bower and Timperley, 1988).

As well as comparing the REE chemistry of the geothermal and surface waters we utilize the relationships between the water and sediment chemistry to further elucidate the influence of rock type on water chemistry. Studies that use the REE chemistry of groundwaters to reconstruct groundwater flow (Johannesson et al., 1996; Johannesson et al., 1997a, b; Johannesson et al., 1999; Dia et al., 2000) focus primarily on the water chemistry with less attention to specific water-rock interactions. The TVZ provides a unique setting for the study of both the relative contributions of REE from geothermal water to surface waters and the degree of REE fractionation during water-rock interaction.

1.1 GEOLOGIC SETTING

In the Taupo Volcanic Zone (TVZ), geothermal and cold spring waters discharge into Lake Taupo in the south as well as along the Waikato River and to the north into Lake Rotorua. The TVZ extends from Mt. Ruapehu (Figure 1A) 300 km northeast to White Island (Cole, 1979). Within the TVZ is a 15-40 km wide band containing 5 volcanic centers. This region has been extensively down-faulted and contains many caldera/crater lakes (Healy, 1975). Andesite volcanoes dominate the southern Tongariro volcanic center. The northern Taupo and Okataina-Rotorua centers are dominated by rhyolite and ignimbrite (welded and ash) (Grange, 1937).

Within the TVZ precipitation provides only ~ 20% of the recharge to the surface waters (Rutherford, 1984; Rutherford et al., 1987). A significant amount to geothermal fluid discharges into TVZ lakes and streams (Timperley and Huser, 1994).

2 Methods

2.1 SAMPLE COLLECTION

Water samples (Table 1) were collected at geothermal springs (T < 110°C) and fresh water sites as part of a broader study to characterize hydrothermal flows and discharges in the TVZ. Water samples were collected using trace metal clean procedures (Hannigan and Basu, 1998; Shiller et al., 2001) and were collected and analyzed in triplicate. Samples were collected using a peristaltic pump equipped with clean Teflon tubing. Water was pumped from at least 0.5 m below the surface directly into a pre-cleaned polycarbonate-filtering device and then filtered through 0.45 μm (Millipore®) membrane filters. Before sampling the filtering device was flushed with 250 mL of sample. In addition, filtered water was used to rinse the sample bottle three times before the filtered sample was collected. Sample splits were acidified in the field to pH <2 using ultra-pure HNO_3 for trace metal and REE analysis. Water samples were then placed in clean plastic bags and then in a clean plastic chest before transportation back to the lab. Unfiltered, unacidified samples were also collected for anion measurements. All samples were stored at 4°C and shipped to the US on blue ice (no more than 1 week after collection). At each site sediment, precipitate, and rock samples were also collected.

2.2 ANALYTICAL METHODS

Sediment and rock samples were prepared for high-resolution inductively coupled plasma mass spectrometry analysis using the procedure outlined in Hannigan and Basu (1998). Sediment REE concentrations were measured using the Finnigan MAT Element 2 magnetic sector inductively couple mass spectrometer (ICP-MS) at Old Dominion University.

SITE	Site description	pH	T °C
LT	LT - Lake Taupo at foreshore station 1 (Environment Waikato monitoring station)	7.38	16.2
WHF	WHF - Waikato River 0.25 km upstream from Huka Falls. Water in contact with rhyolitic ash and welded ignimbrite	6.79	15.8
CP	CP- Wairakei geothermal spring. High total suspended solid content (chocolate).	7.22	52.1
HH	HH - High temperature geothermal spring at Sulphur Bay 0.5 km south of Lake Rotorua. HH is lined by sinter and geothermal sands (realgar and orpiment).	8.43	90.3
CH	CH- lower temperature geothermal spring at Sulphur Bay 0.5 km south of Lake Rotorua and 6 meters from HH. CH is also lined with sinter but contained no geothermal sands.	8.11	62.1
WSP	WSP - Waikato River at Spa Park. Cold geothermal spring discharges into bed of Waikato.	7.48	45.3
WOT	WOT - Wai-O-tapu thermal park champagne pool. Significant sinter, realgar and orpiment deposition.	8.47	74.2

Table 1. Sample description information, pH and temperature data.

All sample preparation was performed in a class 100 clean laboratory at Arkansas State University. Fifty mg of sediment (hand crushed to <60 mesh) was dissolved in 3 mL ultra pure HNO_3 and 1mL ultra pure HF in Teflon bombs on a hot plate for 24 hours. The lids were then removed and samples dried completely at low temperature (40°C) in a laminar flow box. Samples were then brought into solution using 1 mL ultra pure concentrated HNO_3. The liquid was then transferred to 100 mL PTFE bottle and brought to 99 mL in 18.2 MΩ-cm water. Samples were diluted 1:100 prior to the addition of 2 ppb [115]In as an internal standard. Instrument configuration included the MCN100 nebulizer. External precision was calculated using external standardization based on measurements of a range of elemental standards (1 ppb, 10 ppb, 100 ppb, 1 ppm and 10 ppm). Accuracy was determined by measuring the U.S.G.S. shale standard SDO-1 (Ohio Devonian Oil Shale) as an unknown. The measured SDO-1 values are generally within 0.5% of the reference value for this standard (Potts *et al.*, 1992) with no observed systematic differences.

The Cl⁻ and SO_4^{2-} content of the waters was measured by Ion Chromatography (DX 120). The major, trace, and REE composition of the waters were measured using the Finnigan MAT Element magnetic sector inductively coupled mass spectrometry (ICP-MS) at Woods Hole Oceanographic Institution. Prior to spiking with an internal standard (1 ppb ^{115}In), sample splits for major element analysis were diluted 1:100. Potassium was measured at high mass resolution, Ca and Mg were measured at medium mass resolution, and Na and the remaining trace elements and REE were measured at low mass resolution without dilution in both counting and analog modes. Samples were aspirated directly into the magnetic sector ICP-MS using an MCN100 nebulizer in tandem with a guard electrode. The guard electrode enhances signal and decreases oxide interferences, particularly BaO and lanthanide oxides. External precision was calculated using external standardization based on measurements of a range of major, trace and REE standards (1 ppt, 10 ppt, 50 ppt, 100 ppt, 500 ppt and 1 ppb). La, Ce, Pr, Nd, Sm, Eu, and Dy concentrations are determined to better than 1%, while Gd, Tb, Ho, Er, Tm, Yb and Lu are determined to better than 2%. Accuracy was determined based on NIST 1643d (water) run as an unknown. The measured NIST 1643d values are generally within 0.12% of the reference value for this standard (NIST) with no observed systematic differences. No chemistry blanks are associated with the liquid samples aside from that attributable to the filtration process. Filtration blanks were too low to calculate. Analytical blank data are shown in Table 2.

3 Results

Table 2 contains the major element, anion, trace, and rare earth element data for the water samples and sediments. The degree to which the chemistry of the waters is characterized by the underlying bedrock lithology is shown on a tri-linear plot of cation data (Figure 3). Values for geothermal water (Gibbs, 1979), cold springs (Timperley, 1983), and average lake water (White et al., 1980) are also plotted on Figure 3. Overlain on Figure 3 are the fields defined by a variety of andesitic, rhyolitic and geothermal waters (Rosen and Coshell, 1998).

Rocks (Figure 4), which typify the eruptive units within the TVZ, are typical of chondrite normalized island arc volcanics (Houghton and Wilson, 1998). Neither the rhyolite nor andesite has a discernible Eu/Eu*. The scoria possesses a slight negative Eu/Eu* (< 1) indicative of feldspar removal during magma differentiation (Bence et al., 1980). Lake sediment from Taupo (LT-sed) shows slight LREE enrichment relative to chondrite with no Eu/Eu*. Compared to the Taupo rhyolite (TR) the sediment (TR-sed) chondrite-normalized REE abundances are nearly identical. The REE abundances of the Waikato river sediment (WHF-sed) are similar to Lake Taupo, with the exception of a pronounced negative Eu/Eu*. The chondrite-normalized pattern of the Tarawera scoria (TS) from the Okataina-Rotorua region (Figure 1) possesses a nearly flat pattern. Geothermal sediments from sites within this region show significant variability when normalized to

Sample Name	Cl⁻	SO₄²⁻	Na	Mg	K	Ca	Ba	Sr	Y	Zr	U	Th	Pb	La	Ce
LT	9.26	5.47	14.2	2.780	2.01	5.76	14.6	817	0.160	2.5	0.063	0.049	7.64	201	385
WHF	9.15	5.84	16.7	3.120	2.46	6.81	14.1	825	0.100	0.190	0.057	0.021	7.13	5.94	9.29
WSP	15.6	7.40	22.6	0.891	1.31	1.56	147	2395	1.09	0.160	0.053	0.025	15.1	45.2	96.2
CP	821	39.4	648	0.451	23.1	12.4	211	2445	1.035	0.721	0.030	0.027	16.6	1743	5203
HH	1280	47.1	1102	0.221	174	21.1	127	1264	3.44	0.900	0.015	0.011	10.8	260	422
CH	580	29.8	890	0.430	9.16	2.49	113.8	1115	0.548	0.312	0.039	0.018	5.10	84.7	181
WOT	1180	42.1	1115	0.192	74.6	19.4	147	1197	2.29	0.841	0.011	0.009	8.50	894	2714
Water Analytical Blank	nd	nd	0.001	0.012	0.016	0.042	0.031	0.011	0.005	0.005	0.001	nd	0.023	2.00	3.18
NIST 1643d measured			19.20	7.75	2.10	29.2	481.2	259.4					17.42		
NIST 1643d certified			22.07	7.989	2.356	31.04	506.5	294.8					18.15		
LT- sed														41.6	96.4
WHF - sed														23.5	48.7
WSP - sed														14.5	40.8
WSP-soil														20.4	19.5
CP - sed														48.2	91.2
HH - sed														23.9	61.2
CH - sed														5.66	12.4
WOT - sed														8.95	17.5
WOT - wht sand														0.352	0.662
WOT - realgar														21.15	37.68
TR - Taupo Rhyolite														30.9	59.8
RA - Ruapehu Andesite														8.56	20.7
TS - Tarawera Scoria														11.1	25.2
Rock Analytical Blank														0.289	0.443
SDO-1 measured														38.5	79.2
SDO-1 certified														38.5	79.3

SDO-1 measured values average of 15 measurements taken during water and rock runs; certified values Potts et al., 1992.

NIST 1643d measured values average of 15 measurements taken during water run; certified values NIST

Analytical Blanks - 1% ultrapure HNO₃ and 18.6 M   water

Table 2. Major, trace and rare earth element data for water and sediment/rock samples. Mean values of triplicate analyses are shown. RSD values are less than 0.5%. Anion data in g/m³, water trace and REE data in parts-per-trillion, sediment/rock values in parts-per-million. Precip- precipitate from geothermal fluid.

Sample Name	Pr	Nd	Sm	Eu	Gd	Tb	Dy	Ho	Er	Tm	Yb	Lu	Eu/Eu*	Ce/Ce*
LT	47.0	181	36.7	18.5	46.9	5.33	28.8	6.02	15.1	2.11	13.6	2.23	1.35	-0.044
WHF	1.30	5.16	0.963	0.898	1.23	0.214	1.04	0.320	1.23	0.253	2.06	0.576	2.50	-0.128
WSP	13.9	75.4	19.7	15.4	28.1	4.52	30.2	9.01	27.3	4.36	30.9	6.66	1.98	-0.062
CP	1090	5844	1689	424	1622	239	1298	236	648	95.9	635	101	0.778	-0.031
HH	52.8	231	58.0	29.5	88.9	14.2	100	25.7	83.6	12.4	88.7	17.6	1.25	-0.114
CH	21.7	89.0	18.3	6.17	25.3	4.77	27.7	5.94	17.0	2.82	23.1	3.71	0.871	-0.008
WOT	552	2960	854	280	825	122	664	123	338	50.1	333	53.8	1.01	-0.021
Water Analytical Blank	0.308	1.51	0.351	0.086	0.390	0.063	0.306	0.236	0.180	0.053	0.129	0.021		
NIST 1643d measured														
NIST 1643d certified														
LT - sed	8.45	31.8	6.13	2.24	6.45	0.939	5.25	0.965	2.97	0.453	3.28	0.477	1.08	0.053
WHF - sed	4.27	19.4	3.54	0.795	3.78	0.522	2.71	0.597	2.00	0.364	2.54	0.426	0.661	-0.001
WSP - sed	2.53	10.1	2.14	1.16	2.59	0.394	2.12	0.447	1.62	0.268	1.80	0.281	1.49	0.143
WSP-soil	8.93	16.6	1.95	7.59	1.46	0.491	1.63	0.266	1.63	0.350	0.95	0.177	0.006	0.032
CP - sed	6.56	22.7	3.80	0.811	4.49	0.633	3.60	0.790	2.65	0.433	2.91	0.445	0.596	-0.006
HH - sed	4.64	18.4	3.66	0.937	3.93	0.645	3.25	0.766	2.31	0.372	2.60	0.355	0.751	0.096
CH - sed	1.20	5.01	1.20	0.205	1.34	0.230	1.21	0.289	0.912	0.145	1.06	0.172	0.490	0.016
WOT - sed	1.53	6.18	1.34	0.361	1.49	0.268	1.57	0.363	1.10	0.177	1.23	0.189	0.774	-0.014
WOT - wht sand	0.074	0.277	0.072	0.017	0.070	0.012	0.063	0.014	0.040	0.007	0.051	0.009	0.714	-0.040
WOT - realgar	5.11	18.08	3.460	0.70	2.901	0.252	0.933	0.141	0.288	0.042	0.35	0.047	0.667	-0.070
TR - Taupo Rhyolite	7.02	29.7	6.42	1.76	5.63	0.871	5.19	1.23	3.52	0.484	3.34	0.513	0.889	-0.043
RA - Ruapehu Andesite	2.61	11.7	2.52	0.901	2.61	0.487	3.15	0.721	2.07	0.312	2.16	0.326	1.07	0.018
TS - Tarawera Scoria	2.90	12.6	2.90	0.942	2.75	0.478	2.83	0.630	1.81	0.266	1.81	0.267	1.01	0.011
Rock Analytical Blank	0.042	0.217	0.048	0.012	0.064	0.009	0.043	0.033	0.026	0.008	0.018	0.003		
SDO-1 measured	8.88	36.6	7.67	1.60	7.39	1.20	6.01	1.20	3.60	0.449	3.40	0.538	0.644	-0.016
SDO-1 certified	8.90	36.6	7.70	1.60	7.40	1.20	6.00	1.20	3.60	0.450	3.40	0.540	0.644	-0.016

$Eu/Eu* = Eu_N / (Sm_N * Gd_N)^{0.5}$

$Ce/Ce* = 2[Ce]/([La]+[Pr])$

Table 2. Continued

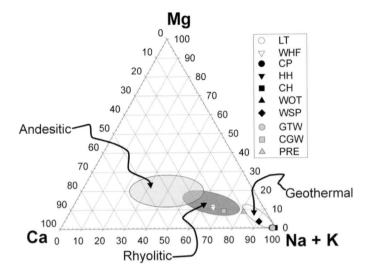

Figure 3. Tri-linear cation diagram showing the major cations in solutions from the TVZ. Also shown are the andesitic, rhyolitic and geothermal fields defined by Rosen and Coshell (1998). None of the TVZ samples in this study fall within the andesitic field. Samples from the Taupo center fall within the rhyolite field with high temperature samples from the Rotorua-Okataina center plotting in the geothermal field.

chondrite, as well as when normalized to the Taupo rhyolite (TR). When these data are normalized to the Tarawera scoria (TS) or to Ruapehu andesite (RA) the variation is minimized, although the significant Eu/Eu* still persist.

Figures 5A and B show the chondrite-normalized patterns of the water and sediment samples. The surface water samples LT and WHF retain the sediment pattern, with the exception of the sign of the Eu/Eu* in the WHF sample that is positive in the water but negative in the sediment. This inverse relation will be addressed below. Geothermal fluids have chondrite-normalized patterns quite unlike the sediment patterns. Specifically, Chocolate Pot (CP), Hot Hole (HH) and Wai-O-tapu (WOT), the three highest temperature samples (Table 1), show significant REE fractionation particularly in the LREE (La to Sm) as well as in the nature of the Eu- and Ce- anomalies (defined as Ce/Ce*=2[Ce]/([La]+[Pr])). Lower temperature geothermal waters also show REE fractionation, particularly in the LREE.

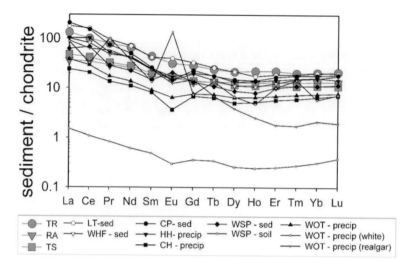

Figure 4. Chondrite-normalized REE abundance patterns for TVZ sediments. Chondritic values from Anders and Grevesse (1989). Rock samples Taupo Rhyolite – TR, Ruapehu Andesite – RA and Tarawera Scoria (TS) have chondrite-normalized patterns typical of island arc volcanics (Houghton and Wilson, 1998). Abundance patterns of sediments collected from the sites shown in Figure 1 show REE abundance patterns similar to the rocks with the exception of variable Eu/Eu* and Ce/Ce* which result from competitive processes of weathering and secondary mineral precipitation.

4 Discussion

In order to assess the relative influence of water-rock interactions on the REE composition of geothermal and surface waters from within the TVZ, the waters are normalized to their associated sediments (Figure 6). Solutions from geothermal fields such as the TVZ show significant REE enrichments relative to average groundwater (Goff and Grisby, 1982; Michard, 1989). Although Lake Taupo (LT) showed LREE fractionation when normalized to chondrite the degree of fractionation disappears when normalized to the associated LT sediment. Overall LT water retains the sediment character similar to that of Taupo rhyolite (TR).

4.1 Eu/Eu* AND Ce/Ce* ANOMALIES

Most interesting is the negative Ce/Ce* (-) and positive Eu/Eu* (> 1) of the water samples and sediments/precipitates (Table 2). These data indicate the effects of secondary mineral precipitation on the REE. The negative Ce/Ce* in Lake Taupo (LT) water and other water samples is indicative of the oxidation of Ce(III) to Ce(IV) with Ce precipitated, potentially, as CeO_2 at the pH range of these samples (Brookins, 1988; deBaar et

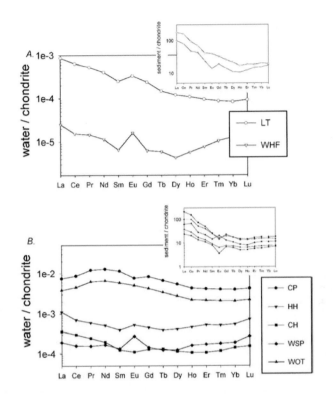

Figure 5. A. Chondrite-normalized patterns for the surface water samples (Table 1). Chondritic values from Anders and Grevesse (1989). Also shown are the sediment patterns. The surface water samples LT and WHF retain the sediment pattern with the exception of the sign of the Eu/Eu* in the WHF sample that is positive in the water but negative in the sediment. This points to a unique relationship between sediment and water Eu/Eu* to be addressed below. B. Chondrite-normalized patterns for the geothermal fluids (Table 1). Chondritic values from Anders and Grevesse (1989). Also shown are the sediment patterns. Lower temperature geothermal waters also show REE fractionation particularly in the LREE. The three highest temperature samples (Table 1) show significant REE fractionation particularly in the LREE (La to Sm) as well as in the nature of the Eu- and Ce- anomalies.

al., 1988; Brookins, 1989; Braun et al., 1990). Several water samples show positive Eu/Eu* and their sediment/precipitate counterparts possess negative Eu/Eu*. Unlike the Ce/Ce*, the Eu/Eu* is related to the dominance of Eu^{3+} across Eh-pH ranges. The addition of Eu^{3+} to the water samples and the resultant positive Eu/Eu* is related to hydrothermal alteration of minerals from the aquifer material or sediment and the consequent increase in concentration of aqueous Eu^{3+}.

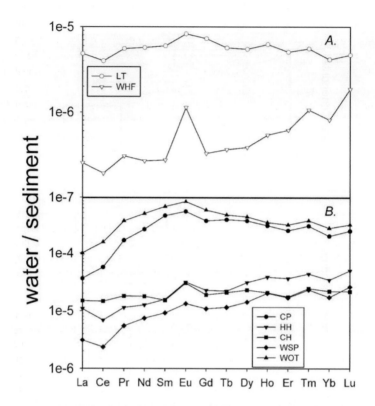

Figure 6. Sediment normalized REE patterns of the surface waters (A) and geothermal fluids (B). A. LT showed LREE fractionation when normalized to chondrite (Figure 6A). The extent of fractionation is lessened when normalized to the associated sediment. Overall surface waters retain the sediment character, which is similar in REE abundances to Taupo rhyolite. B. Several samples show positive Eu/Eu* in spite of the fact that their sediment counterparts possess negative Eu/Eu*. Solutions from geothermal fields such as the TVZ show significant REE enrichments relative to average groundwater (Goff and Grisby, 1982; Michard, 1989).

The relation between Ce/Ce* and Eu/Eu* change dramatically between the sediments and waters (Figure 7). The water samples show an inverse relationship between Ce/Ce* and Eu/Eu*. None of the geothermal fluids analyzed in this study represent deep reducing geothermal waters and are likely to be well oxidized. Therefore, the negative Ce/Ce* values of the geothermal fluids may reflect the oxidized character of the fluids.

At Lake Taupo (LT) and Spa Park (WSP), the sediments (LT-sed and WSP-sed) possess positive Ce/Ce*, which may reflect an inherited pattern from aquifer materials. These sediments are a mixture of volcanic material. Although the dominant rock types (RA, TS and TR) do not possess positive Ce/Ce*, the soils developed on top of the Taupo ash at Spa Park (WSP-soil) do have positive Ce/Ce* (average = 0.032). The contribution of surface run-off to the composition of the sediments at LT and WSP may account for the sediment having a positive Ce/Ce*. In the case of the precipitates in geothermal pools, several of the precipitates have negative Ce/Ce* values. Precipitates from Hot Hole (HH) and Cold Hole have slightly positive Ce/Ce*, which are, like those found in the waters of Spa Park (WSP), a combination of retention of aquifer signature and mixing of soil as well as precipitation of Ce-enriched phases.

The positive Eu/Eu* in sediments from the surface water sites (LT and WHF) is an inherited characteristic from the bedrock lithology (mixtures of TR, TS and RA). The negative Eu/Eu* in the geothermal precipitates may be related to hydrothermal alteration and removal of Eu from the fluids during precipitation of secondary minerals. It is possible that the negative Eu/Eu* is imparted by the geothermal fluids discharging into the bed of the Waikato, however, where geothermal fluid discharges into the Waikato at Spa Park (WSP-sed) the sediment has a positive Eu/Eu*.

Although the chemistry of Eu and Ce and the development of anomalies are relatively well understood (e.g., Taylor and McLennan, 1985; Brookins, 1989; and references therein), the specifics of the fractionation of the REE during hydrothermal alteration is less well understood. The inverse relationship of Eu/Eu* and Ce/Ce* between the waters and sediments, as well as the changes in REE abundances relative to chondrite between the sediments and waters, suggests competitive chemical processes resulting in REE fractionation. Mineral fractionation, either by hydrothermal alteration or precipitation, can lead to progressive changes in the REE abundances across the entire series. The fluids analyzed in this study are highly evolved with higher anomalies accompanying more fractionated REE patterns. The anomalous behavior of Eu has been noted in relation to hydrothermal alteration (Haas et al., 1995; Moeller, 1998; Irber, 1999, and references therein). Although the fractionation in the TVZ samples is not as extreme as those discussed by Irber (1999), the inverse relationship between the anomalies required we explore the possibility that aquifer wall rock type and inherited geochemistry are not solely responsible for the observed variations in REE chemistry of the TVZ fluids.

The majority of water samples have depleted or enriched Sr/Eu ratios relative to chondrite (~139; Anders and Grevesse, 1989). Sr/Eu values below that of chondrite indicate higher degrees of differentiation of the fluids related to the leaching of minerals from the wall rock and the precipitation of secondary minerals. Increasing Sr/Eu correlates with higher Eu/Eu* in the fluids, which suggests that, although Sr and Eu may decouple during fractionation of the parent magma (cf. Irber, 1999), the fluid Sr and Eu remain coupled. This indicates the importance of water-rock interactions (leaching of feldspars) in controlling the REE chemistry of the geothermal fluids.

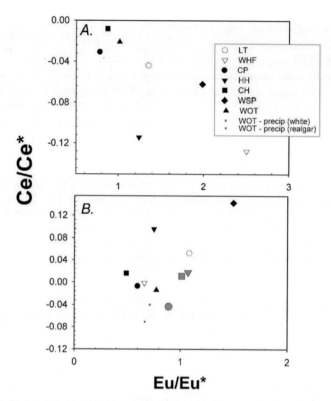

Figure 7. Ce/Ce* and Eu/Eu* in (A) water and (B) sed/precip. The water samples show an inverse relationship between Ce/Ce* and Eu/Eu*. Grey Circle – Taupo Rhyolite, Grey square – Ruapehu Andesite, Grey Triangle – Tarawera Scoria. The development of negative Ce/Ce* values of the waters reflects the oxidized character of the fluids. Several sediments possess positive Ce/Ce* which could result from precipitation of minerals from deeper fluids at an earlier time. As mentioned previously, the negative Eu/Eu* in the sediments is related to the removal of feldspar from the source magma and the inheritance of this pattern by river and lake sediment. Geothermal precipitates also possess negative Eu/Eu* that may be related to the hydrothermal alteration and removal of Eu from mineral fractions.

4.2 MIXING AND WEATHERING

The geothermal fluid influences on surface water and shallow groundwater composition are broad in nature and are idealized in Figure 8. Of the samples analyzed for this study we did not sample deep geothermal fluids, which are characterized by temperatures in excess of 250°C. The relations presented in Figures 3 and 8 are used to assess which elements will be most useful for assessing the relative contributions from water-rock interactions. The balance of metals within surface waters reflects weathering of aquifer wall rock by geothermal water discharging into river and lakebeds. Timperley (1987) demonstrated that geothermal waters contributed to the sulfate and chloride budgets of surface waters within the TVZ (Figure 9). The relationship between Ca and Cl in the waters indicates water-rock interactions occurring in the predominantly geothermal samples. The samples fall on a simple mixing line indicative of water-rock interactions with concentrations of Ca, K and Na increasing and Mg decreasing as temperature of the water increases (Table 3). As a result of water-rock interaction, Ca increases and this leads to precipitation of Ca-bearing minerals (Chon et al., 1998). This relationship also suggests the leaching of Ca-plagioclase and Ca-Mg minerals within the aquifer wall rock. The high concentration of K is explained by the weathering of K-feldspar, also common in the rocks found within the TVZ (Nördstrom et al., 1989).

The chemistry of the TVZ fluids can be described as mixtures of fresh and geothermal fluids. The chemistry of the geothermal fluids is, in turn, influenced by chemical weathering of minerals from the aquifer wall rock and secondary mineral precipitation from geothermal fluids. To further explore the geochemical variation between sample sites in the context of mixing we used the relationships between Ce and Pb (Figure 10). As Pb remains within the mineral during feldspar weathering preferentially to Ce, the weathering of plagioclase produce a negative correlation between Ce/Pb and Eu/Eu*. The data follow a general trend from low temperature fresh waters to higher temperature geothermal waters. This relationship suggests that the majority of geochemical variation in the water composition is related to mixing of fresh and geothermal end-members. Because the data do not fall on a discrete hyperbolic mixing curve, however, mixing is not the sole cause of geochemical variation between the sites. As suggested above, the differentiation in REE patterns between sites is related to a combination of mixing, precipitation, and water-rock interaction. The case of Ce and/or Pb conservation during mixing was further explored by comparing the Ce/Pb and Eu/Eu* values (Figure 10B). The relation in these data provides further evidence of the importance of water-rock interactions with increasing control on REE fractionation in the warmer geothermal samples.

5 Conclusions

The anomalous behavior of Ce and Eu as well as the degree of fractionation in REE patterns between waters and sediments, precipitates, and aquifer material indicate that the geothermal water composition results from competing processes with surface waters being less fractionated than higher temperature geothermal waters. The variations in Ce, Eu, and total REE concentrations are the result of inherited source rock signatures and water-rock interaction (sorption and precipitation). The negative Ce/Ce* are the result of the oxidation state of the waters rather than inherited from the source rock, which do not possess a cerium anomaly (Figure 5). Positive Eu/Eu* are the result of differential leaching of minerals from the aquifer wall-rocks. The negative Eu/Eu* in some of the geothermal fluids and associated precipitates (Table 3) may be the result of

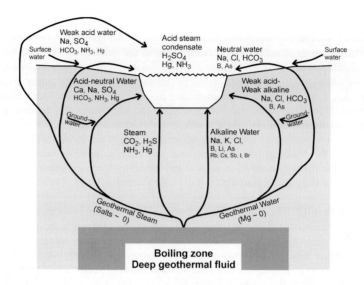

Figure 8. Idealized relationships between geothermal fluids and surface waters. Increasing type size indicates increasing relative abundances in particular fluids. The higher temperature fluids (Table 3) represent samples of "Alkaline Water". Cooler groundwater samples and surface water samples have chemistries similar to the "Neutral Water". After Timperley (1987).

precipitation of Ca minerals under high temperatures at depth and the development of negative Eu/Eu* in the remaining fluids (Moeller, 1998; Moeller, 2000).

In order to adequately address the competitive effects of inheritance and water-rock interactions, higher temperature reducing geothermal fluids must be sampled. In addition, a more thorough study of the surface-groundwater mixing is required. Incompatible element mixing trajectories, such as those shown in Figure 10, demonstrate mixing between high temperature geothermal fluids and low temperature surface waters. The degree to which the mixing relationships define the observed REE variations is, however, compounded by the extent of water-rock interaction.

In summary, the REE patterns of the waters, sediments, precipitates and aquifer material show varying degrees of fractionation. The different REE signatures between the cooler shallow groundwaters and surface waters and the deeper warm groundwaters are the results of a combination of processes such as the oxidizing conditions of water-rock interactions which prevents the mobilization of Ce over the other REE resulting in a negative Ce anomaly. The fractionation of the REE is also attributable to secondary mineral precipitation and complexation of REE by sulfate and chloride (Haas et al., 1995; Wood 1990a, b).

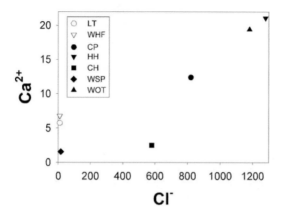

Figure 9. Ca *versus* Cl in the waters indicates water-rock interactions dominate the chemistry of the geothermal samples. The samples plot close to a simple mixing line with concentrations of Ca and Cl increasing as water temperature increases.

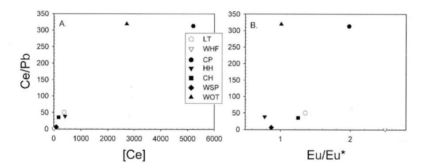

Figure 10. (A) Ce/Pb – Ce and (B) Ce/Pb – Eu/Eu* relationships of the TVZ samples. The relationships shown suggest that the majority of geochemical variation in the water composition is related to mixing of fresh and geothermal end-members, however, the data do not fall on a discrete hyperbolic mixing line. Competitive processes such as secondary mineral precipitation also fractionate the REE (B). As Pb remains within the mineral during feldspar weathering preferentially to Ce, the weathering of plagioclase produce a negative relation between Ce/Pb and Eu/Eu*.

The inverse relationship between Ce/Ce* and Eu/Eu* between the water and solid samples is indicative of the chemistry of the fluids and the nature of their interaction with the surrounding aquifer wall rock. With further study and the inclusion of deep high temperature (> 250°C) samples we may be able to accurately assess the importance of magmatic processes on the development of REE signatures in the geothermal fluids and the processes which govern the retention and differentiation of these patterns across the TVZ. The combination of incompatible element geochemistry with REE geochemistry in the TVZ illuminated the unique geochemical character of this hydrothermal region.

The trace elements and REE are useful as hydrological tools as well as for teasing apart the chemical processes governing the chemical evolution of geothermal fluids. Future studies will continue to assess the nature of water-rock interaction from both an aqueous and petrogenetic perspective to better understand the nature of groundwater flow within the region. By understanding the factors contributing the REE variation in hydrothermal regions such as the TVZ we will be able to use the REE as a non-conservative groundwater tracer by fingerprinting the various controls on REE fractionation, specifically the inherited versus water-rock induced variation.

Acknowledgements

Many thanks to L. Ball of the WHOI-ICP-MS facility and Z. Chen of LITER at ODU. Drs. R. Murray, K. Johannesson, and an anonymous reviewer are thanked for thorough and helpful reviews. This research was supported, in part, by a grant from the ACS Petroleum Research Fund, PRF #36568-GB2.

References

Anders, E. and Grevesse, N. 1989. Abundances of the elements: meteoric and solar. *Geochim. Cosmochim. Acta*, **53**, 197-214.
Bence, A.E., Grove, T.L. and Papike, J.J. 1980. Basalts as probes of planetary interiors; constraints on the chemistry and mineralogy of their source regions. *Precambrian Res.*,**10**, 249-279.
Bower, J.E. and Timperley, M.H. 1988. Lithium and rubidium in the Waikato River, New Zealand. *NZ J. Mar. Fresh. Res.*, **22**, 201-214.
Braun, J.J., Pagel, M., Muller, J.P., Bilong, P., Michard, A. and Guillet, B. 1990. Cerium anomalies in lateritic profiles. *Geochim. Cosmochim. Acta*, **54**,781-795.
Brookins, D.G. 1989. *Eh-pH diagrams for Geochemists*. Springer-Verlag Pubs., NY, 176p.
Brookins, D.G. 1989. Aqueous geochemistry of rare earth elements. In: P.H. Ribbe (ed.) Geochemistry and Mineralogy of Rare Earth Elements. Reviews in Mineralogy, Vol. 21. Mineralogical Society of America (Washington, DC), pp. 201-225.
Cole, J.W., 1979. Structure, petrology and genesis of Cenozoic volcanism, Taupo Volcanic Zone, New Zealand – a review. *NZ J. Geol. Geophys.*, **22**, 631-657.
deBaar, H.J.W., German, C.R., Elderfield, H., and Van Gaans, P. 1988. Rare earth element distributions in anoxic waters of the Cariaco Trench. *Geochim. Cosmochim. Acta*, **52**, 1203-1220.

Dia, A., Gruau, G., Olivie-Lauquet, G., Rious, C, Molenat, J. and Curmi, P. 2000. The distribution of rare earth elements in groundwaters: Assessing the roe of source-rock composition, redox changes and colloidal particles. *Geochim. Cosmochim. Acta*, **64**, 4131-4151.

Ding, Z.L., Sun, J.M., Yang, S.L. and Liu, T.S. 2001. Geochemistry of the Pliocene red clay formation in the Chinese Loess Plateau and implications for its origin, source provenance and paleoclimate change. *Geochim. Cosmochim. Acta*, **65**, 901-913.

Elderfield, H.R., Upstill-Goddard, R. and Sholkovitz, E.R., 1990. The rare earth elements in rivers, estuaries and coastal sea waters: processes affecting crustal input of elements to the ocean and their significance to the composition of sea water. *Geochim. Cosmochim. Acta*, **54**, 971-991.

Eser, P. and Rosen, M.R. 1999. The influence of groundwater hydrology and stratigraphy on the hydrochemistry of Stump Bay, South Taupo Wetland, New Zealand, *J. Hydrol.*, **220**, 27-47.

Gibbs, M.M. 1979. Groundwater input to Lake Taupo, New Zealand: Nitrogen and phosphorus inputs from Taupo township. *NZ J. Sci.*, **22**, 235-243.

Giggenbach, W.F., Sheppard, D.S., Robinson, B.W., Stewart, M.K. and Lyon, G.L. 1994. Geochemical structure and position of the Waiotapu geothermal field, New Zealand. *Geothermics*, **23**, 599-644.

Giff, F. and Grisby, C.O. 1982. Valles caldera geothermal systems, New Mexico, USA. *J. Hydrol.*, **56**, 119-136.

Goldstein, S.J. and Jacobsen, S.B., 1987. The Nd and Sr isotopic systematics of river water dissolved material: implications for the sources of Nd and Sr in seawater. *Chem. Geol.*, **66**, 245-272.

Goldstein, S.J. and Jacobsen, S.B., 1988a. The Nd and Sr isotopic systematics of river water suspended material: implications for crustal evolution. *Earth Planet. Sci. Lett.*, **87**, 249-265.

Goldstein, S.J. and Jacobsen, S.B., 1988b. Rare earth elements in river waters. *Earth Planet. Sci. Lett.*, **89**, 35-47.

Graham, I.J. 1992. Strontium isotope composition of Rotorua geochemical waters. *Geothermics*, **21**, 165-180.

Grange, L.I. 1937. The geology of the Rotorua-Taupo subdivision, Rotorua and Kaimanawa Divisions. *Geol. Surv. NZ Bull.*, **37**, 138 p.

Haas, J.R., Shock, E.L. and Sassani, D.C. 1995. Rare earth elements in hydrothermal systems: Estimates of standard partial molal thermodynamic properties of aqueous complexes of the rare earth elements at high pressures and temperatures. *Geochim. Cosmochim. Acta*, **59**, 4329-4350.

Hannigan, Robyn E. and Basu, Asish R. 1998. Late Diagenetic Trace Element Remobilization in Organic-Rich Black Shales of the Taconic Foreland Basin of Québec, Ontario and New York. *In*, J. Schieber, W. Zimmerle and P. Sethi (eds.) *Mudstones and Shales: Recent Progress in Shale Research*, Schweizerbart'sche Verlagsbuchhandlung. 209-234.

Hannigan, R.E. and Sholkovitz, E.R. 2001. Rare Earth Element Chemistry of Natural Waters: Chemical Weathering and Dissolved REE Contents in Major River Systems. *Chem. Geol.*, **175**, 495-508.

Healy, J. 1975. Volcanic lakes. In, V.H. Jolly and J.M.A. Brown (eds.) *New Zealand lakes*. New Zealand, Auckland University Press. 388 p.

Houghton, B.F., and Wilson, C.J.N. 1998. Fire and water: Physical roles of water in large eruptions at Taupo and Okataina calderas. In, G.B. Arehart and J.R. Hulston (eds.) *Proc. Water-Rock Interaction* **9**, 25-30.

Irber, W. 1999. The lanthanide tetrad effect and its correlation with K/Rb, Eu/Eu*, Sr/Eu, Y/Ho, and Zr/Hf of evolving peraluminum granite suites. *Geochim. Cosmochim. Acta*, **63**, 489-508.

Johannesson, K.H., Lyons, W.B., Yelken, M.A., Gaudette, H.E. and Stetzenbach, K.J., 1996. Geochemistry of the rare-earth elements in hypersaline and dilute acidic natural terrestrial waters: Complexation behavior and middle rare-earth element enrichments. *Chem. Geol.*, **133**, 125-144.

Johannesson, K.H., Stetzenbach, K.J., and Hodge, V.F. 1997a. Rare earth elements as geochemical tracers of regional groundwater mixing. *Geochim. Cosmochim. Acta*, **61**, 3605-3618.

Johannesson, K.H., Stetzenbach, K.J., Hodge, V.F., and Kreamer, D.K. 1997b. Delineation of groundwater flow systems in the southern Great Basin using aqueous rare earth element distributions. *Ground Water*, **35**, 807-819.

Johannesson, K.H., Farnham, I.M., Guo, C. and Stetzenbach, K.J. 1999. Rare earth element fractionation and concentration variations along a groundwater flow path within a shallow, basin-fill aquifer, southern Nevada, USA. *Geochim. Cosmochim. Acta*, **60**, 1695-1707.

Land, M., Ohlander, B., Ingri, J. and Thunberg, J. 1999. Solid speciation and fractionation of rare earth elements in a spodosol profile from northern Sweden as revealed by sequential extraction. *Chem. Geol.*, **160**, 121-138.

Michard, A., 1989. Rare earth element systematics in hydrothermal fluids. *Geochim. Cosmochim. Acta*, **53**, 745-750.\

Moeller, P. 1998. Eu anomalies in hydrothermal minerals: Kinetic versus thermodynamic interpretation. *Proc. Quad. IAGOD Symp.*, **9**, 239-246.

Morey, G.B. and Setterholm, D.R. Rare earth elements in weathering profiles and sediments of Minnesota: Implications for provenance studies. *J. Sed. Res.*, **67**, 105-115.]

Nördstrom, D.K., Ball, J.W., Donahoe, R.J. and Whittemore, D. 1989. Groundwater chemistry and water-rock interactions at Stripa. *Geochim. Cosmochim. Acta*, **53**, 1727-1740.

Nyakairu, G.W.A. and Koeberl, C. 2001. Mineralogical and chemical composition and distribution of rare earth elements in clay-rich sediments from central Uganda. *Geochem. J.*, **35**, 13-28.

Potts, P.J., Tindle, A.G. and Webb, P.C., 1992. Geochemical reference material compositions: rocks, minerals, sediments, soils, carbonates, refractories & ores used in research & industry. CRC Press, Boca Raton, 126 pp.

Rosen, M.R. and Coshell, L. 1998. Influence of eruptive volcanic lithologies on surface and groundwater chemical compositions, Lake Taupo, New Zealand. In, G.B. Arehart and J.R. Hulston (eds.) *Proc. Water-Rock Interaction* **9**, 181-184.

Rutherford, J.C. 1984. Trends in Lake Rotorua water quality. *NZ J. Mar. Fresh. Res.*, **18**, 355-365.

Rutherford, J.C., Williamson, R.B. and Cooper, A.B. 1987. Nitrogen, phosphorus and oxygen dynamics in rivers. In, A.B. Viner (ed.) Inland Waters of New Zealand. *DSIR Bulletin*, **241**, 139-165.

Sholkovitz, E.R., 1992. Chemical evolution of rare earth elements: fractionation between colloidal and solution phases of filtered river water. *Earth Planet. Sci. Lett.*, **114**, 77-84.

Sholkovitz, E.R., 1995. The Aquatic Chemistry of Rare Earth Elements in Rivers and Estuaries. *Aquat. Geochem.*, **1**, 1-34.

Schouten, S.G. 1983. Budget of water and its constituents for Lake Taupo. In: Dissolved Loads of Rivers and surface water quality relationships. *Inter. Assoc. Hydro. Sci. Pub.*, **141**, 277-297.

Shiller, A.M., Chen, Z. and Hannigan, R. In Press. A time series of dissolved rare earth elements in the Lower Mississippi River. *Proc. Water-Rock Interaction* **10**, 1005-1008.

Simmons, S.F., Stewart, M.K., Robinson, B.W. and Glover, R.B. 1994. The chemical and isotopic compositions of thermal waters at Waimangu, New Zealand. *Geothermics,* **23**, 539-553.

Smedley, P.L. 1991. The geochemistry of rare earth elements in groundwater from the Carnmennellis area, southwest England. *Geochim. Cosmochim. Acta*, **55**, 2767-2779.

Taylor, S.R. and McLennan, S.M., 1985. *The Continental Crust: its Composition and Evolution*. Blackwell Scientific Publications, Boston, 312 pp.

Taylor, S.R. and McLennan, S.M. 1995. The geochemical evolution of the continental crust. *J. Rev. Geophys.*, **33**, 241-265.

Timperley, M.H. 1983. Phosphorus in spring waters of the Taupo Volcanic Zone, North Island, New Zealand. *Chem. Geol.*, **38**, 287-306.

Timperley, M.H. 1987. Regional influences on lake water chemistry. *DSIR Bull.*, **241**, 97-111.

Timperley, M.H. and Huser, B.A. 1994. Natural geothermal inflows to the Waikato River. In, S. Soengkono and K.C. Lee (eds.), *Proceedings of the 16th New Zealand geothermal workshop*, 57-63.

Viers, J., Dupre, B., Polve, M., Schott, J., Dandurand, J-L., and Braun, J-J. 1997. Chemical weathering in the drainage basin of a tropical watershed (Nsimi-Zoetele site, Cameroon): Comparison between organic-poor and organic-rich waters. *Chem. Geol.*, **140**, 181-206.

White, E., Downes, M., Gibbs, L., Kemp, L., Mackenzie, L. and Payne, G. 1980. Aspects of the physics, chemistry and phytoplankton biology of Lake Taupo, *NZ J. Mar. Fresh. Res.*, **14**, 139-148.

Wombacher, F. and Muenker, C. 2000. Pb, Nd, and Sr isotopes and REE systematics of Cambrian sediments from New Zealand: Implications for the reconstruction of the early Paleozoic Gondwana margin along Australia and Antarctica. *J. Geol.*, **108**, 663-686.

Wood, S.A. 1990a. The aqueous geochemistry of the rare-earth elements and yttrium, 1: Review of available low temperature data for inorganic complexes and the inorganic REE speciation of natural waters. *Chem. Geol.*, **82**, 159-186.

Wood, S.A. 1990b. The aqueous geochemistry of the rare-earth elements and yttrium, Part 2: Theoretical predictions of speciation in hydrothermal solutions to 350°C at saturated water vapor pressure. *Chem. Geol.*, **88**, 99-125.

Chapter 4

THE AQUEOUS GEOCHEMISTRY OF THE RARE EARTH ELEMENTS AND YTTRIUM. PART 13: REE GEOCHEMISTRY OF MINE DRAINAGE FROM THE PINE CREEK AREA, COEUR D'ALENE RIVER VALLEY, IDAHO, USA

SCOTT A. WOOD, WILLIAM M. SHANNON & LESLIE BAKER

Department of Geological Sciences, Box 443022, University of Idaho, Moscow, ID, 83844-3022, USA

1. Introduction

The production of acidity by the oxidation of sulfide minerals, whether occurring naturally or as a result of mining operations, is a serious environmental concern owing to the potential for acidification of lakes, streams and ground waters, and the concomitant increase in mobility of toxic heavy metals (see Alpers and Blowes, 1994; Jambor and Blowes, 1994). Acidified waters draining mine workings are usually referred to as acid-mine drainage (AMD). However, a more general term which also covers generation of acid by unexploited sulfide mineralization is acid-rock drainage (ARD). Recently, the geochemistry of REE in acid-rock drainage has received much attention (Carlson-Foscz et al., 1991; Miekeley et al., 1992; Webb et al., 1993; Nordstrom et al., 1995; Gimeno et al., 1996; Verplanck et al., 1997; Leybourne et al., 1998; 2000; Pearce et al., 1998; White et al., 1998; Elbaz-Poulichet and Dupuy, 1999; Hollings et al., 1999; Verplanck et al., 1999; Gimeno et al., 2000; Åström, 2001; Worrall and Pearson, 2001a,b; Gammons et al., 2003). The motivation for these studies is the possibility that the REE might be useful tracers in furthering our understanding of the processes controlling ARD.

The present study focuses on drainage from a number of abandoned mine sites (including adits and seeps from tailings piles) in the Pine Creek area (Figures 1 and 2) of Idaho, at the southern edge of the Bunker Hill Superfund Site, about 10 km south of Smelterville. With one exception, the mine drainage is neutral to slightly alkaline in pH. This is a result of the neutralization of acid produced during sulfide oxidation by the carbonate gangue minerals present in the mine workings. In spite of the generally near-neutral pH values, all the waters contain relatively high concentrations of zinc and some waters contain high concentrations of other heavy metals. We have analyzed these waters to improve our knowledge of the behavior of the REE during sulfide oxidation followed by neutralization via carbonate dissolution.

2. Study area

The mineralization of the Pine Creek area has been described in numerous theses and reports (e.g., Jones, 1919; Forrester and Nelson, 1944; Campbell, 1949; Gin, 1953; Silverman, 1957). Sulfide-bearing veins occur in argillites and interbedded quartzite bands and lenses of the Prichard formation of the Proterozoic Belt Series. Primary (post-metamorphic) minerals of the Lower Prichard Formation include graphite, ilmenite, magnetite, pyrite, pyrrhotite, quartz, white mica and/or illite, tourmaline, and

K.H. Johannesson, (ed), Rare Earth Elements in Groundwater Flow Systems, 89-110.

Figure 1. Location of the Pine Creek study area within the Coeur d'Alene mining district.

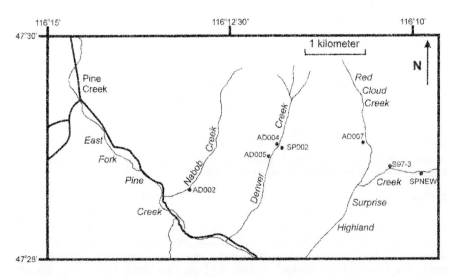

Figure 2. Location of the various sampling sites in the Pine Creek area. The heavy lines represent roads, whereas the lighter lines represent creeks. Identification numbers with an AD prefix denote sites where waters emerge from the adits of abandoned mines. Identification numbers with S or SP prefixes refer to seeps from tailings piles.

zircon. Hydrothermal alteration of the metasediment resulted in formation of biotite, chalcopyrite, clinochlore, hematite, leucoxene, limonite, muscovite, orthoclase, pyrite, pyrrhotite, quartz and siderite (Silverman, 1957). Sphalerite, galena and pyrrhotite are the main sulfide minerals in the mineralized veins, but arsenopyrite, chalcopyrite, pyrite, stibnite and tetrahedrite also occur in various combinations in some veins. Gangue minerals are andesine, ankerite, chlorite, dolomite, magnesite, muscovite, white mica, siderite and quartz. Anglesite, azurite, bornite, cerussite, covellite, limonite, malachite, massicot and stibiconite occur as secondary minerals within oxidized zones of the veins. The mines have been inactive since the mid 1950's to mid 1970's. Total production of ore from the area during the period 1910-1951 was about 2,500,000 tons (Gin, 1953). The grade of the deposits was highly variable, with an average combined Pb and Zn of 10% (Gin, 1953). In addition to Zn and Pb, some mines also produced significant Ag and much smaller quantities of Sb, W, Cu and Au. Details of the individual sample sites are given in Wood et al. (in prep.).

3. Methodology

3.1. SAMPLING

Samples for REE analysis were obtained in June 1999 and April 2000, although we have analyzed the same mine waters for major elements and transition metals over the course of nearly ten years. All sample containers were constructed of high-density polyethylene. Those containers destined to hold samples for metal determinations were soaked in 10% nitric acid overnight and rinsed with deionized (DI) water, labeled, closed and sealed with Parafilm. However, before closing and sealing, a sufficient quantity of high-purity, concentrated nitric acid was added such that the final acid concentration after filling the container with sample was 5 volume %. Containers to be used for storing samples for anion analyses were soaked in DI water for several days, rinsed again with DI water, labeled and sealed with Parafilm. Both unfiltered and filtered aliquots of each mine water were taken for metal determinations. Filtered aliquots were obtained by forcing the sample through Gelman Supor 450^{TM} 0.45-µm polysulfone membranes in the field using a hand-operated peristaltic pump.

3.2. FIELD MEASUREMENTS

Temperatures were measured with an Omega 871A digital thermometer equipped with a chromel-alumel (type K) thermocouple. The pH was determined using a Radiometer PHM80 portable pH meter and a combination glass pH electrode. The electrode was calibrated using at least three of five NIST buffer solutions (pH = 2, 4, 7, 9 and 10). Conductivity was determined using a portable conductivity meter (YSI Model 33). The Eh was determined with a platinum indicator electrode and a saturated calomel reference electrode, calibrated against two redox buffers. Additional details of these field measurements are given in Wood et al. (in prep.).

3.3. LABORATORY MEASUREMENTS

Immediately on return to the laboratory, aliquots of the unfiltered, unacidified samples were subjected to alkalinity titration using standardized HCl solutions. The titration data were interpreted using the Gran method (cf. Stumm and Morgan, 1981).

Unacidified mine-drainage samples were analyzed using a Dionex AI-450 ion chromatography (IC) workstation to determine the concentrations of the anions fluoride, chloride, bromide, nitrate, sulfate, and phosphate. An isocratic elution with a sodium bicarbonate-sodium carbonate eluent was employed together with a Dionex AS4-SC column. No additional peaks were observed on chromatograms that would indicate the presence of major anionic species other than those listed above.

Concentrations of Al, Ba, Ca, Cd, Cu, Fe, K, Mg, Mn, Na, Pb, Si, Sr, Ti, and Zn in acidified filtered and unfiltered samples were determined using a Perkin-Elmer Optima 3000-XL inductively coupled plasma-atomic emission spectrometer (ICP-AES) equipped with an axial torch. The concentrations of many other elements (e.g., As, Sb, Co, Ni, W, Mo) were below detection. A conservative limit of detection for these metals is about 5-10 µg/L. All calibration standards, blanks, and quality-control standards were matrix-matched to the samples with respect to acid concentration. More details are available in Wood et al. (in prep.).

The July 1999 samples were preconcentrated for REE analysis using ferric hydroxide co-precipitation (Buchanan and Hannaker, 1984; Weisel et al., 1984; Greaves et al., 1989; Welch et al., 1990; Johannesson and Lyons, 1995; van Middlesworth and Wood, 1998; Shannon and Wood, this volume). An aliquot of ultrapure Fe solution is added to the sample and thoroughly mixed. Then, high-purity ammonium hydroxide is added to cause amorphous, reddish-brown to yellow Fe-hydroxide to precipitate. The precipitate was collected by filtration onto acid-washed, quartz-fiber filters (Whatman P/N 1851-047, with an effective pore size of 2.2 µm) and redissolved in 20-40 mL of Seastar Baseline 4% nitric-1% hydrochloric acid solution. The quartz-fiber filters employed are depth-type filters and are considered to be ideal for the collection of precipitated iron hydroxides. The ferric hydroxide co-precipitation procedure results in a concentration factor of 25-50 times. It has been shown previously that, when the pH is adjusted to between 8 and 9 during the Fe-hydroxide co-precipitation, recovery of each REE is greater than 90% (Buchanan and Hannaker, 1984; van Middlesworth and Wood, 1998; Shannon and Wood, this volume).

For the April 2000 samples, a different method of separation and preconcentration of the REE was employed. The method is based on the extraction of the REE from the acidified aqueous sample into a mixture of mono- and bis-2-ethylhexyl phosphate esters dissolved in n-heptane (Shabani et al, 1990; Aggarwal et al, 1996). The REE are back-extracted into an aqueous phase by addition of n-octanol and 6-N hydrochloric acid. For further details see Shannon and Wood (this volume) and references therein.

The Fe co-precipitated samples were analyzed for REE using a Perkin Elmer model Elan 250 ICP-MS which was retrofitted with an EPT multiplier detector, argon mass-flow controllers for coolant, auxiliary, and nebulizer flows, a water-cooled spray chamber to reduce oxide formation, and a high-solids Meinhard-style nebulizer. Transport, ionization, and space-charge interferences were corrected by use of Co, In, and Bi as internal standards. Typically oxide formation was near or below 10% as measured from $^{248}[ThO]/^{232}Th$ ratios. Correction for isobaric oxide interference were made for BaO on Eu and Sm, PrO on Gd, and NdO on Tb. Correction was made by periodically running both a 10-mg/L Ba standard and a mixed Pr-Nd standard. Quality control was monitored by periodic analysis of two or more mid-level calibration standards and the calibration blank. The internal standards were used to compensate for instrument drift. All data reduction was accomplished off-line using a standardized computer spreadsheet. Rare earth elements in the extracts from the April 2000 samples were determined using a Hewlett Packard model 4500 ICP-MS, housed at Washington State University. For this instrument, oxide formation as measured by $^{156}[CeO]/^{140}Ce$ was typically less than 0.5%, equivalent to $^{248}[ThO]/^{232}Th$ of less than 1%. In this case we used Ru, In, and Re as internal standards, but the correction for isobaric interferences and quality-control procedures were similar to those employed with the Perkin Elmer instrument.

Analysis of selected samples for REE using both analytical protocols yielded results that were identical within the precision of the analysis. Thus, we are confident that any observed variations in REE behavior between the two sampling dates are real and not artifacts of having used two different analytical methods. For presentation, all REE concentrations were normalized to NASC (North American Shale Composite; Gromet et al., 1984), which is a good proxy for average North American crustal abundances of REE. The Fe-coprecipitation method resulted in detection limits $\leq 10^{-7}$ times NASC for the light REE (La-Nd) and 10^{-6} times NASC for the middle and heavy REE. The solvent extraction method yielded detection limits up to one order of magnitude lower than those of the Fe co-precipitation method for all REE. The improvement in sensitivity of the latter protocol is almost entirely due to the fact that the newer Hewlett Packard instrument is more sensitive than the older Perkin Elmer instrument. For comparable concentration factors, the Fe-hydroxide co-precipitation and the solvent extraction technique yield nearly identical sensitivities when REE determination is by the same instrument (see also Shannon and Wood, this volume).

For the April 2000 samples, we also analyzed the solid material left behind on the filter membranes for REE. This was accomplished by placing each membrane with its load of solid material into a solution of Seastar Baseline 4% nitric-1% hydrochloric acid until most of the solid residue dissolved. The resulting solution was then directly analyzed via ICP-MS using the Hewlett Packard model 4500 instrument and the correction procedures outlined above for the April 2000 water samples. The digestion procedure described above dissolves any Fe oxyhydroxide or carbonate material, but does not dissolve most silicates or refractory REE minerals (e.g., monazite). No attempt was made to determine the weight of solid material dissolved. Thus, the absolute concentrations of REE in the extracts reflect an unknown and variable dilution factor, and only the relative REE concentrations are significant.

3.4. SPECIATION AND SOLUBILITY CALCULATIONS

In order to gain some insight into the likely predominant dissolved REE species in the mine drainage analyzed in this study, we calculated the distribution of La^{3+} and Yb^{3+} among the following species: Ln^{3+}, $LnSO_4^+$, $LnHCO_3^{2+}$, $LnCO_3^+$, $LnOH^{2+}$, $Ln(OH)_2^+$, $Ln(OH)_3^0$, $Ln(OH)_4^-$, $LnCl^+$, $LnCl_2^+$, $LnCl_3^0$, $LnCl_4^-$, LnF^+, LnF_2^+, LnF_3^0, and LnF_4^-, where $Ln = La^{3+}$ or Yb^{3+}. Speciation calculations were conducted only for the July 1999 samples. The required stability constants for the REE complexes were taken from the compilation of Haas et al. (1995). To accomplish these calculations, the activities of all REE-complexing ligands were calculated using SOLMINEQ88 (Kharaka et al., 1988) with the compositional data shown in Table 1 as input. The activities of the free anions SO_4^{2-}, HCO_3^-, OH^-, Cl^-, and F^- given as output from SOLMINEQ88 were then used to calculate the speciation of the REE using a spreadsheet, the assumption being that the REE are present in too low a concentration to affect the activities of the free ligands. To determine whether certain solid phases might control the measured REE concentrations, the latter were compared with concentrations of REE calculated assuming saturation with respect to $Ln(OH)_3(s)$, $Ln_2(SO_4)_3(s)$ or $Ln_2(CO_3)_3(s)$. Solubility products reported by Diakonov et al. (1998a,b), Rard (1988), and Firsching and Mohammadzadel (1986), respectively, were employed in these calculations.

4. Results

Major features of the chemical composition of the waters analyzed are given in Tables 1 and 2. With the exception of AD004, the waters range in pH from 6.94 to 8.44, i.e., from neutral to slightly alkaline (Tables 1, 2). The water issuing from AD004 is acidic (2.30-3.3). In most cases, the waters were more acidic in April 2000 than in June 1999. Based on sampling these waters over many years we find that, in general, the drainage is more acidic during the spring when discharge rates are high, than in summer when they are comparatively low. The differences in pH values between July and April are relatively small for most of the waters (<0.5 pH units), whereas the waters draining AD004 and AD005 show an entire unit pH difference between the two sampling campaigns. Sulfate and bicarbonate are the dominant anions in these waters, with sulfate greatly predominating in the one acidic water (AD004). Calcium, magnesium, and zinc are the dominant metals. Zinc concentrations are relatively high in all the waters, ranging from 1 to 240 mg L^{-1}.

Table 1. Chemical characteristics of Pine Creek mine drainage (June 1999). Note: All analyses refer to samples passed through a 0.45-μm filter. ND – not detected.

Site	T(°C)	Cond (μmhos)	pH	Eh (mV vs. SHE)	HCO_3^- (mg L^{-1})	SO_4^{2-} (mg L^{-1})	Cl^- (mg L^{-1})	F^- (mg L^{-1})	NO_3^- (mg L^{-1})
AD002	10.5	700	8.44	470	150	500	0.45	0.01	0.38
AD004	19.1	1180	3.27	730	0	700	0.26	0.20	0.01
AD005	7.1	500	7.84	550	38	360	0.33	0.10	0.39
AD007	10.8	290	7.95	510	83	170	0.31	0.06	0.46
SP002	18.2	298	7.05	470	135	120	0.38	0.18	0.17
SPNEW	16.6	126	7.83	520	38	44	0.33	ND	0.10
ADNEW	11.2	115	7.76	500	180	81	0.33	0.02	0.18
S97-3	14.5	470	7.26	610	150	260	0.34	0.02	0.09

Site	Al (mg L^{-1})	Ca (mg L^{-1})	Fe (mg L^{-1})	K (mg L^{-1})	Mg (mg L^{-1})	Mn (mg L^{-1})	Na (mg L^{-1})	Zn (mg L^{-1})
AD002	0.01	120	0.18	0.75	54	1.0	7.8	9.1
AD004	5.0	55	12	0.58	33	5.4	3.9	79
AD005	ND	49	0.18	0.46	28	2.8	3.8	65
AD007	ND	38	0.17	0.36	18	0.14	4.2	22
SP002	ND	26	2.5	0.62	30	1.0	6.1	1.0
SPNEW	0.01	12	0.18	0.82	6.2	0.04	2.6	5.5
ADNEW	ND	42	0.17	0.77	22	0.29	3.1	1.9
S97-3	ND	72	0.17	0.51	34	0.62	5.7	3.1

The results for REE concentrations in the mine waters are given in the appendix, and displayed in NASC-normalized format in Figures 3a-c. Almost all the waters share the characteristic of being middle rare earth element (MREE)-enriched, with a maximum at Gd (or less commonly Eu). The REE heavier than Gd exhibit slightly decreasing abundances relative to NASC with increasing atomic number, whereas the REE lighter than Gd show a stronger decrease in abundance relative to NASC with decreasing atomic number. In many cases the NASC-normalized abundance of La is greater than that of Ce, so that the patterns take on sinusoidal shape, and subtle negative Ce anoma-

Table 2. Chemical characteristics of Pine Creek mine drainage (April 2000). Note: All analyses refer to samples passed through a 0.45-μm filter. ND – not detected.

Site	T(°C)	Cond (μmhos)	pH	Eh (mV vs. SHE)	HCO_3^- (mg L^{-1})	SO_4^{2-} (mg L^{-1})	Cl$^-$ (mg L^{-1})	F$^-$ (mg L^{-1})	NO_3^- (mg L^{-1})
AD002	7.3	348	8.35	580	113	350	0.71	0.08	0.16
AD004	8.9	910	2.30	720	0	370	0.47	0.13	0.30
AD005	5.7	890	6.94	640	13	580	0.48	0.18	0.45
AD007	8.2	480	7.53	670	20	92	0.86	0.09	0.02
SP002	8.3	445	7.30	540	170	130	0.51	0.19	0.06
SPNEW	6.6	235	7.58	550	85	61	0.36	0.03	0.11
ADNEW	9.8	600	7.92	---	180	110	0.35	0.05	0.05
S97-3	10.4	315	7.02	---	170	230	0.46	0.08	0.10

Site	Al (mg L^{-1})	Ca (mg L^{-1})	Fe (mg L^{-1})	K (mg L^{-1})	Mg (mg L^{-1})	Mn (mg L^{-1})	Na (mg L^{-1})	Zn (mg L^{-1})
AD002	ND	97	ND	0.76	46	0.78	7.1	12
AD004	3.0	29	9.8	0.58	18	2.4	3.0	60
AD005	1.3	77	0.48	0.66	47	5.3	4.4	240
AD007	0.02	14	ND	0.41	6.6	0.53	2.2	24
SP002	0.09	28	7.4	0.74	35	2.0	7.3	1.6
SPNEW	0.01	11	ND	0.86	6.0	0.19	2.6	8.3
ADNEW	0.02	49	ND	0.98	26	0.51	3.7	4.7
S97-3	0.02	70	ND	0.59	34	0.64	6.9	2.5

lies are evident. Exceptions to the general pattern described above only occur for the waters with the lowest REE concentrations, where the abundance is approximately 10^{-6} times NASC or lower. In these cases, the LREE show a trend very similar to the other waters, but the MREE and HREE are relatively constant or increase slightly from Gd to Lu (rather than showing a maximum at Eu-Gd). It is not clear whether this slight difference in abundance patterns for the samples with lower REE concentrations is real or an artifact of having concentrations near the analytical detection limit. However, the lower-abundance patterns are relatively smooth, suggesting that proximity to the detection limit is not a problem and the observed difference may be real.

It is remarkable that the MREE-enriched pattern is found both in mine waters with low pH and relatively little visible Fe oxyhydroxide precipitation (e.g., AD004), and mine waters with near-neutral pH and abundant Fe oxyhydroxide precipitation (e.g.,SPNEW, AD005, SP002). It is possible that abundant Fe oxyhydroxides have precipitated from the AD004 water at depth. However, dissolved oxygen measurements show that all the mine drainages are typically near saturation with respect to atmospheric oxygen (Wood et al., in prep.), and the measured Eh values (Tables 1 and 2) are consistent with those expected for waters in contact with the atmosphere (cf. Garrels and Christ, 1965). The Eh value of the acidic water is somewhat higher than those of the near-neutral to alkaline waters, but this is expected as Eh increases with decreasing pH at constant partial pressure of oxygen. Thus, the higher Eh does not indicate intrinsically more oxidizing conditions in the water emanating from AD004, and given its low pH, precipitation of abundant Fe-oxyhydroxides at depth seems unlikely. The similarity in REE patterns in waters from which Fe-oxyhydroxides are precipitating in abundance with those from waters in which Fe-oxyhydroxides are sparse may therefore place some constraints on the role of such phases in controlling the REE patterns.

The highest REE concentrations generally were found in the one acidic mine water (AD004; see Figure 3). Moreover, the concentrations of REE in the filtered aliquots of AD004 are generally only slightly less than those in the unfiltered aliquots, suggesting that the REE are present predominantly in a form (probably in true solution) that passes through a 0.45-μm filter. Finally, the absolute concentrations of REE in mine water AD004 were very similar in both 1999 and 2000.

Although the highest REE concentrations typically were found in the most acidic water, relatively high concentrations were also found in the slightly alkaline mine water AD005 (Figure 4). In fact, in April 2000 the REE concentrations in AD005 were essentially the same as those in AD004. On the other hand, in June 1999, the concentration of REE in AD005 was almost an order of magnitude lower than in April 2000. In April 2000, the filtered and unfiltered aliquots of AD005 had nearly identical REE concentrations. However, in June 1999, the filtered aliquot had a slightly lower REE content than the unfiltered aliquot. Zinc concentration is also typically high (60-235 mg L^{-1}) in both AD004 and AD005, and in April 2000 the Zn concentration of neutral AD005 was almost four times higher than that in the acidic AD004. Moreover, like the REE, Zn concentrations in AD005 were much higher in April 2000 than in June 1999, and the pH was lower in April 2000. These data indicate an increase in acid production and metal leaching in spring compared to the summer. It is interesting to note that water AD005 emerges from an adit only a few hundred meters down Denver Creek from AD004. The similarity in Zn and REE concentrations may suggest that the waters draining AD004 and AD005 may be related in some manner.

In all adits other than AD004 and AD005, and in all seeps, the unfiltered aliquot had higher REE concentrations than the filtered aliquot irrespective of the sampling date. In some cases, the differences between filtered and unfiltered samples were greater than an order of magnitude. For SP002, the unfiltered samples contain nearly as much REE as both filtered and unfiltered samples from AD004, but the filtered samples from SP002 contain orders of magnitude less REE. At SP002, abundant iron oxyhydroxides are precipitating, and it seems likely that they are removing substantial amounts of the REE from solution.

The NASC-normalized patterns for filtered and unfiltered aliquots of the same sample generally are parallel (Figure 3). This is true even in samples where there is a large difference in the absolute REE concentrations between filtered and unfiltered aliquots. This finding suggests that there is no significant fractionation of REE between the suspended solids and solution in these mine waters. To further test for the possible existence of fractionation, we analyzed the solids remaining on the membranes after

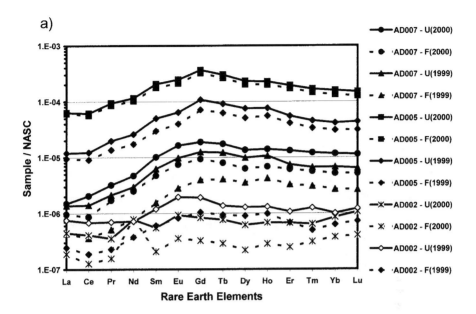

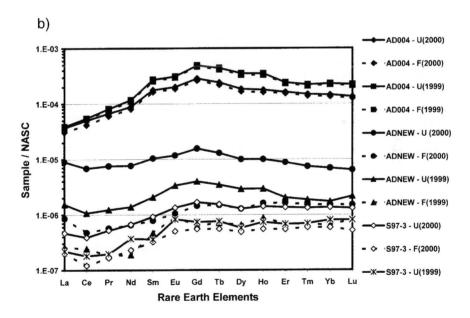

Figure 3. NASC-normalized REE concentrations of the rare earth elements in waters draining mine workings from the Pine Creek area. Data were obtained in July 1999 and April 2000. Solid lines connecting REE data represent unfiltered water samples, and dashed lines represent samples filtered through a 0.45-μm membrane. Samples are grouped for clarity of presentation: a) AD002, AD005, and AD007; b) AD004, ADNEW, and S97-3; c) SPNEW and SP002.

c)

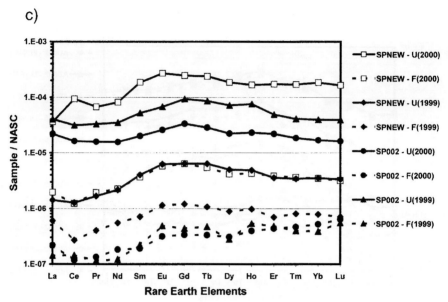

Figure 3. Continued.

filtration of the April 2000 samples. The data are plotted as the ratio of REE concentration in the filtered water relative to that in the solid digest in Figure 4. In interpreting the data in Figure 4 it is important to recall that the absolute values of the ratio depicted are a function of unknown dilution factors (see §3.3), and so only the REE patterns (not their absolute abundances) have significance. Figure 4 shows that there is no consistent pattern of fractionation of the REE between water and solid. Patterns for S97-3 and SPNEW, where Fe-oxyhydroxides are clearly precipitating in abundance and form a significant fraction of the suspended material trapped by the filter, are essentially flat, showing no fractionation whatsoever. Slight LREE-depletion is exhibited by AD004 and AD007, and slight LREE-enrichment is exhibited by AD005. In both these cases, the patterns are essentially flat from Sm to Lu, fractionation only being apparent from La to Nd. The pattern for SP002 is flat from La to Dy, but shows HREE-enrichment from Ho to Lu. Finally, the pattern for AD002 shows a slight MREE-depletion (the apparent Nd anomaly is probably due to minor contamination). The lack of any consistent, significant fractionation trends agrees with the parallelism of REE patterns for filtered and unfiltered aliquots as shown in Figure 3. Particularly interesting is the general lack of MREE-enriched or –depleted patterns in Figure 4.

The results of the speciation calculations are shown in Figures 5 and 6, where La^{3+} and Yb^{3+} are used to represent light and heavy REE, respectively. In the case of both La^{3+} and Yb^{3+}, only three species make contributions of more than 2%: $LnCO_3^+$, $LnSO_4^+$, and Ln^{3+}. The only other species making contributions greater than 0.5% was $LnHCO_3^{2+}$. In all the near-neutral to alkaline waters, except that draining ADNEW, $LnCO_3^+$ predominates, accounting for from ~100% (Yb^{3+} in SPNEW) to ~55% (La^{3+} in AD005).

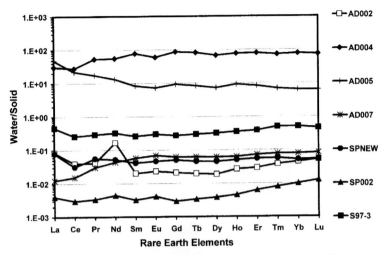

Figure 4. Results of the analysis of filter residues from the April 2000 sampling campaign. The data are presented as the concentration of REE in the water divided by the concentration of REE in the solution extract resulting from acid digestion of the corresponding filter residue.

Typically, Yb^{3+} shows a greater dominance than La^{3+} of the species $LnCO_3^+$, a consequence of the fact that the stability constants for carbonate complexes increase from light to heavy REE (cf. Cantrell and Byrne, 1987; Wood, 1990). In the waters in which $LnCO_3^+$ predominates, $LnSO_4^+$ is modeled as being present in proportions ranging from negligible to ~35%, and Ln^{3+} is modeled as accounting for negligible percentages up to ~12%. In the water draining ADNEW (pH = 7.8), sulfate complexes predominate for both La and Yb, with lesser contributions from the carbonate complex. In the only acidic drainage (AD004), La is distributed as 16% La^{3+} and 84% $LaSO_4^+$, and Yb is distributed as 18% Yb^{3+} and 82% $YbSO_4^+$, with negligible contributions from other species including carbonate complexes. The database of Haas et al. (1995) does not contain thermodynamic data for higher sulfate complexes, e.g., $Ln(SO_4)_2^-$ and $Ln(SO_4)_3^{3-}$, and so it is possible that the contribution of sulfate complexes in these waters has been underestimated. However, the highest activity of SO_4^{2-} in these waters is approximately 3×10^{-3} M (AD004), a value at which higher sulfate complexes make less than a 20% contribution to Ln speciation even in the absence of other ligands, according to calculations summarized by Wood (1990). In all other waters with lower activities of sulfate and higher activities of carbonate, the contribution of higher sulfate complexes is likely to be very small.

Gimeno et al. (2000) have demonstrated that competition by Al with REE for ligands can affect speciation calculations significantly for the REE in acidic waters. In the near-neutral waters investigated here, dissolved Al concentrations are low (< 0.4 mg L^{-1}) and the effect of competition is minimal. Only in the acidic water, AD004, where Al con-

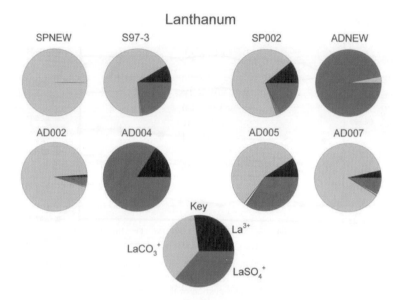

Figure 5. Results of lanthanum speciation calculations for the June 1999 samples. The key shows the shadings for the three main species. Other species account for less than 2% of total La.

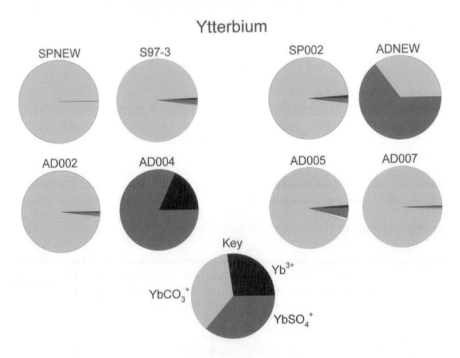

Figure 6. Results of ytterbium speciation calculations for the June 1999 samples. The key shows the shadings for the three main species. Other species account for less than 2% of total Yb.

centrations were measured to be ~ 3 mg L^{-1}, is competition from Al likely to be significant. Nevertheless, competition was taken into account when the activities of the free ligands were calculated using SOLMINEQ.88.

Solubility calculations indicate that the waters are strongly undersaturated (a minimum of a factor of 1000 undersaturated and typically significantly more) with respect to pure endmember phases of the type $Ln(OH)_3(s)$, $Ln_2(SO_4)_3(s)$ and $Ln_2(CO_3)_3(s)$. Phosphate was at or below the detection limit (50-100 μg L^{-1}) of our IC method in all of these waters, so neither saturation state with respect to REE phosphate phases (e.g., Byrne and Kim, 1993; Johannesson et al., 1995; Liu and Byrne, 1997) nor complexation of REE by phosphate (e.g., Bingler and Byrne, 1989; Byrne et al., 1991) can be evaluated. Thermodynamic data for other potential REE phases, e.g., $LnOHCO_3(s)$ are unavailable.

5. Discussion

Shale-normalized MREE-enrichment appears to be a common feature of ARD and has been previously reported by Nordstrom et al. (1995), Gimeno et al. (1996), Leybourne et al. (1998), Elbaz-Poulichet (1999), Verplanck et al. (1997; 1999), Gimeno et al. (2000), Worrall and Pearson (2001a), and Gammons et al. (2003). However, although MREE-enrichment is common in mine drainage, it is not invariably present. For example, Leybourne et al. (1998, 2000) also report mine waters with HREE-enriched shale-normalized patterns. Our study shows that MREE-enriched patterns also occur in ARD that has been neutralized by carbonate gangue minerals. It seems clear that the MREE-enrichment is not due solely to preferential complexation of REE in solution. The dominant REE species in ARD that has not been neutralized are the simple, hydrated ion Ln^{3+} and sulfate complexes. As pointed out by Wood (1990), among others, the stabilities of sulfate complexes are similar across the entire REE series, and therefore sulfate complexation alone cannot result in significant fractionation. Johannesson et al. (1996) pointed out that the stability constants for MREE-sulfate complexes actually are slightly higher than those for LREE or HREE, but also concluded that sulfate complexation was not the sole reason for MREE-enrichment in naturally acidic ground waters. In any case, in most of the near-neutral mine waters studied here, carbonate complexes are predominant over sulfate complexes, yet the MREE-enrichment persists. This finding suggests that dominance of sulfate complexes is not a necessary condition to achieve MREE-enrichment. Furthermore, the stabilities of carbonate complexes increase strongly and steadily from LREE to HREE (Cantrell and Byrne, 1987; Wood, 1990). Hence, carbonate complexation could be responsible for HREE-enrichment or -depletion, but by itself cannot yield MREE-enrichment.

The commonly observed precipitation of Fe oxyhydroxides from ARD and their known affinity for REE (as attested by the efficacy of the ferric hydroxide co-precipitation technique employed in this study) raises the possibility that precipitation of these phases is somehow responsible for MREE-enrichment. In our study, the fact that REE concentrations in near-neutral mine drainage are significantly higher in unfiltered aliquots than in filtered aliquots, especially in samples where Fe oxyhydroxides are clearly precipitating, suggest that the latter have removed REE from solution. However, as described above in the results section, our data are inconsistent with a significant role for iron oxyhydroxide precipitation in the production of the observed MREE-enrichment. If iron oxyhydroxides were preferentially removing LREE and HREE from solution to form MREE-enriched solutions, then the patterns shown in Figure 4 should be consistently MREE-depleted. Moreover, the NASC-normalized patterns for unfiltered and filtered aliquots would not be expected to be parallel. Finally, the acidic water, AD004, from which only small amounts of Fe-oxyhydroxides were seen to be precipitating at the point of discharge, has a similar MREE-enriched pattern to those waters from which Fe oxyhydroxides were clearly seen to have precipitated in abundance. Because neither complexation nor fractionation by precipitating Fe oxyhydroxides appears to be responsible for the MREE-enriched patterns, we tentatively suggest that these patterns are inherited from the minerals from which the REE were leached. Unfortunately, detailed information on the REE content of minerals in rocks through which the mine waters may have passed is not available.

Elbaz-Poulichet and Dupuy (1999) showed that little fractionation occurred between suspended particulate matter (including Fe oxyhydroxides) and water in the ARD-impacted Tinto and Odiel Rivers in Spain. In a study of REE geochemistry of acid waters from hard-rock mines in Colorado and Montana, Verplanck et al. (1999) concluded that the NASC-normalized REE pattern of the waters was source-related. Although precipitation of Fe oxyhydroxides removed some of the REE, no significant fractionation was found to occur. Worrall and Pearson (2001a,b) similarly concluded that precipitation of Fe ochres does not result in REE fractionation of waters draining coal mines in England. These conclusions are in agreement with the findings of this study.

On the other hand, Grandjean-Lécuyer et al. (1993) concluded that progressive extraction of LREE from river water by Fe oxyhydroxides could result in the precipitation of biogenic apatites with MREE-enrichment. Johannesson et al. (1996) invoked a combination of sulfate complexation and either dissolution of MREE-enriched Fe-Mn oxyhydroxides or sorption equilibrium with the latter phases to explain MREE-enrichment in some naturally acidic ground waters. Fractionation of the REE between iron ochres and mine waters was observed by Pearce et al. (1998), with La being preferentially partitioned into the water compared to Sm. Johannesson and Zhou (1999) showed that dissolution of MREE-enriched oxides/oxyhydroxides was a probable source of MREE-enrichment in the waters of a naturally acidic Arctic lake (Johannesson and Lyons, 1995). Åström (2001) called upon preferential scavenging of HREE by Fe oxyhydroxides to explain LREE/MREE-enrichment in streams and soil solutions impacted by oxidation of sulfides in marine sediments in Finland.

Experimental studies by Bau (1999) and Ohta and Kawabe (2001) show that the LREE (La-Nd) are relatively depleted in Fe oxyhydroxides compared to coexisting aqueous solutions. These findings contrast with the preferential extraction of LREE by Fe oxyhydroxides assumed by Grandjean-Lécuyer et al. (1993). In the experimental studies, cerium exhibits different behavior from the other REE owing to its ability to exist in the tetravalent state. The partition coefficient for Ce between Fe oxyhydroxide and solution appears to increase with time, reflecting slow oxidation and removal of Ce from solution with time. We observed no strong Ce anomalies in any of the waters from Pine Creek, even though all the waters studied were near saturation with respect to atmospheric oxygen. Thus, the lack of strong Ce anomalies suggests that precipitation of Fe-oxyhydroxides and consequent removal of REE is rapid at Pine Creek, such that there is insufficient time for Ce in solution to become oxidized and removed preferentially. Relatively rapid precipitation of Fe-oxyhydroxides might also help explain the apparent lack of fractionation of the trivalent REE between the mine waters and Fe-oxyhydroxides.

In a study of the REE geochemistry of the Berkeley Pit in Butte, Montana, Gammons et al. (2003) showed that precipitates (likely consisting of gypsum, schwertmannite, and possibly strengite), which form from the acidic Berkley Pit water, remove the LREE preferentially to the HREE (with a maximum K_d actually occurring at Nd). However, it was concluded that because the K_d values were relatively low, very large amounts of solid would have to precipitate to produce the NASC-normalized REE pattern observed in the Berkeley Pit waters.

Thus, there seems to be a discrepancy between the results of experimental and some field studies that suggest that precipitation of Fe oxyhydroxides should result in fractionation of the REE in the coexisting waters, and other field studies (including ours) in which no evidence for such fractionation exists. The reasons for this discrepancy are not clear, but may include differences in solution compositions (pH, ligand composition, etc.), differences in the rate of precipitation of the Fe oxyhydroxides, presence or absence of other REE-fractionating phases (e.g., Mn oxyhydroxides, schwertmannite), the relative mass of Fe oxyhydroxides precipitated from solution, and the influence of microbial activity.

As noted in the Results section, the only waters in which the concentrations of REE were similar in both filtered and unfiltered aliquots were the acidic AD004 and the near-neutral AD005. The similarity in REE concentrations in filtered and unfiltered aliquots suggests, but does not prove, that the REE are present in dissolved form. Although 0.45 μm is often taken as an operational division between dissolved ions and suspended particles, small colloidal particles can pass through a 0.45-μm filter. Nevertheless, it seems reasonable that the REE in the acidic waters draining AD004 are indeed in dissolved form, because low-pH waters are known typically to contain much higher dissolved REE concentrations than higher-pH waters. On the other hand, the neutral waters from AD005 contain similar REE concentrations as those from AD004, which is somewhat unusual. Moreover, the filtered and unfiltered aliquots of AD005 also have similar REE concentrations. This is not the case with any of the other near-neutral to alkaline mine waters in this study. This similarity in REE concentrations in filtered and unfiltered aliquots of waters from AD005 is particularly interesting given that yellow- to buff-colored oxyhydroxides precipitate from the water as it discharges from the adit. These oxides appear to be slightly different from those observed elsewhere, which are orange in color. Unfortunately, the precipitates observed in the field have not been analyzed. Examination of Figures 4 and 5 fails to reveal any significant differences in REE speciation between AD005 and the other near-neutral mine waters that could account for high REE solubility in spite of high pH. It is possible that the REE load in AD005 is predominantly present as very fine colloidal particles that pass through the 0.45-μm filter employed, which would explain both the higher overall REE concentrations and the similarity in filtered and unfiltered aliquots. The physical proximity of AD004 and AD005, and their similar Zn and REE concentrations may suggest some kind of direct relationship, e.g., a hydrological connection, between them. It is possible that initially the two mine waters have similar compositions, but that AD005 is rapidly neutralized (probably via contact with carbonates), resulting in strong oversaturation with respect to some mineral or minerals that contain Zn and REE, and the formation of colloidal particles of these minerals. In any event, it seems likely that the fluids emanating from AD005 were originally acidic, allowing them to attain high

REE concentrations, and then subsequently neutralized. Initially near-neutral fluids would not be capable of leaching REE from minerals in the rocks hosting the mineralization.

Of additional interest is the fact that, although the pH in AD004 varied from 3.3 in June 1999 to 2.3 in April 2000, there was very little difference in the REE concentrations, especially those of the LREE. Surprisingly, the HREE concentrations were slightly higher in June 1999 when the pH was higher, contrary to the expected trend. Also, Zn concentrations only varied from 80 to 60 mg L^{-1}, and were slightly higher at the higher pH. The variations in REE and Zn concentrations in AD004 observed between June 1999 and April 2000 are small considering the unit pH difference, and probably can be considered to be the same within analytical and sampling uncertainty. This finding may suggest that neither the REE nor the Zn concentrations are being controlled by equilibrium with any solid phases in AD004. Indeed, calculations show that no Zn phases have attained saturation (Wood et al., in prep.), and no REE phases for which thermodynamic data are available have attained saturation either. Finally, as noted above, there is little evidence at AD004 for the presence of solids, including Fe oxyhydroxides, which could exert a sorption control on the concentrations of the REE or Zn.

The above situation contrasts with that of AD005, in which both Zn and the REE are significantly more concentrated in the lower pH samples from April 2000. The pH varies from 7.8 to 6.9 and Zn concentrations vary from 65 to 235 mg L^{-1} in June 1999 and April 2000, respectively. REE concentrations vary over nearly an order of magnitude between the two sampling dates. Interestingly, the difference in REE concentrations between filtered and unfiltered aliquots was larger (but still comparatively small) in June 1999 when the pH was higher, than in April 2000 when the pH was lower. These findings are all consistent with some kind of solubility or sorption control of Zn and REE concentrations in AD005.

6. Conclusions

The concentrations of REE in mine waters from the Pine Creek Pb-Zn district, Idaho, in June 1999 and April 2000 have been determined. All but one of these mine waters are near-neutral to slightly alkaline owing to the presence of carbonate gangue minerals. All the mine waters exhibit MREE-enriched, NASC-normalized patterns, most with a fairly steep increase in normalized abundance from La to Ga, followed by a shallower decrease from Gd to Lu. The highest REE concentrations were found in the most acidic (pH = 2.3-3.3) mine water. In this water, the REE contents of the filtered and unfiltered aliquots were identical, and there was no significant difference between concentrations measured in the two sampling campaigns. The REE in the acidic water appear to be present in true solution with no solubility or sorption control. One near-neutral mine water, issuing from an adit in close proximity to the adit from which the acidic mine water emerged, had similarly high REE concentrations and relatively little difference between filtered and unfiltered aliquots. This water may have originally been acidic, but then neutralized and supersaturated relatively rapidly such that small colloidal particles containing REE and Zn were formed. However, all other near-neutral to slightly alkaline mine waters exhibit lower REE concentrations, greater differences between filtered and unfiltered aliquots, and greater temporal variations.

The results suggest that, in the near-neutral to slightly alkaline mine waters, the REE are being removed by co-precipitation with or sorption onto Fe oxyhydroxides. However, the formation of Fe oxyhydroxides upon emergence of these mine waters does not appear to be causing fractionation responsible for the observed MREE-enrichment of the waters. Complexation also does not appear to be responsible for MREE-enrichment. In the acidic water, the predominant REE species are the uncomplexed, hydrated ions and the sulfate complexes. However, in the near-neutral to alkaline waters carbonate complexes most often predominate, with a lesser contribution from sulfate complexes. The reason for MREE-enrichment in these mine waters is not clear, but we suspect that it may represent a pattern inherited from minerals in the source rocks. All of the waters are strongly undersaturated with respect to pure endmember REE hydroxide, carbonate and sulfate phases.

The behavior of Zn, the heavy metal present in highest concentration in these mine waters, shows some parallels with that of the REE. In particular, neither REE nor Zn concentrations appear to be controlled by equilibria with any solid phases in the one acidic mine water, but may be so controlled in all the other near-neutral to alkaline waters. Moreover, both Zn and REE are unusually high in concentration in the neutral mine water emanating from AD005. The results of this study show that aqueous REE geochemistry can provide useful insights into processes occurring in waters draining former mining districts.

Acknowledgments

We would like to thank the many students in the course GEOL 478/578 who, since 1992, helped sample and analyze mine waters from the Pine Creek district. We also acknowledge the assistance of Jennifer Gustafson and Tom Williams in obtaining samples in 1999 and 2000. Dr. Charles Knaack granted access to and provided technical assistance with ICP-MS analyses at the Washington State University Geoanalytical Laboratories.

References

Aggarwal, J.K., Shabani, M.B., Palmer, M.R., and Ragnarsdottir, K.V. 1996 Determination of the rare earth elements in aqueous samples at sub-ppt levels by inductively coupled plasma mass spectrometry and flow injection ICPMS. *Analytical Chemistry*, **68**, 4418-4423.

Alpers, C.N. and Blowes, D.W., eds. 1994 Environmental geochemistry of sulfide oxidation. *American Chemical Society Symposium Series*, **550**, 681p.

Åström, M. 2001 Abundance and fractionation patterns of rare earth elements in streams affected by acid sulphate soils. *Chemical Geology*, **175**, 249-258.

Bau, M. 1999 Scavenging of dissolved yttrium and rare earth elements by precipitating iron oxyhydroxide: experimental evidence for Ce-oxidation, Y-Ho fractionation and lanthanide tetrad effect. *Geochimica et Cosmochimica Acta*, **63**, 67-77.

Bingler, L.S. and Byrne, R.H. 1989 Phosphate complexation of gadolinium(III) in aqueous solution. *Polyhedron*, **8**, 1315-1320.

Buchanan, A.S. and Hannaker, P. 1984 Inductively coupled plasma spectrometric determination of minor elements in concentrated brines following precipitation. *Analytical Chemistry*, **56**, 1379-1382.

Byrne, R.H. and Kim, K.-H. 1993 Rare earth precipitation and coprecipitation behavior: The limiting role of PO_4^{3-} on dissolved rare earth concentrations in sewater. *Geochimica et Cosmochimica Acta*, **57**, 519-526.

Byrne, R.H., Lee, J.H., and Bingler, L.S. 1991 Rare earth element complexation by PO_4^{3-} ions in aqueous solution. *Geochimica et Cosmochimica Acta*, **55**, 2729-2735.

Campbell, A.B. 1949 The paragenesis of the lead-zinc ores, Pine Creek district, Shoshone County, Idaho. Unpublished M.A. thesis, Washington University, 64p.

Cantrell, K.J. and Byrne, R.H. 1987 Rare earth element complexation by carbonate and oxalate ions. *Geochimica et Cosmochimica Acta*, **51**, 597-605.

Carlson-Foscz, V.L., Oreskes, N., and Nordstrom, D.K. 1991 Mobility of rare earth elements in the Ophir region, San Juan Mountains, Colorado. *EOS Transactions of the American Geophysical Union*, **72**, 308.

Diakonov, I.I., Tagirov, B.R., and Ragnarsdottir, K.V. 1998a Standard thermodynamic properties and heat capacity equations for rare earth element hydroxides. I. $La(OH)_3$ and $Nd(OH)_3$. Comparison of thermochemical and solubility data. *Radiochimica Acta*, **81**, 107-116.

Diakonov, I.I., Ragnarsdottir, K.V., and Tagirov, B.R. 1998b Standard thermodynamic properties and heat capacity equations of rare earth hydroxides: II. Ce(III)-, Pr-, Sm-, Eu(III)-, Gd-, Tb-, Dy-, Ho-, Er-, Tm-, Yb-, and Y-hydroxides. Comparison of thermochemical and solubility data. *Chemical Geology*, **151**, 327-347.

Elbaz-Poulichet, F. and Dupuy, C. 1999 Behaviour of rare earth elements at the freshwater-seawater interface of two acid mine rivers: the Tinto and Odiel (Andalucia, Spain). *Applied Geochemistry*, **14**, 1063-1072.

Firsching, F.H., and Mohammadzadel, J. 1986 Solubility products of the rare earth carbonates. *Jour. Chem. Eng. Data* **31**, 40-42.

Forrester, J.D. and Nelson, V.E. 1944 Lead-zinc deposits of the Pine Creek area, Coeur D'Alene mining region, Shoshone County, Idaho. U.S. Geol. Surv. Strategic Minerals Investigation.

Gammons, C.H., Wood, S.A., Jonas, J.P., and Madison, J.P. 2003 Geochemistry of the rare earth elements and uranium in the acidic Berkeley Pit lake, Butte, Montana. *Chemical Geology* **198**, 269-288.

Garrels, R.M. and Christ, C.L. 1965 Solutions, Minerals, and Equilibria. Harper & Row, New York, 450p.

Gimeno, M.J., Auqué, L.F., Nordstrom, D.K., and Bruno, J. 1996 Rare earth element (REE) geochemistry and the tetrad effect in the naturally acidic waters of Arroyo del Val, northeastern Spain. *Geological Society of America Annual Meeting Abstracts with Program*, **28**, 468.

Gimeno, M.J., Auqué, L.F., and Nordstrom, D.K. 2000 REE speciation in low-temperature acidic waters and the competitive effects of aluminum. *Chemical Geology*, **165**, 167-180.

Gin, T.T. 1953 Mineralization in the Pine Creek area, Coeur D'Alene mining region, Idaho. Unpublished Ph.D. thesis, University of Utah, 73 pp.

Grandjean-Lécuyer, P., Feist, R., and Albarède, F. 1993 Rare earth elements in biogenic apatites. *Geochimica et Cosmochimica Acta*, **57**, 2507-2514.

Greaves, M.J., Elderfield, H. and Klinkhammer, G.P. 1989 Determination of the rare earth elements in natural waters by isotope-dilution mass spectrometry. *Analytica Chimica Acta*, **218**, 265-280.

Gromet, L.P., Dymek, R.F., Haskin, L.A. and Korotev, R.L. 1984 The "North American shale composite": Its compilation, major and trace element characteristics. *Geochimica et Cosmochimica Acta*, **48**, 2469-2482.

Haas, J. R., Shock, E. L., and Sassani, D. C. 1995 Rare earth elements in hydrothermal systems: Estimates of standard partial molal thermodynamic properties of aqueous complexes of the rare earth elements at high pressures and temperatures: *Geochimica et Cosmochimica Acta*, **59**, 4329-4350.

Hollings, P., Hendry, M.J., and Kerrich, R. 1999 Sequential filtration of surface and ground waters from the Rabbit Lake uranium mine, northern Saskatchewan, Canada. *Water Quality Research Journal of Canada*, **34**, 221-247.

Jambor, J.L. and Blowes, D.W., eds. 1994 Environmental geochemistry of sulfide mine-wastes. *Mineralogical Assocation of Canada Short Course Handbook*, **22**, 438p.

Johannesson, K.H. and Lyons, W.B. 1995 Rare earth element geochemistry of Colour Lake, an acidic freshwater lake on Axel Heiberg Island, Northwest Territories, Canada. *Chemical Geology*, **119**, 209-223.

Johannesson, K.H., Lyons, W.B., Stetzenbach, K.J., and Byrne, R.H. 1995 The solubility control of rare earth elements in natural terrestrial waters and the significance of PO_4^{3-} and CO_3^{2-} in limiting dissolved rare earth concentrations; a review of recent information. *Aquatic Geochemistry*, **1**, 157-173.

Johannesson, K.H., Lyons, W.B., Yelken, M.A., Gaudette, H.E., and Stetzenbach, K.J. 1996 Geochemistry of the rare earth elements in hypersaline and dilute acidic natural terrestrial waters: complexation behaviour and middle rare earth element enrichments. *Chemical Geology*, **133**, 125-144.

Johannesson, K.H., and Zhou, X. 1999 Origin of middle rare earth element enrichments in acid waters of a Canadian High Arctic lake. *Geochimica et Cosmochimica Acta*, **63**, 153-165.

Jones, E.J., Jr. 1919 A reconnaissance of the Pine Creek district. *U.S. Geological Survey Bulletin*, **710**, 1-36.

Kharaka, Y.K., Gunter, W.D., Aggarwal, P.K., Perkins, E.H. and DeBraal, J.D. 1988 SOLMINEQ88: A computer program for geochemical modeling of water-rock interactions. *U.S.G.S. Water Resources Investigations Report*, **88-4227**, 420 p.

Leybourne, M.I., Goodfellow, W.D., and Boyle, D.R. 1998 Hydrogeochemical, isotopic, and rare earth element evidence for contrasting water-rock interactions at two undisturbed Zn-Pb massive sulphide deposits, Bathurst Mining Camp, N.B., Canada. *Journal of Geochemical Exploration*, **64**, 237-261.

Leybourne, M.I., Goodfellow, W.D., Boyle, D.R., and Hall, G.M. 2000 Rapid development of negative Ce anomalies in surface waters and contrasting REE patterns in groundwaters associated with Zn-Pb massive sulphide deposits. *Applied Geochemistry*, **15**, 695-723.

Liu, X. and Byrne, R.H. 1997 Rare earth and yttrium phosphate solubilities in aqueous solution. *Geochimica et Cosmochimica Acta*, **61**, 1625-1633.

Miekeley, N., Couthino de Jesus, H., Porto da Silveira, C.L., Linsalata, P., and Morse, R. 1992 Rare-earth elements in groundwaters from the Osamu Utsumi mine and Morro do Ferro analogue study sites, Poços de Caldas, Brazil. *Journal of Geochemical Exploration*, **45**, 365-387.

Nordstrom, D.K., Carlson-Foscz, V., and Oreskes, N. 1995 Rare earth element (REE) fractionation during acidic weathering of San Juan Tuff, Colorado. *Geological Society of America Annual Meeting Abstracts with Program*, **27**, 199.

Ohta, A. and Kawabe, I. 2001 REE(III) adsorption onto Mn dioxide (δ-MnO_2) and Fe oxyhydroxide: Ce(III) oxidation by δ-MnO_2 . *Geochimica et Cosmochimica Acta*, **65**, 695-703.

Pearce, N.J.G., White, R.A., Fuge, R. 1998 Behaviour of rare earth elements and heavy metals in ochreous drainage; analogue elements for the behaviour of the trans-uranic metals. *Geological Society of America Annual Meeting Abstracts with Program*, **30**, 128.

Rard, J.A. 1988 Aqueous solubilities of praseodymium, europium, and lutetium sulfates. *Journal of Solution Chemistry*, **17**, 499-517.

Shabani, M.B., Akagi, T., Shimizu, H., and Masuda, A. 1990 Determination of trace lanthanides and yttrium in seawater by inductively coupled plasma mass spectrometry after preconcentration with solvent extraction and back-extraction. *Analytical Chemistry*, **62**, 2709-2714.

Silverman, A.J. 1957 Structural terminations of vein type ore deposits. Part I. The Sidney Mine, Pine Creek area, Coeur D'Alene mining region, Idaho. Unpublished M.A. thesis, Columbia University, 52p.

Stumm, W. and Morgan, J.J. 1981 Aquatic chemistry - An introduction emphasizing chemical equilibria in natural waters. John Wiley & Sons, New York, 780p.

van Middlesworth, P.E. and Wood, S.A. 1998 The aqueous geochemistry of the rare earth elements and yttrium: Part 7: REE, Th and U contents in thermal springs associated with the Idaho Batholith. *Applied Geochemistry*, **13**, 861-884.

Verplanck, P.L., Nordstrom, D.K., Taylor, H.E., and Wright, W.G. 1997 Nonconservative nature of rare earth elements in an acidic alpine stream, upper Animas River basin, Colorado. *Geological Society of America Annual Meeting Abstracts with Program*, **29**, 152.

Verplanck, P.L., Nordstrom, D.K., and Taylor, H.E. 1999 Overview of rare earth element investigations in acid waters of U.S. Geological Survey abandoned mine lands watersheds. In: D.W. Morganwalp and H.T. Buxton (eds.) U.S. Geological Survey Toxic Substances Program: Proceedings of the Technical Meeting. Water Resources Report of Investigations, WRI 99-4018-A, pp. 83-92.

Webb, C., Davis, A.D., and Hodge, V.F. 1993 Geochemical characterization of acidic waters from uranium mines in the Black Hills of South Dakota. *Geological Society of America Annual Meeting Abstracts with Program*, **25**, 323.

Weisel, C.P., Duce, R.A. and Fasching, J.L. 1984 Determination of aluminum, lead and vanadium in North Atlantic Seawater after coprecipitation with ferric hydroxide. *Analytical Chemistry*, **56**, 1050-1052.

Welch, S., Lyons, W.B. and Kling, C.A. 1990 A co-precipitation technique for determining trace metal concentrations in iron-rich saline solutions. *Environmental Science and Technology*, **11**, 141-144.

White, R.A., Pearce, N.J.G., and Fuge, R. 1998 Behaviour of rare earth elements and other metals in synthetic and natural acid mine drainage, a laboratory study. *Geological Society of America Annual Meeting Abstracts with Program*, **30**, 254.

Wood, S.A. 1990 The aqueous geochemistry of the rare earth elements and yttrium. Part I. Review of available low temperature data for inorganic complexes and the inorganic REE speciation of natural waters. *Chemical Geology*, **82**, 159-186.

Wood, S.A., Baker, L.L, Shannon, W.M. in prep. Heavy metal concentrations in acidic and near-neutral mine drainage from the Pine Creek area, Coeur d'Alene River valley, Idaho.

Worrall, F. and Pearson, D.G. 2001a The development of acidic groundwaters in coal-bearing strata: Part I. Rare earth element fingerprinting. *Applied Geochemistry*, **16**, 1465-1480.

Worrall, F. and Pearson, D.G. 2001b Water-rock interaction in an acidic mine discharge as indicated by rare earth element patterns. *Geochimica et Cosmochimica Acta*, **65**, 3027-3040.

Appendix

Table A1. Concentrations of REE (ng/L) of Pine Creek mine drainage (June 1999). U – unfiltered; F – filtered through a 0.45-µm membrane.

Sample	La	Ce	Pr	Nd	Sm	Eu	Gd	Tb	Dy	Ho	Er	Tm	Yb	Lu
AD002 – U	23	47	5.2	25	2.3	6.4	9.2	1.1	7.0	1.3	3.4	0.6	2.9	0.5
AD002 – F	7.5	13	1.7	14	1.0	3.3	6.3	0.8	4.9	1.0	2.2	0.2	1.9	0.3
AD004 – U	1200	3800	610	3500	370	1500	2300	36	1900	350	780	110	680	100
AD004 – F	1100	3600	550	3300	340	13040	2100	320	1700	310	710	95	620	93
AD005 – U	360	830	140	790	74	270	510	72	400	74	170	21	120	20
AD005 – F	290	620	100	540	46	160	340	49	270	54	130	16	89	14
AD007 – U	41	95	16	92	11	33	59	9.6	53	10	23	3.1	19	2.9
AD007 – F	13	25	3.8	24	3.3	8.3	18	3.2	19	4.1	10	1.4	7.8	1.2
ADNEW – U	45	72	9.0	42	3.8	11	18	2.7	16	2.8	6.4	0.9	4.9	1.0
ADNEW – F	7.6	16	1.3	6.9	1.1	2.5	3.1	0.6	3.5	0.8	2.1	0.3	1.9	0.4
SP002 – U	1300	2200	250	1100	81	290	450	70	400	76	160	20	120	18
SP002 – F	4.2	10	0.9	5.3	0.6	1.2	2.0	0.4	1.5	0.5	1.5	0.2	1.1	0.2
SPNEW - U	44	91	13	71	7.7	23	33	5.3	29	5.0	12	1.7	11	1.6
SPNEW - F	17	18	2.9	16	1.3	3.7	5.1	0.8	4.7	0.9	2.1	0.4	2.2	0.3
S97-3 – U	6.5	12	1.4	11	1.0	1.9	4.0	0.6	3.1	0.7	2.1	0.3	2.3	0.4
S97-3 – F	2.7	20	0.4	2.6	0.5	0.5	0.9	0.1	0.7	0.2	0.5	ND	0.6	0.1

Table A2. Concentrations of REE (ng/L) of Pine Creek mine drainage (April 2000). U – unfiltered; F – filtered through a 0.45-µm membrane.

Sample	La	Ce	Pr	Nd	Sm	Eu	Gd	Tb	Dy	Ho	Er	Tm	Yb	Lu
AD002 – U	14	30	2.8	22	3.1	1.2	4.4	0.7	3.6	0.7	2.3	0.3	2.6	0.5
AD002 – F	6.0	9.2	1.2	22	1.2	0.4	1.7	0.2	1.3	0.3	0.8	0.2	1.2	0.2
AD004 – U	1200	3700	530	2800	1000	250	1500	210	1100	190	550	74	450	64
AD004 – F	9908	3100	480	2600	9301	230	1300	190	9707	170	520	69	420	60
AD005 – U	2000	4500	740	3600	1200	310	1900	250	1300	240	660	83	490	72
AD005 – F	2000	4200	670	3300	1000	260	1600	220	1200	210	580	72	410	60
AD007 – U	481	150	25	1401	58	20	100	15	78	15	45	6.1	36	5.5
AD007 – F	31	63	14	75	27	9.3	48	6.8	37	7.1	21	2.8	16	2.5
ADNEW - U	290	500	60	230	59	15	81	11	58	10	30	3.7	21	3.1
ADNEW – F	28	35	4.5	19	4.5	1.3	7.7	1.3	7.4	1.7	5.5	0.8	4.6	0.7
SP002 – U	710	1200	130	480	120	32	180	24	130	24	75	9.3	52	7.8
SP002 – F	6.9	8.6	1.1	5.0	1.1	0.4	1.8	0.3	1.8	0.4	1.5	0.2	1.6	0.3
SPNEW – U	1200	6800	530	2600	1100	330	1300	200	1100	170	580	85	570	79
SPNEW – F	62	90	15	68	21	7.2	33	4.6	24	4.4	13	1.8	11	1.5
S97-3 – U	15	28	4.1	19	5.2	1.7	8.9	1.3	7.4	1.5	4.6	0.7	4.2	0.6
S97-3 – F	6.4	8.8	1.3	6.8	1.8	0.6	3.1	0.5	2.8	0.6	1.9	0.3	1.8	0.2

Chapter 5

LANTHANIDE ELEMENTS IN ALTERATION MINERALS AS A GUIDE TO TRANSPORT PHENOMENA IN THE VADOSE ZONES OF CONTINENTAL TUFFS

DAVID VANIMAN, DAVID BISH, & STEVE CHIPERA

Group EES-6, Geology, Geochemistry, and Hydrology, MS D462, Los Alamos National Laboratory, Los Alamos, NM 87545 USA

1. Introduction: The utility of REE data from groundwater-deposited minerals in vadose-zone rocks

Other chapters in this book focus primarily on direct analysis and interpretation of rare-earth elements (REE) in groundwaters, or on the determination of factors relating groundwater REE data to correlative data from suspended sediments or host rock. In this chapter we address the interpretation of REE data from groundwater-deposited minerals in vadose-zone rocks where the REE compositions of the source waters are unknown. The goal here is to illustrate methods by which useful information can be extracted from such minerals on the transport of REE and related heavy metals by groundwater. Although disequilibrium processes and the lack of specific data on water/mineral partition coefficients for the REE prevent estimation of concentrations of these elements in groundwater from mineral data, REE patterns in vadose-zone minerals can be very useful. The data discussed in this chapter illustrate applications developed from common vadose-zone precipitates or products of groundwater alteration (calcite, Mn oxides, and clays) in a widespread continental volcanic lithology (tuff). This lithology provides a leachable or readily altered source material that in many instances presents a traceable REE signature, particularly when the tuff has cooled quickly and provides a glassy source material for groundwater leaching. Two localities are considered in this chapter (Figure 1). One is Yucca Mountain in Nevada, where there is a thick sequence of predominantly rhyolitic Miocene (12-13 Ma) tuffs with rarer quartz-latitic tuffs in a thick (~500 m) vadose zone. The other is at Pajarito Mesa, part of the Pajarito Plateau in New Mexico where the early Pleistocene Bandelier Tuff (1.2-1.6 Ma and ~250 m thick) is entirely within the vadose zone and overlies fanglomerates and basaltic lavas that place it ~100 m above the regional water table.

The precipitated or altered phases emphasized here range from structurally and chemically simple calcite to complex hydrous alteration products. Clay minerals with sufficient REE data for consideration include smectites, generally interstratified with lesser amounts of illite, and kaolinites or halloysites. The Mn oxides include pyrolusite, cryptomelane, birnessite, and rancieite but these "identifications" as definitive mineral types often belie an association with amorphous Mn oxides that may contain a high percentage of the REE. Many other phases have potential significance but are not covered (e.g., zeolites) for a lack of sufficient data.

The REE, particularly the lanthanides, play a significant role in understanding the origins of rocks and minerals. Many summaries of the use of REE in geochemistry are

K.H. Johannesson,(ed), Rare Earth Elements in Groundwater Flow Systems , 111-140.

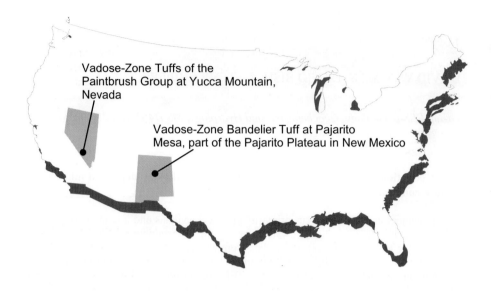

Figure 1. Locations of Yucca Mountain in Nevada and the Pajarito Plateau (including the Pajarito Mesa) in New Mexico.

available in book form (e.g., Henderson, 1984; Lipin and McKay, 1989) or as chapters within compendia dealing with the REE (e.g., Taylor and McLennan, 1988). The bulk of the research in this area, however, is focused on the geochemical partitioning of the REE in igneous and metamorphic systems. The behavior of REE in sedimentology, weathering processes, and aqueous geochemistry has received less coverage. In part this unequal coverage is a result of the very systematic mineral-mineral and mineral-melt partitioning properties of the REE at high temperature, allowing their use in the solution of problems in metamorphic and especially in igneous systems. Although comparable understanding of water-mineral REE partitioning is compromised by kinetic factors, disequilibrium, complexation with other species in solution, and lack of data, there is a growing body of work, particularly on seawater-mineral systems, that illustrates the very useful information to be gained (see for example de Baar et al., 1991).

The vadose-zone products of groundwater interaction with tuff described here illustrate two aspects of REE composition that are common threads within this book. First, the shapes of REE patterns and the magnitudes of Eu anomalies are both useful tracers of source rocks for REE in groundwaters. This does not discount possible effects of incongruent dissolution in host rocks (i.e., preferential dissolution of minerals with REE patterns that differ from the whole rock), but recognizes that such effects need to be accounted for in data interpretations. Second, the development of Ce anomalies is far more prevalent in groundwater/mineral systems than it is in the better-studied igneous and metamorphic systems. In groundwater systems, Ce stands out as the REE most susceptible to anomalous separation. Europium separation may also occur but under very different circumstances; to be separated from the other REE in natural aqueous processes, Eu^{3+} must be reduced to Eu^{2+} at conditions of low Eh (<-0.35 v) that can only occur in very reducing groundwater. In contrast, separation of Ce^{3+} from the other trivalent REE is attained by oxidation to Ce^{4+} at conditions more readily attained in aqueous systems of moderately elevated pH or Eh (Brookins, 1989). Because Eh and pH both play a role in Ce^{4+}/Ce^{3+} speciation, interpretations of these parameters from groundwater alteration and precipitation products can be ambiguous, but other data can be used to reduce uncertainties in interpretation. This is particularly the case where redox reactions can be deduced for minerals that accumulate Ce.

The groundwater-formed minerals used as examples in this chapter include calcite, clays, and Mn oxides. These minerals were selected based on (1) evidence of a significant role in accumulation of REE from groundwater (calcite, clays) or (2) evidence of a controlling effect in Ce^{4+}/Ce^{3+} speciation (Mn oxides). Other alteration minerals (e.g., opal) are common but contain REE in low abundance or are too uncommon in tuffs for practical use (e.g., fluorite). For the chosen minerals, comparisons with the same or similar minerals from other groundwater environments are drawn to illustrate a small part of the range of REE compositions that can be found. The magnitude of variation and the rich diversity of REE patterns in these minerals reveal how little we know and how much is to be gained by further study of minerals formed from groundwater.

2. Methods

The focus of this chapter is on the subset of REE that comprise the lanthanide elements. The systems examined focus on groundwater-deposited alteration minerals of the vadose zone from continental tuffs, but the widespread distribution of these alteration minerals (especially calcite and clays) makes the methods described here more broadly applicable. There are many methods for analyzing the lanthanide elements in minerals. Where groundwater-formed minerals are abundant enough to be separable by hand picking (e.g., calcite) or by sample crushing and sequential sedimentation of the finest material (e.g., clays), enough material can be collected for analysis by well-established methods such as instrumental neutron activation analysis (INAA) or induction-coupled plasma mass spectrometry (ICP-MS). Electron microbeam (microprobe or scanning electron microscope) methods can be used where lanthanide elements occur in concentrations of a fraction of a weight percent or more. Ion probe methods are very effective when the lanthanide elements occur in concentrations of a few to tens of ppm and an element of relatively invariant concentration (e.g., Ca in calcite) is present for reference intensity. Laser ablation ICP-MS holds considerable promise but requires further development to be more generally applicable. In this chapter the focus is on INAA data obtained from mineral separates, either hand-picked (calcite) or sedimented (clays).

2.1   ANALYSIS OF MINERAL SEPARATES

Minerals provide the media for REE accumulation from groundwater. In order to obtain useful information about mineral/water controls on lanthanide element distributions, analyses of pure minerals are needed. One method of obtaining these analyses is through mineral separates. Where the minerals are coarse enough and readily visible, hand separation is possible. Calcite crystals in fractures and voids can be concentrated in quantities necessary for analysis by several bulk methods, including INAA and solution ICP-MS. Where intergrown with other clear minerals such as opal, differences in short-wave UV fluorescence can be used to pick out the opal, which will fluoresce yellow-green in many occurrences. For a simple phase such as calcite with a limited range of Ca content, the analytical results for Ca can be used as an aid in assessing the purity of the material separated. For greater confidence, it is highly recommended that the mineral separate be hand-crushed in an agate mortar and analyzed by X-ray diffraction (XRD) to determine whether other minerals are present.

For clay minerals, some samples can be hand-picked where the clays are localized in fractures and voids, but crushing and powdering of the whole-rock sample is more generally useful. Crushing may be done in a W-carbide shatterbox, followed by sample disruption with a sonic probe, sequential sedimentation in deionized water, and centrifugation to separate the 0.1-0.5 µm size fraction (Chipera et al., 1993). In many samples this treatment is sufficient to obtain clay separates of >95% purity from tuff matrices, with impurities consisting principally of feldspar and silica minerals. Evaporation of the centrifuged supernatant can be used to obtain what is generally purer clay as a <0.1-µm size fraction left in suspension. In some cases, however, impurities that are very important in REE studies (e.g., phosphate minerals such as apatite and crandallite) may increase in the finer clay fractions. This problem is discussed in section 4.3.

The idea of mineral "purity" is fairly straightforward in calcite but more complex in clay minerals. The purity of clay separates must be considered in terms of the common intergrowth of many different clay minerals in a single "clay" sample. Although clay separates may consist of >99% clay by weight, this measure may include several clay-family phyllosilicates such as interstratified illite/smectite (I/S), kaolinite, and chloritic components, often with remnant mica that may provide nuclei for clay-mineral growth. Such complex intergrowths are so inseparable that they must of necessity be treated as a single component in lanthanide element analysis. Some additional information can be gleaned where multiple clay size fractions or multiple separates from the same parent sample have different proportions of the clay minerals, but there is always a concern about small-scale variability in the parent sample and possible geochemical differences between different crystallite sizes of the same clay-mineral constituent.

2.2   MICROBEAM METHODS

Analysis of minerals in polished mounts or thin sections can be accomplished by electron microprobe or ion probe. Most groundwater-deposited minerals have lanthanide element abundances too low for analysis by microprobe, but exceptions such as Mn oxides can provide important information about vadose-zone lanthanide element accumulation (e.g., Carlos et al., 1993). Practical application of electron microprobe analysis is generally limited to the abundant light lanthanide elements (e.g., La and Ce). Ion probe analysis is far more sensitive but requires an index element of relatively fixed abundance for quantitative analysis; however, relative lanthanide element abundance patterns can often be determined where no index element is present. Calcite, when relatively pure and close to ideal $CaCO_3$ composition, provides Ca as a useful index element (40.04% Ca in ideal calcite) for ion probe analysis of lanthanide elements. Denniston et al. (1997) used this method to examine zoned REE profiles in calcites from Yucca Mountain, obtaining useful data for La, Ce, Nd, Sm, Eu, Dy, Er, and Yb with supporting data on Mn, Fe, and Sr from the same analytical points. Laser ablation ICP-MS also holds considerable promise for the future in calcite analysis for REE content (e.g., Ionov and Harmer, 2002).

3. Examples of mineral systems that reflect vadose-zone water compositions

A large number of minerals deposited by groundwater are amenable to determination of lanthanide elements. The examples emphasized here illustrate common minerals that may be found in many different localities and in many different rock types. Certain REE-rich minerals deposited in small amounts can be important in groundwater/mineral systems. The Mn oxides are described here because of their ubiquity in altered tuffs, but other REE-rich minerals in other systems (e.g., cerianite and florencite in laterites, Valeton et al., 1997) can play comparable roles and must be considered in those systems.

3.1 CALCITE: A PASSIVE AND RELATIVELY SIMPLE RECORD

The record of calcite deposition in the vadose zone at Yucca Mountain has received considerable attention because of the potential development of a repository for high-level nuclear waste at this site (Peterman et al., 1992; Whelan and Stuckless, 1992; Vaniman and Chipera, 1996). A simplified stratigraphy of tuffs in the vadose zone at Yucca Mountain with average unit thicknesses is shown in Figure 2. This stratigraphy is dominated by devitrified tuffs (crystallized during cooling to feldspars and silica minerals), with the exception of basal upper rhyolites plus the upper quartz-latitic horizon (stratigraphic levels 2 and 3 in Figure 2) and the vitric-to-zeolitic base of the Topopah Spring Tuff (stratigraphic level 8 in Figure 2) that are either glassy or partially to completely altered to clay and/or zeolites (clinoptilolite ± mordenite). Most important for the data described in this chapter is the difference in lanthanide composition between rhyolitic and quartz-latitic compositions, the former characterized by relatively low rise in chondrite-normalized patterns of the light lanthanides with strong negative Eu anomalies and the latter by steep rise in the light lanthanides and essentially no Eu anomaly (Figure 2).

Questions of vadose-zone hydrology, paleoclimate influence on recharge, and vadose-zone mass transfer are all linked to understanding calcite deposition. For a more comprehensive study of calcite at Yucca Mountain that considers the data from stable and radiogenic isotopes (see Stuckless et al., 1991). For the purposes of this chapter we

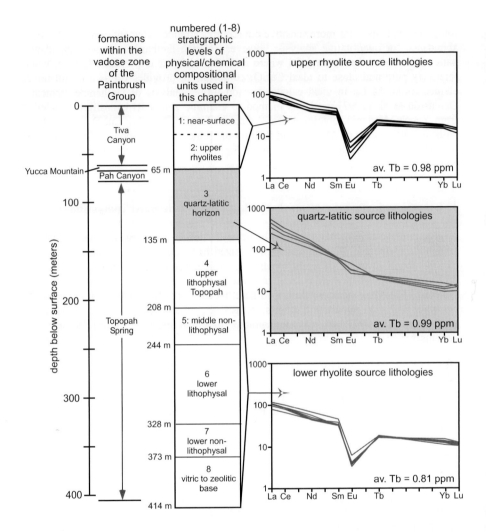

Figure 2. Simplified stratigraphy of vadose-zone Paintbrush Group tuffs at Yucca Mountain. Depositional formations are shown on the left (Tiva Canyon Tuff, Yucca Mountain Tuff, Pah Canyon Tuff, and Topopah Spring Tuff), at their average thicknesses for the portion of Yucca Mountain considered in this chapter. The middle column summarizes physical/chemical stratigraphic levels (based on either cooling-unit features or rhyolite versus quartz-latitic composition). Thicknesses for both the formations and the physical/chemical stratigraphic levels are averages based on stratigraphic information from six drill holes. The physical/chemical stratigraphic levels are numbered in sequence from the top (1 through 8) for cross-reference with Figure 3 and Figure 5. Representative whole-rock chondrite-normalized lanthanide patterns are illustrated for upper rhyolite, quartz-latitic, and lower rhyolite lithologies. Average Tb abundances are shown for comparison because Tb in the Yucca Mountain samples is least affected by depth of Eu anomaly or by changes in chondrite-normalized lanthanide slope.

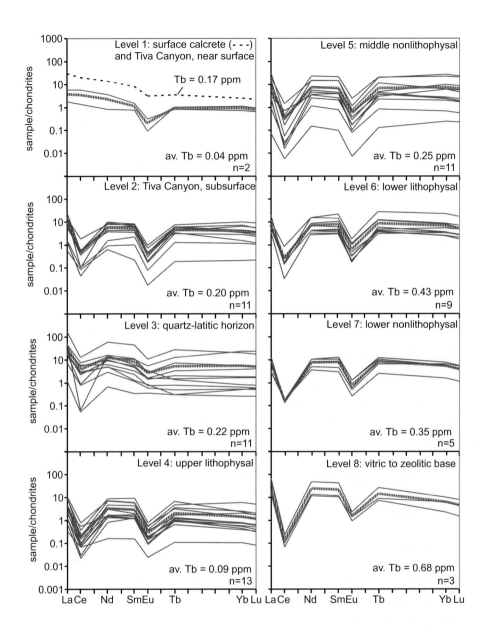

Figure 3. Chondrite-normalized lanthanide patterns for calcites from the eight physical/chemical stratigraphic levels shown in Figure 2. The average calcite lanthanide pattern (vertical shading) is shown for each level, as is the average Tb content. Note that Unit 1 includes a dashed pattern for calcrete, consisting of 78% calcite and 22% opal, from a surface trench.

focus solely on the information that bears on vadose-zone groundwater REE sources and transport.

Calcite deposited in fractures and voids of the vadose zone at Yucca Mountain has surprisingly high REE content. Light lanthanide contents in these calcites tend to rise with depth, from ~5 times chondritic close to the surface to values up to 100 times chondritic at depth, although there is not a simple correlation between depth and abundance (Figure 3). These abundances contrast greatly with calcites from soil carbonate horizons, from travertine mounds in springs of the region, and from calcite-lined veins exposed in fossil spring systems of the region. All of these other depositional systems have calcites with lanthanides generally below chondritic abundances (Vaniman et al., 1995).

The octahedral or six-fold coordinated Ca site in calcite can readily accept divalent cations in the size range of 0.7 to 1 Å. Lanthanide elements are ~0.85 to 1.03 Å in six-fold coordination, with the lightest lanthanides (La, Ce^{3+}) too large to fit in the Ca site. All lanthanides carry too high a charge (except for Eu^{2+} at low fO_2) for replacement of Ca without compensating vacancy defects or defects that allow monovalent substitutions, and the large size of the light lanthanides is an added liability if they are to be incorporated into calcite. As a result, one might expect the lanthanide patterns of calcites to have lower light-to-heavy lanthanide ratios than groundwaters from which they precipitated. This appears to be the case in comparing the relatively "flat" lanthanide patterns of Yucca Mountain calcites (Figure 3) with the light-rare-earth enriched patterns of host tuffs (Figure 2).

3.1.1. Pervasive Negative Ce Anomalies in Calcite

Chondrite-normalized lanthanide patterns of Yucca Mountain calcites show a striking and pervasive development of negative Ce anomalies. In many samples where the other REE determined by INAA are well above detection limits, Ce is excluded from the calcites to the extent that it cannot be detected. The pervasive negative Ce anomaly in the Yucca Mountain calcites is reminiscent of the common negative Ce anomalies of shallow, oxygenated marine waters, where Ce depletion from water can be attributed to oxidation of Ce^{3+} to Ce^{4+} with fixation onto and sedimentation of Fe, Mn oxide particles (e.g., de Baar et al., 1991). This similarity is not superficial, for a comparable process is occurring in the very different continental vadose system at Yucca Mountain.

In the marine system, extraction of Ce by Fe, Mn oxides from shallow seawater is inferred from the complementary positive Ce anomalies of Fe, Mn oxides dredged from the seafloor (Piper, 1974). The link to Mn oxides is more direct and obvious at Yucca Mountain, because these oxides are closely associated with calcite deposits. Figure 4 shows a splay of pyrolusite crystals growing on the surface of a fracture-lining calcite from 265 m depth in the vadose-zone tuff at Yucca Mountain. The pyrolusite crystals have high Ce content in energy-dispersive SEM analysis and the underlying calcite crystal is strongly Ce depleted (INAA analysis). Carlos et al. (1993) made a comprehensive survey of the Mn oxides at Yucca Mountain by electron microprobe and found that rancieite and pyrolusite both contained Ce considerably in excess of La. The prime importance of rancieite is suggested by its ubiquitous distribution in the vadose zone and high average Ce content (27,000 ppm; Carlos et al., 1993). The average Ce

Figure 4. Pyrolusite crystals formed on the surface of a calcite crystal in the lower lithophysal zone, Yucca Mountain.

content calculated for Yucca Mountain calcite with no Ce anomaly is 13 ppm, whereas the actual average Ce content is <0.8 ppm. To reach this diminished concentration, a unit mass of rancieite can account for the Ce depletion in about 2.3×10^3 equivalent masses of calcite. Leaching studies indicate an average weight abundance of about 1500 ppm calcite in the Yucca Mountain tuffs, whereas the Mn oxides are less abundant (~100 ppm Mn oxide in Yucca Mountain tuff; Zielinski, 1983). Even if the Mn oxides are <1% efficient in removing Ce from solution, they are capable of completely removing Ce from the groundwaters that precipitate calcite in the vadose zone.

The mechanism of Ce removal from solution by Mn oxides is linked to redox effects at the Mn oxide surface. The surface-mediated oxidation of Ce^{3+} to Ce^{4+} by synthetic and natural oxide phases, particularly Mn oxides, has been observed using X-ray absorption near-edge structure (XANES) techniques (Bidoglio et al., 1992). Many Mn oxide minerals have large surface areas and effective sorption sites for dissolved cations such as Ba^{2+}, Ni^{2+} and Pb^{2+} (Hem 1978, McKenzie 1989). More importantly, the most common oxidation state of Mn in these minerals is Mn^{4+}, a highly oxidizing species that can facilitate transformations in the oxidation states of sorbed species such as Co, Ce, Cr and Pu (Manceau and Charlet, 1992; Manceau et al., 1997; Duff et al., 1999). In the vadose zone at Yucca Mountain, groundwater Ce^{3+} is probably removed from solution and oxidized to Ce^{4+} upon sorption to the predominantly Mn^{4+} oxides rancieite and pyrolusite.

The source rocks for REE in vadose-zone waters at Yucca Mountain are soils and tuffs that have no Ce anomaly. Figure 2 shows the range of chondrite-normalized REE patterns in the tuff sources. The soils include thin A horizons and both clay-rich and calcite/opal-enriched B horizons. Studies of C and O isotopes in the calcites of the vadose zone suggest that dissolution of calcites from the B horizon is a significant factor in providing the C and O structural constituents for deeper calcites throughout the vadose zone (Whelan and Stuckless, 1992; Whelan et al., 1994). Because Ca is not abundant in the tuffs, it is likely that much of the Ca is also derived from soil sources. However, Sr isotopic studies indicate that some of the trace elements in the vadose-zone calcites are derived from dissolution of tuff by groundwater (Johnson and DePaolo, 1994). This conclusion is supported by an analysis of the slopes of light-lanthanide patterns and the development of Ce and Eu anomalies in the calcites.

For comparison with the range of chondrite-normalized REE patterns in source soils and rocks at Yucca Mountain (Figure 2), Figure 3 shows the chondrite-normalized lanthanide patterns in calcites from several horizons in the vadose zone at Yucca Mountain. None of the sources, either soils or tuffs, possess significant Ce anomalies; the negative Ce anomalies in the calcites are distinctive and require Ce fractionation from the source compositions.

The negative Ce anomalies in vadose-zone calcites at Yucca Mountain have significance for understanding the migration of heavy metals toward the water table. Figure 3 shows that in the vadose zone, only near-surface calcites lack negative Ce anomalies. Within ~10-50 m of the surface, smooth chondrite-normalized patterns through the light lanthanides vanish and all calcites acquire negative Ce anomalies (Ce/Ce^* of 0.2 or less, where Ce is the measured cerium abundance and Ce^* is the abundance estimated for a smooth chondrite-normalized pattern between La and Nd). However, the range of Ce anomalies becomes less negative in the quartz-latitic horizon where glass is abundant; below this interval the Ce anomalies again become increasingly negative with depth, reaching values of Ce/Ce^* of ~0.01 and lower in the deepest vadose-zone calcite samples. These profiles show that the addition of fresh lanthanides in solution, from sources without Ce depletion, is overwhelmed by the losses of Ce from solution to Mn oxides with the exception of those calcites that are within a few tens of meters of soil sources. Even dissolution of fresh glass in the quartz-latitic horizon, which adds more fresh lanthanides than does dissolution of less soluble devitrified tuff, cannot regenerate lanthanide patterns without Ce anomalies, although the maximum calcite Ce/Ce^* rises to ~0.45 at this horizon (Figure 5a). The values of Ce/Ce^* in calcites diminish progressively below the quartz-latitic horizon, where glass sources are absent. The stratigraphy of Ce/Ce^* values in vadose-zone calcites thus points to the locations of the principal sources for lanthanides in solution (soils and the quartz-latitic horizon) and reveals the transport distance required for vadose-zone waters to lose most of their Ce to Mn oxides (~20-100 m).

3.1.2. Tracking REE Sources with Eu Anomalies and La/Sm ratios
The tuffs and soils at Yucca Mountain represent a limited number of source lithologies with distinctive REE compositions. The quartz-latitic horizon in Figure 5a (level 3), where the Ce anomalies in the calcites are diminished (see above), is also an interval where the light-to-medium lanthanide ratio (La/Sm) of the calcites is higher than

elsewhere and the Eu anomalies may be absent or diminished (Figure 3). Both of these features represent dissolution of significant amounts of the local quartz-latitic glass. Figure 5b shows almost complete reversion to deeper Eu anomalies ratios in the next lowest horizon (compositional unit 4, the upper lithophysal horizon of the Topopah Spring Tuff). This sharp difference indicates how limited the influence of local glass dissolution can be. Although the process of Ce removal from solution by Mn oxides is spread over ranges of many tens of meters, the dissolution signatures of lanthanide sources are strongly influenced by local differences in tuff composition. If the Ce/Ce* data were not considered, the Eu/Eu* and La/Sm data might be interpreted to indicate no vertical communication between the different stratigraphic levels.

3.2 CLAYS: AN INTERACTIVE AND COMPLEX RECORD

Calcite lining the fractures in tuff requires addition of carbonate and calcium to a rock type that virtually lacks the former and is deficient in the latter. In contrast, clay minerals can form directly from most rock types containing silicon, aluminum, and a few transition metals plus alkalis or alkaline earths with essentially no geochemical gain or loss, except for the addition of water. If this were always the case, then the genetic interpretation of clays would be straightforward, but there are many situations where the less mobile constituents of the clay (especially aluminum) are derived in place but augmented by influx of more soluble constituents. Clays are also prone to translocation after formation, most clearly by eolian, fluvial, and soil eluviation/illuviation processes but also as fine particulates and even finer colloids that can migrate for considerable distances through fractures and porous media. Clay structure and composition thus holds tremendous potential for unraveling the solute, colloid, and even Eh/pH characteristics of complex groundwater systems, but the complexity of clay minerals can make this a difficult task.

Consider the differences between clay minerals formed in the vadose-zone tuffs at Yucca Mountain and at Pajarito Mesa. In the tuffs at Yucca Mountain, clays separated from either rhyolitic or quartz-latitic horizons have iron and lanthanide compositions that reflect those of the host rock (Figure 6a,b). In the Bandelier Tuff of Pajarito Mesa, clays separated from the tuff matrix are strikingly different from the host tuff in iron and lanthanide content (Figure 7). The reason for differing clay formation at Yucca Mountain and at Pajarito Mesa can be found in the petrography of clay occurrences at these two sites. The vadose-zone clays at Yucca Mountain formed principally by alteration of glass in the nonwelded margins of major cooling units, where saturation in the porous vitric rock has promoted alteration well above the regional water table. The vadose-zone clays of the Bandelier Tuff formed principally in units that are completely devitrified, without any glass and in a crystalline system that has abundant Fe-rich minerals (especially pyroxene and olivine) that are rare or absent in the tuffs of Yucca Mountain. The Bandelier clays form on a smaller scale and can be seen in thin section as alteration rims localized around these Fe-rich minerals (Stimac et al., 1996; however, this is not always the case, for there are deeper vitric units in the Bandelier Tuff that can support perched water and are clay-altered, but the clays from these units are largely unstudied). The first challenge in studying clay systems is thus to determine initially whether the petrographic setting of the clay supports formation in place and if so, whether the conditions of formation favor complete alteration of the host (usually the

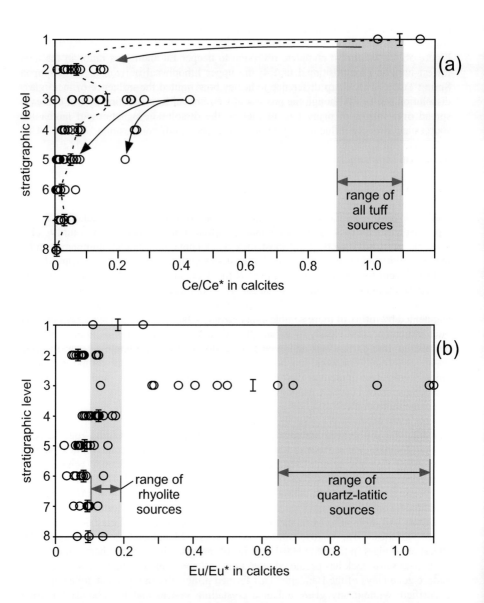

Figure 5. Plots of (a) Ce anomalies (Ce/Ce*, where Ce* represents predicted content at no anomaly) and (b) Eu anomalies (Eu/Eu*, where Eu* represents predicted content at no anomaly) in calcites from the eight physical/chemical stratigraphic levels illustrated in Figure 2. The average for each level is shown as a vertical bar and the corresponding ranges for the local tuffs are shaded. Dashed curve in (a) connects averages for each level, and arrows illustrate Ce depletion phenomena beneath major lanthanide inputs from the surface (level 1) and from the quartz-latitic vitric horizon (level 3). Panel (b) shows prominent input from quartz-latitic glass with high Eu/Eu* that characterizes level 3.

case in formation from glass) or incongruent alteration (possibly the case if specific minerals are altered).

The REE play an important role in interpretation of clay alteration, for the resistance of the REE (particularly lanthanides) to fractionation permits recognition of specific source lithologies. With appropriate caution, the REE can also serve as surrogates for studies of other heavy metals such as the actinides. Clays, particularly smectites, have a reputation for effective sorption of heavy metals from solution. However, one of the problems in trace-element studies of clay/groundwater systems is the determination of how much of the reaction with groundwater is attributable to the clay minerals and how much to the minor but often very active phases that are commonly associated with clays. In the case of clays that occur in the shallow vadose zone of tuffs that form Pajarito Mesa, association with intergrown Mn oxides has a determinative effect.

3.2.1. *Acquisition of both Negative and Positive Ce Anomalies in Clays of Pajarito Mesa*

Figure 8 summarizes the chondrite-normalized lanthanide patterns of clays from Pajarito Mesa. In contrast to the fracture-filling calcites at Yucca Mountain, the fracture clays of Pajarito Mesa have both positive and negative Ce anomalies. Clays from the soils may have small Ce anomalies, but these are minor in comparison with Ce anomalies of the fracture clays. The large negative and positive Ce anomalies in fracture clays are an important factor in evaluating lanthanide fractionation in the clay system. As discussed above in section 3.1.1, consistent negative Ce anomalies in the calcites of Yucca Mountain result from stripping of Ce from downward-flowing groundwater by Mn oxides that occur in the same fracture system but are seldom intergrown with the calcites. In contrast, occurrence of both negative and positive Ce anomalies within clays at Pajarito Mesa requires a different explanation, although the mechanism involved still requires participation of Mn oxides.

Figure 9 is an SEM backscattered-electron image of a representative intergrowth of Mn oxide with one of the shallow fracture clays from Pajarito Mesa. The clay has bright high-atomic-weight regions of Mn oxide that are rich in Ce and Ba (energy-dispersive SEM data) with associated accumulations of Ni and Pb (synchrotron X-ray fluorescence data; Vaniman et al., 2002). X-ray diffraction analysis of hand-picked clay separates with abundant dark spots of Mn oxide shows that these spots are composed principally of birnessite. Localization of Mn oxides along clay laminae, with highest concentration along walls of microfractures within clay bodies, indicates migration of Mn and fixation on clay particles at a late stage in illuviation or after the clay body had formed.

INAA data for clay concentrates, summarized in Vaniman et al. (2002), show that the fracture clays of Pajarito Mesa have a trend of Ce-Mn correlation with approximately 0.1 to 0.3 Ce cations for each Mn. Combined with the direct evidence of Ce-rich Mn oxides within the clays, it is evident that the Mn oxides define the Ce concentrations in this clay system. It is the content of other lanthanides in the host clay that determines whether an individual sample will have either a positive or a negative Ce anomaly.

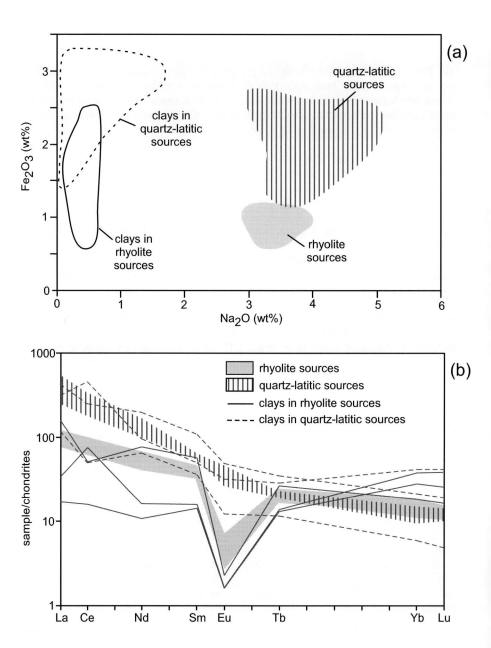

Figure 6. Fe₂O₃-Na₂O relations (a) and chondrite-normalized lanthanide patterns (b) in host vitric tuffs and in clay separates from those tuffs at Yucca Mountain. In both panels the clays reflect the compositional differences between rhyolite versus quartz-latitic sources, although the relations are affected by significant Na loss and development of positive and negative Ce anomalies in formation of clay from the vitric tuffs.

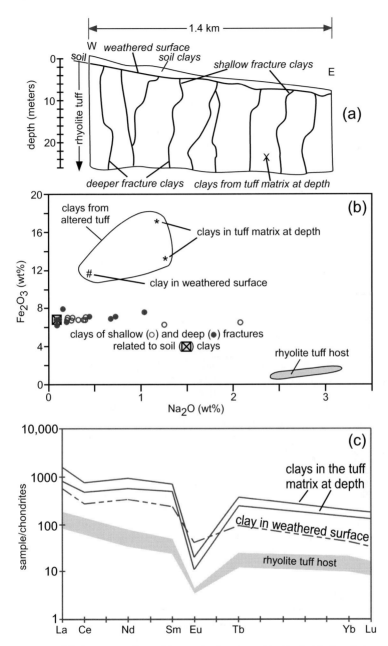

Figure 7. Simplified cross-section with sample locations (a), Fe_2O_3-Na_2O relations (b), and lanthanide patterns (c) for a portion of the Bandelier Tuff on Pajarito Mesa. Data for clays from the surface soil, from fracture-fillings, and from altered tuff (weathered at surface or altered at depth) are shown in (b); note the very high Fe content of clays from altered tuff (compare Figure 6a). In (c) the range of chondrite-normalized lanthanide patterns for the rhyolite tuff host is contrasted with the elevated patterns of clays weathered from tuff at the surface or formed by alteration concentrated around Fe-rich minerals in the tuff matrix at depth.

Textural evidence of illuviation at Pajarito Mesa leaves little doubt that clays in the fractures can be derived from surface soils (Davenport et al., 1995; Vaniman et al., 2002). Illuviation of clay particles may be facilitated along openings created by live or decaying roots. Although on the scale of individual samples (~5 g) the clays may have either a positive Ce anomaly or a negative Ce anomaly, the combined mode of multiple samples from either the shallow fractures or the deeper fractures has no Ce anomaly. As seen in Figure 8, the Ce content in either the shallow or the deeper fracture environment is relatively limited and the expression of a negative or positive Ce anomaly depends on whether the other lanthanides are more or less abundant in each individual clay sample.

There is an increase in all lanthanides in clays of Pajarito Mesa in the transition from shallow to deeper fractures. Although there are fewer data from soil clays, the two samples analyzed (<1 ka age and 30 ka; Figure 8) suggest that unless sources of eolian input have varied significantly from late Pleistocene to Holocene time, there is also an increase in lanthanide content of clays with soil maturation. Taken together, the data from soil and fracture clays indicate that both time and transport distance lead to accumulation of lanthanides by clays.

Average lanthanide content in clays from the deeper fractures is approximately three times that of the shallow fracture clays. This rise in lanthanide content is evidence that the fracture system is not closed but is in fact open to the transport and accumulation of lanthanide elements. However, at both shallow and deep levels within the fracture system, the chondrite-normalized lanthanide data (Figure 8) indicate that there is a correlation between high lanthanide content and negative Ce anomalies, or low lanthanide content and positive Ce anomalies ($R^2 = 0.79$). This correlation is a result of the small amount of Ce variation between clay samples in each environment, contrasted with large variability in local abundance of other lanthanide elements. The Mn oxides are distributed within the clays uniformly enough that the Ce abundance of each clay sample from a given depth in the fracture system is relatively constant. Mn and Ce both vary by a factor of only 2 within either shallow or deep fracture clays, whereas lanthanide elements other than Ce vary by factors of 5 in shallow fracture clays and 13 in deep fracture clays. Uniformity in Ce content would not occur if Mn oxides were not ubiquitous in fracture clays or were inaccessible to fluids, but Mn oxides are widely distributed and textural evidence (Figure 9) indicates that Mn oxides occur along pervasive microfractures in clay bodies. In contrast to relatively fixed Ce concentration controlled by these Mn oxides, other lanthanides are accumulated by clays in highly variable amounts.

3.2.2. Tracking REE Sources with Eu Anomalies in Clays
The chondrite-normalized lanthanide patterns in Figure 8 show a wide variety of negative Eu anomalies in clays from the four principal environments at Pajarito Mesa. Variety in Eu anomalies is better displayed in Figure 10, where relative Eu depletion (Eu/Eu*, where Eu is the measured europium abundance and Eu* is the abundance estimated for a smooth chondrite-normalized pattern between Sm and Tb) is plotted

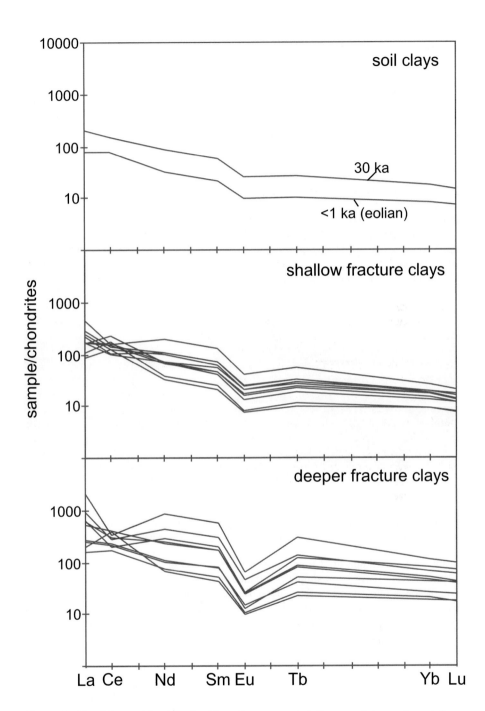

Figure 8. Chondrite-nomalized lanthanide patterns separated from overlying soils and from factures in tuff of Pajarito Mesa. For sample locations refer to Figure 7a.

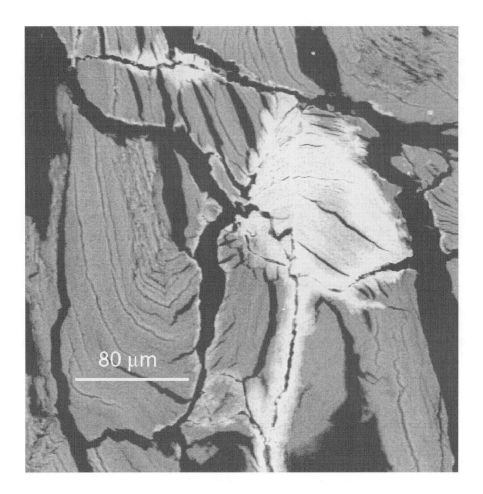

Figure 9. SEM image of Mn oxide (birnessite) concentrated along desiccation fractures in a clay mass and permeating clay laminae. Sample is from a near-surface fracture in the Bandelier Tuff.

against the magnitude of Ce enrichment (Ce/Ce* >1) or depletion (Ce/Ce* <1; see section 3.2.1). There is no simple correlation between Eu and Ce anomalies, although within the group of clays from deep fractures (solid circles) there is a possible correlation between shallower negative Eu anomalies (higher Eu/Eu*) and higher positive Ce anomalies (Ce/Ce* increasing). This tentative correlation may reflect real differences in lanthanide sources or modest Eu enrichment in the higher Ce/Ce* samples where the overall lanthanide pattern is more strongly dominated by proton-donor (or electron-acceptor) surfaces of Mn oxides. Such surfaces may also favor Eu (Erel and Stolper, 1993), but not to the same extent as Ce. The principal inference from this figure is not the relatively minor Ce anomaly variations within classes of Eu anomalies representing different locations of clay formation, but the much greater differences between locations.

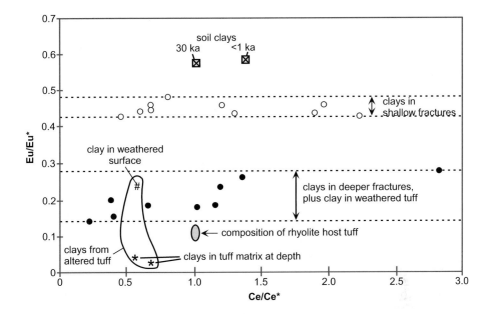

Figure 10. Eu anomalies (Eu/Eu*) versus Ce anomalies (Ce/Ce*) in clay separates and rhyolite host tuff from Pajarito Mesa. Open circles represent clays in a shallow fracture; solid dots represent clays in deep fractures.

Soil clays have the smallest negative Eu anomalies among all clays at Pajarito Mesa. The small magnitude of this Eu anomaly (Eu/Eu* = 0.58) and the overall chondrite-normalized lanthanide pattern are similar to eolian materials that represent typical crustal-average compositions (Taylor and McLennan, 1988). The small negative Eu anomalies of soil clays are distinct from greater negative Eu anomalies in the Bandelier rhyolite host tuff (Figure 10). Among all clays, the largest negative Eu anomalies occur in tuff matrix clays from samples at depth. These clays are exceptionally Fe-rich (Figure 7b) and formed largely by alteration of mafic minerals in the tuff matrix, leaving unaltered the feldspars that retain much of the rock's initial Eu. Between the extremes of soil and tuff-matrix clays fall the other clays, with shallow fracture clays closer to soil clays in Eu/Eu* and deep fracture clays more similar to local tuffs.

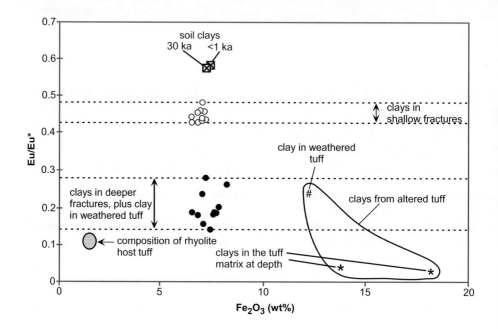

Figure 11. Eu anomalies (Eu/Eu*) versus Fe_2O_3 in clay separates and rhyolite host tuff from Pajarito Mesa. Note the restricted range of Fe_2O_3 content common only to soil and fracture clays.

Small negative Eu anomalies in soil clays reflect their eolian origins, approximating average crustal sources. Exceptionally deep negative Eu anomalies (low Eu/Eu*) of Fe-rich tuff-matrix clays are a product of selective alteration of mafic components in the host rock. These are the only "primary" types of Eu anomalies among clays at the site, and there are few low-temperature geochemical processes that can lead directly to intermediate Eu anomalies. It is possible that total weathering of an upper stratigraphic level in the tuff could produce clays with the composition of deep fracture clays, but the deep fracture clays occur entirely within lower strata that have significantly deeper negative Eu anomalies (shown as "rhyolite host tuff" in Figure 10). It would be more likely that selective weathering of feldspars within the deeper strata could produce clays with a slightly diminished Eu anomaly, but if fracture clays formed by this process they should be less mafic and more aluminous as Eu/Eu* increases, an effect that is not evident. Alternatively, direct formation of diminished negative Eu anomalies could occur through surface-oxidation processes as documented by Erel and Stolper (1993), but this is a small effect that cannot account for the large range of Eu anomalies between soil and deep tuff matrix clays. The intermediate Eu anomalies result from other processes.

Intermediate Eu anomalies in fracture clays at Pajarito Mesa are superimposed on other chemical parameters that reflect transport from the soil environment. Figure 11 illustrates one of these parameters. In Figure 11 the vertical axis shows the magnitude of the negative Eu anomaly as in Figure 10, but the horizontal axis plots weight percent Fe_2O_3 in clay samples and in local tuffs. Figure 11 shows that the Fe_2O_3 contents of soil and fracture clays are similar and much lower than clays in altered tuff. Clays in altered Bandelier tuff are Fe-rich, in contrast to those at Yucca Mountain (Figure 6a). This difference may depend on several factors, but the key factor is found in petrographic evidence of bulk-sample alteration of vitric tuff at Yucca Mountain versus targeted alteration of Fe-silicates (particularly clinopyroxene) in the devitrified Bandelier Tuff. The lack of low-Fe clays in the devitrified Bandelier Tuff precludes mixing of low-Fe clays with the high-Fe tuff alteration clays to produce the 7% Fe_2O_3 clays in the fractures. Transport of clay from soils into the fractures is thus strongly indicated by both petrography and Fe-composition information. This does not, however, explain the variability in fracture-clay lanthanide abundances.

Since lanthanides are not readily accommodated within the clay structures, it is likely that defects, particle size, and precipitation or growth rate control accumulation of lanthanides by clays. The deepening negative Eu anomalies as lanthanides are accumulated in fracture clays are evidence that dissolution of local tuff host provides the additional lanthanides. Lanthanides in the shallow fracture clays have no correlations with elements other than themselves, but the clays of the deeper fractures have fair to strong correlations of lanthanide elements (except Ce) with P (R^2 values against P are La 0.94, Ce 0.35, Nd 0.95, Sm 0.77, Eu 0.96, Tb 0.92, Yb 0.84, and Lu 0.80). This suggests that leaching of trace phosphate phases in the tuff provides some of the lanthanide accumulation in clays of the deeper fractures. Although the trace mineral with highest known lanthanide concentration in the Bandelier Tuff is not a phosphate but the silicate mineral chevkinite (33% lanthanides by weight; Stimac et al., 1996), apatite is more common and provides an alternative source. Chevkinite is also likely to be less soluble than apatite.

4. Cautions for the analysis of mineral separates

Analyses of REE in groundwater-deposited minerals can provide information on water-rock interactions in the vadose zone, as illustrated for calcite and clays in this chapter. However, there are aspects of mineral-separate analysis that require caution in both the preparation of materials for analysis and in the interpretation of results. These aspects include averaging of any compositional zonation finer than the sample size, potential for different grain-sizes to have different compositions, and possible inclusion of mineral impurities with the target mineral. It may seem that these are all potential problems that would be overcome simply by analysis using microbeam methods, especially ion probe analysis, but this is not necessarily the case. For example, when clays are analyzed, microbeam methods are commonly problematic because of beam damage and lack of a useful reference element in predictable content (such as Ca in calcite). Moreover, there are certain advantages analyzing different clay size fractions to determine whether the clay behaves as a homogeneous mass at all size ranges. This last factor can be especially important when there is a desire to know how clays perform geochemically as colloidal size fractions are approached.

4.1 MINERAL ZONATION OR MULTIPLE EPISODES OF GROWTH

The discussion of Yucca Mountain calcites in this chapter is focused on INAA data from hand-picked mineral separates, but parts of this sample suite have also been analyzed by cathodoluminescence and ion microprobe (Denniston et al., 1997). In that study it was found that there was not a simple correlation between lanthanide abundances and cathodoluminescence, but there was a tendency for overall decrease in lanthanide abundance from initial growth (at the fracture wall) toward the rim within single calcite crystals. This spatial information is valuable in evaluating trends with time in calcite precipitation from groundwater. However, it is uncertain what the time scale of crystal growth is and whether the core-to-rim trend in individual crystals represents single-stage evolution or a more profound shift in vadose-zone water compositions over time. Moreover, the trend of decreasing lanthanide content with successive growth layers was not uniform and the few data available suggest that the trend may be more consistent in the northern part of Yucca Mountain than in the central and southern parts. The potential for detailed growth history is there, but is yet to be realized.

4.2 GRAIN-SIZE EFFECTS

Grain-size effects are of much greater consequence in clay analysis than in the analysis of calcites. Calcites, although commonly zoned, can be readily hand separated and analyzed as pure minerals uncontaminated by the host rock. Clays rarely occur within the subsurface vadose zone in masses large enough and pure enough for hand separation and analysis without further processing. This processing typically involves crushing, disaggregation, and sedimentation to purify the clay-mineral components. Such work is more tedious and prone to contamination (see below) but also provides important mineralogical and geochemical information specific to different clay size fractions. There are several advantages to analysis of clay size separates that are not present in the more straightforward analysis of calcites.

First, consideration should be given to whether different clay size fractions are similar in composition. Figure 12 shows data for clay size fractions from both Yucca Mountain and the Pajarito Plateau. Included with the data for the Pajarito Plateau are analyses of clay size fractions from a basalt that is in the vadose zone but hosts perched groundwater.

The two clay size fractions from a shallow fracture fill on Pajarito Mesa (Figure 12a) are very similar, with only a 15% decrease in lanthanide abundance of the finer fraction and parallel chondrite-normalized patterns. The parallel patterns include Ce and Eu anomalies that are identical in the two size fractions. As noted above in section 3.2, these patterns are largely inherited from eolian clays (source-soil clay in Figure 12a) that have been eluviated from overlying soils into the shallow fractures.

The two clay size fractions from vesicles in the flow base of a basalt of the Pajarito Plateau vadose zone (Figure 12b) have lanthanide patterns that parallel their host rock but differ in development of Eu and especially of Ce anomalies. It is not certain that these clays are derived entirely by alteration of the host basalt without geochemical additions or translocation, but parallel lanthanide patterns suggest that they have acquired much of their character from alteration of this glass-rich basalt. The coarser clay separate is 50% depleted in lanthanide elements relative to the host rock and the finer clay separate is further depleted by ~78% relative to the coarser clay. Notable in these patterns is the appearance of a negative Eu anomaly in the clays that does not occur in the bulk rock and the development of a positive Ce anomaly in the clays that increases as clay size diminishes. The negative Eu anomalies in the clays can be attributed to alteration of basalt matrix glass, leaving Eu-enriched plagioclase relatively unaltered. The positive Ce anomalies in the clays may be associated with Mn oxides that are observed in these samples, but further work is needed to test this assumption.

The clay size separates from Yucca Mountain, as noted above (section 3.2), form in vitric tuffs and tend to mimic the composition of the rhyolitic or quartz-latitic host rock (Figure 12c, d). Size fractions of the quartz-latitic clays differ more in the heavy lanthanides than in the light lanthanides; deviations in rhyolitic clay size-fractions are more complex. In both cases, however, the lanthanide abundances of the finer clay fraction are generally diminished relative to the coarser clay. Development of a strongly positive Ce anomaly in the finer rhyolitic clay may again be associated with Mn oxides observed in this sample but not in the quartz-latitic sample, but further work is also needed in this instance.

When information from the different samples presented in Figure 12 is considered together, it is evident that the available data from clay separates are too few to draw many well-supported conclusions. However, the data are sufficient to indicate that congruent versus incongruent alteration of the bulk rock can be recognized and that Ce anomalies can be radically dependent on clay size (whether or not the Ce retention is linked to intergrown Mn oxides). Perhaps the most consistent feature of the lanthanide patterns of clay size fractions shown here is the diminished overall lanthanide abundance in the finest size fractions. This observation indicates that clay particles approaching colloid size are relatively poor hosts for heavy metals. The possible excep-

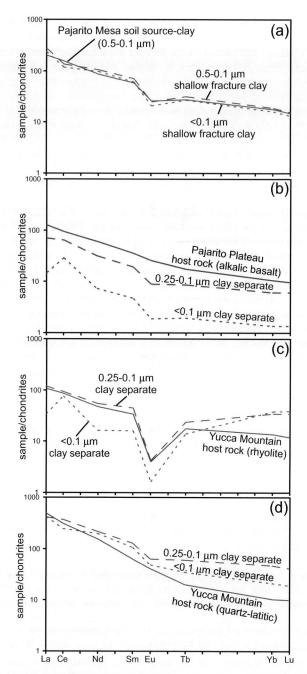

Figure 12. Chondrite-normalized lanthanide patterns of different clay size separates compared with source-clay (a) or host-rock (b, c, d) patterns. Samples are from Pajarito Plateau, which includes Pajarito Mesa (a, b), or from Yucca Mountain (c, d).

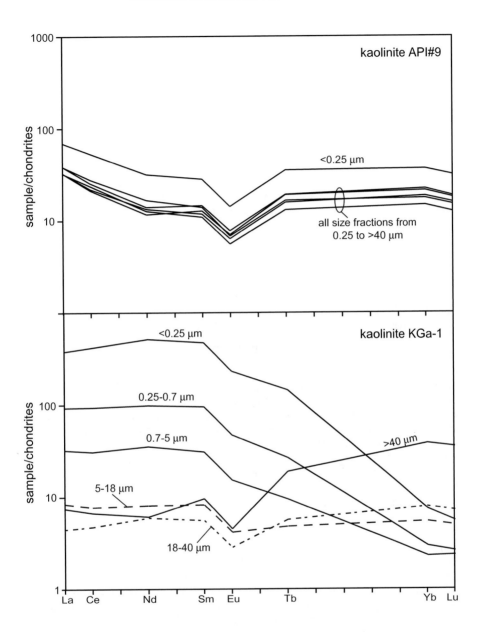

Figure 13. Chondrite-normalized lanthanide patterns in size separates of two kaolinite standard reference materials, American Petroleum Institute (API)#9 and Clay Minerals Society source clay KGa-1.

tion may be those heavy metals susceptible to redox fixation by Mn oxides associated with the clays. Whether those centers are minute equivalents of the intergrown Mn oxides as seen in Figure 9 or Mn resident as part of the clay structure, or some other mechanism entirely, is still beyond the scope of present knowledge.

4.3 PURITY OF MINERAL SEPARATES

For any mineral separate to be analyzed for REE, purity must be a critical consideration. Impurities can be tolerated if it is known what they are and their approximate abundance, but interpretation of data is invariably more straightforward if the mineral analyzed is as pure as possible. For calcite, evaluation of purity is relatively straightforward if the specimen is coarse enough for hand separation and preparation guidelines discussed in section 2.1 are taken into consideration. Clay minerals are a different matter.

Much has been said in this chapter concerning the intimate relationships between Mn oxides and clays. In practice these intergrowths are so intimate that it is impossible to physically separate the two. This is part of the problem in interpreting the varied Ce anomalies in different size fractions (Figure 12) but also part of the story to be read from the analyses. In addition to the role played by Mn oxides, there are other impurities that can occur in clay separates and deserve special attention in REE studies.

High Zr and Hf in analyses of clay separates from the matrix of the Bandelier Tuff indicate that zircon may be a contaminant in the clay-size fractions. The tuff-matrix clays form preferentially around mafic minerals and are associated with vapor-phase needles of zircon that are a few tens of μm in length and <1 μm in diameter (Stimac et al., 1996). In crushing and processing matrix samples, such small and fragile zircons are difficult to separate from clays. Retention of only ~0.1 wt% of such zircons would account for the high Hf and Zr content of these tuff-matrix clay samples (Vaniman et al, 2002). These small amounts of zircon are sufficient to affect chemical analyses but are not evident in XRD analyses. Most importantly for REE studies, the REE content of the zircon is small enough (Stimac et al., 1996) that it has little impact on lanthanide analyses of the clay separates.

The inclusion of minute needles of zircon in clay separates is not a problem of concern for typical clay analyses, because this fragile vapor-phase habit is not common among zircons. A more general problem in processing separations of smectites is the common observation of increasing apatite abundance with diminishing clay size. The apatite concentration may reach several percent in a <0.1 μm clay size separate. A consequence of this apatite association can be a rise in lanthanide abundance that contradicts the general loss of lanthanides that otherwise occurs in most samples as smectite size decreases (Figure 12). At this time, little is known about the apatite association with fine smectite fractions, but the nature of the apatite-clay association should be resolvable with appropriate study by transmission electron microscopy.

A separate study of kaolinites by Bish et al. (1999) provides a remarkable case where another phosphate, crandallite, has a profound effect on lanthanide composition. Figure 13 illustrates the chondrite-normalized lanthanide data for multiple size fractions from two kaolinite samples that are standard clay-mineral reference materials, American Petroleum Institute (API)#9 and the Clay Minerals Society source clay KGa-1. Size fractions of API#9 show little variation in chondrite-normalized pattern except for a significant rise in lanthanide abundance (~110%) in the <0.25 μm size fraction relative to the average of all other size fractions. X-ray diffraction data for the different size fractions show no contaminant except for minor amounts of quartz. The result is strongly parallel chondrite-normalized lanthanide patterns. It is not known whether an increase in REE with smaller kaolinite size is common, an effect opposite to that noted with smectites (Figure 12).

In contrast to the API#9 results, the size fractions of KGa-1 show extreme lanthanide fractionation with light-lanthanide enrichments correlated to the amount of crandallite [CaAl$_3$(PO$_4$)$_2$(OH)$_5$·H$_2$O] in each size fraction. Here is a contaminant worth serious consideration. X-ray diffraction data show that crandallite increases in abundance in the finest three size fractions of KGa-1. Small amounts of anatase are also present, and illite in the >40 μm fraction indicates a detrital phyllosilicate source with higher Tb-Lu content, but crandallite has the greatest impact on lanthanide abundances. The rise in La-Tb with crandallite contamination illustrates a situation where lanthanide abundances in the contaminant overshadow those of the host clay.

The impact of contaminants on clay-mineral REE studies is serious but should not be viewed as a block to useful interpretation. The single most important step to be taken to ensure accurate interpretation of clay-mineral REE data is the analysis of several size fractions. Where differences in REE abundances are found, especially relative differences between light and heavy lanthanides, contaminant minerals should be suspected. Splits of the different size separates analyzed for REE should be retained, or analyzed splits preserved if not consumed, in order to permit XRD analysis for contaminant minerals. In many cases this will not be possible for most samples because the amount of clay that can be collected may be very small and multiple size separates cannot be processed. This situation can be partially remedied if many samples are collected and those that are most abundant processed to obtain at least two size separates. The effort can be great, but the additional information gained from different size fractions can be an unexpected and sometimes most-significant source of information (e.g., kaolinite KGa-1 in Figure 13).

5. Conclusions and suggestions for further work

Studies of lanthanide elements in alteration minerals can be useful in determining transport distances and processes in unsaturated rocks. In many instances, variations in lanthanide abundances and ratios, particularly the development of Ce anomalies, can be used to infer mechanisms of heavy-metal accumulation by specific alteration minerals such as Mn oxides. Alteration minerals such as calcite and clays tend to be less selective in lanthanide accumulation and may reflect the more active role of associated Mn oxides. Caution should be exercised in interpretations of lanthanide information from clays, where intergrown trace minerals (particularly phosphates) may be exceptionally enriched in lanthanide elements. The best constraints on interpretations of transport systems are obtained where several different alteration minerals can be extracted and analyzed.

Based on data from two tuff systems in different settings (Yucca Mountain, Nevada, and Pajarito Mesa, New Mexico), it is possible to identify features of these systems that are most important to the successful application of this method. The most important of these is the occurrence of stratigraphically distinctive source-rock properties, such as differing magnitudes of Eu anomalies or differences in light-to-heavy lanthanide fractionation between source-rock strata. These differences provide markers that can be tracked in alteration phases that precipitate along flow paths, particularly where flow paths pass into strata with different properties. Information on source rocks and on accumulation by alteration minerals is obtained by comparing lanthanide features characteristic of specific strata, such as Eu anomalies, with those features such as Ce anomalies that are altered as lanthanides are accumulated. The evidence from unsaturated tuff systems indicates that relative lanthanide abundances in solution can reflect local tuff compositions. However, modification of lanthanide abundances and ratios can also occur, beginning with incongruent dissolution of the source rock (e.g., dissolution of Fe-rich minerals in the Bandelier Tuff) and progressing further through groundwater interactions with other alteration minerals. Groundwater modification through redox interaction with Mn oxides is observed within transport distances of just a few tens of meters at Yucca Mountain.

Future work will benefit tremendously from advances in high-precision spot analysis. Advances in laser-ablation ICP-MS and in secondary-ion mass spectrometry leading to REE data from small analytical points will improve studies of both zoned calcite crystals and small clay bodies. Particularly revealing would be such spot analyses of Mn oxides, which commonly have very complex associations with Ce measurable by electron microprobe on the scale of a few micrometers but seldom yield reasonable data for the other REE. However, one important source of information that would be lost in such analyses would be the differences in REE abundances and patterns for different clay size fractions. These will still need to be extracted and separated much as they are now. Finally, there are other analytical possibilities that were not pursued here but are well within the range of current methods. Examples include standard ICP-MS analysis of stepwise acid leachates of calcite deposits or similar stepwise dissolution and analysis (e.g., dithionate leaching) of clay-Mn oxide associations. The possibilities rapidly expand beyond the time and capacity available to any one research group and open up a wide range of research topics for understanding the REE records locked in minerals precipitated from groundwater systems.

References

Bidoglio G., Gibson P. N., Haltier E., Omenetto N., and Lipponen M., XANES and laser fluorescence spectroscopy for rare earth speciation at mineral-water interfaces. *Radiochim. Acta* **58-59**, 1992, 191-197.

Bish D. L., Vaniman D. T., and Chipera S. J., Effects of particle size and trace-mineral content on kaolin trace element chemistry (abst.). In *Clay Min. Soc. 36th Ann. Mtg.*, Purdue Univ., 1999, p.15.

Brookins D. G., Aqueous geochemistry of rare earth elements. In *Geochemistry and Mineralogy of Rare Earth Elements* (eds. B. R. Lipin and G. A. McKay), Chap. 8, 1989, pp. 201-225. Min. Soc. Amer.

Carlos B. A., Chipera S. J., Bish D. L., and Craven S. J., Fracture-lining manganese oxide minerals in silicic tuff, Yucca Mountain, Nevada, U.S.A. *Chem. Geol.* **107**, 1993, 47-69.

Chipera S. J., Guthrie G. D., Jr., and Bish D. L., Preparation and purification of mineral dusts. In *Health Effects of Mineral Dusts* (eds. G. D. Guthrie and B. T. Mossman), Chap. 6, 1993, pp. 235-249. Min. Soc. Amer.

Davenport D. W., Wilcox B. P., and Allen B. L., Micromorphology of pedogenically derived fracture fills in Bandelier Tuff, New Mexico. *Soil Sci. Soc. Amer. J.* **59**, 1995, 1672-1683.

de Baar H. J. W., Schijf J., and Byrne R. H., Solution chemistry of the rare-earth elements in seawater. *Euro. Jour. Solid-State and Inorg. Chem.* **28**, 1991, 357-373.

Denniston R. F., Shearer C. K., Layne G. D., and Vaniman D. T., SIMS analyses of minor and trace element distributions in fracture calcite from Yucca Mountain, Nevada, USA. *Geochim. Cosmochim. Acta* **61**, 1997, 1803-1818.

Duff M. C., Hunter D. B., Triay I. R., Reed D. T., Sutton S. R., Bertsch P. M., Shea-McCarthy G., Kitten J., Eng P., Chipera S. J., and Vaniman D. T., Mineral associations and average oxidation states of sorbed Pu on tuff. *Env. Sci. Tech.* **33**, 1999, 2163-2169.

Erel Y. and Stolper E. M., Modeling of rare-earth element partitioning between particles and solution in aqueous environments. *Geochim. Cosmochim. Acta* **57**, 1993, 513-518.

Hem J. D., Redox processes at surfaces of manganese oxide and their effects on aqueous metal ions. *Chem. Geol.* **21**, 1978, 199-218.

Henderson P. (ed.), Rare Earth Element Geochemistry. Elsevier, N. Y., 1984, 466 pp.

Ionov D. and Harmer R. E., Trace element distribution in calcite-dolomite carbonatites from Spitskop: inferences for differentiation of carbonatite magmas and the origin of carbonates in mantle xenoliths. *Earth Planet. Sci. Lett.* **198**, 2002, 495-510.

Johnson T. M. and DePaolo D. J., Interpretation of isotopic data in groundwater-rock systems: Model development and application to Sr isotope data from Yucca Mountain. *Water Resources Res.* **30**, 1994, 1571-1587.

Lipin B.R. and McKay G. A. (eds.), Geochemistry and mineralogy of rare earth elements. *Min. Soc. Amer. Rev. Mineralogy*, vol. 21, 1989, 348 pp.

Manceau A. and Charlet L., X-ray absorption spectroscopic study of the sorption of Cr(III) at the oxide/water interface: I. Molecular mechanism of Cr(III) oxidation on Mn oxides. *J. Colloid Interface Sci.* **148**, 1992, 443-458.

Manceau A., Drits V. A., Silvester E., Bartoli C. and Lanson B., Structural mechanism of Co^{2+} oxidation by the phyllomanganate buserite. *Am. Min.* **82**, 1997, 1150-1175.

McKenzie R. M., Manganese oxides and hydroxides. In *Minerals in Soil Environments* (eds. J. B. Dixon and S. B. Weed), 1989, pp. 439-465. Soil Sci. Soc. Amer.

Peterman Z. E., Stuckless J. S., Marshall B. D., Mahan S. A., and Futa K., Strontium isotope geochemistry of calcite fracture fillings in deep core, Yucca Mountain, Nevada - A progress report. *Proc. 3rd Int. Conf. High Level Rad. Waste Mgmt.*, 1992, 1582-1586.

Piper D. Z., Rare earth elements in ferromanganese nodules and other marine phases. *Geochim. Cosmochim. Acta* **38**, 1974, 1007-1022.

Stimac J. D., Hickmott D., Abell R., Larocque A. C. L., Broxton D., Gardner J., Chipera S., Wolff J., and Gauerke E. Redistribution of Pb and other volatile trace metals during eruption, devitrification, and vapor-phase crystallization of the Bandelier Tuff, New Mexico. *J. Volc. Geotherm. Res.* **73**, 1996, 245-266.

Stuckless J. S., Peterman Z. E., and Muhs D. R. U and Sr isotopes in ground-water and calcite, Yucca Mountain, Nevada: Evidence against upwelling water. *Science* **254**, 1991, 551-554.

Taylor S. R. and McLennan S. M., The significance of the rare earths in geochemistry and cosmochemistry. Chapter 79 in *Handbook on the Physics and Chemistry of Rare Earths* (Gschneider K. A., Jr., and Eyring, L., eds.), vol. 11, Elsevier, N. Y., 1988, 485-578.

Valeton I., Schumann A., Vinx R., and Wienke M., Supergene alteration since the upper Cretaceous on alkaline igneous and metasomatic rocks of the Poços de Caldas ring complex, Minas Gerais, Brazil. *Appl. Geochem.* **12**, 1997, 133-154.

Vaniman D. T. and Chipera S. J., Paleotransport of lanthanides and strontium recorded in calcite compositions from tuffs at Yucca Mountain, Nevada, USA. *Geochim. Cosmochim. Acta* **60**, 1996, 4417-4433.

Vaniman D. T., Chipera S. J., and Bish D. L., Petrography, mineralogy, and chemistry of calcite-silica deposits at Exile Hill, Nevada, compared with local spring deposits. *Los Alamos Nat. Lab. Rept.* LA-13096-MS, 1995, 70 pp.

Vaniman D. T., Chipera S. J., Bish D. L., Duff M. C., and Hunter D. B., Crystal chemistry of clay-Mn oxide associations in soils, fractures, and matrix of the Bandelier Tuff, Pajarito Mesa, New Mexico. *Geochim. Cosmochim. Acta.* **66**, 2002, 1349-1374.

Whelan J. F. and Stuckless J. S., Paleohydrologic implications of the stable isotopic composition of secondary calcite within the Tertiary volcanic rocks of Yucca Mountain, Nevada. *Proc. 3rd Int. Conf. High Level Rad. Waste Mgmt.*, 1992, 1572-1581.

Whelan J. F., Vaniman D. T., Stuckless J. S., and Moscati R. J., Paleoclimatic and paleohydrologic records from secondary calcite, Yucca Mountain, Nevada. *Proc. 5th Int. Conf. High Level Rad. Waste Mgmt.*, 1994, 2738-2745.

Zielenski R. A., Evaluation of ash-flow tuffs as hosts for radioactive waste: Criteria based on selective leaching of manganese oxides. *U. S. Geol. Survey Open-file Rept.* 83-480, 1983, 21 pp.

Chapter 6

ORIGIN OF RARE EARTH ELEMENT SIGNATURES IN GROUNDWATERS OF SOUTH NEVADA, USA: IMPLICATIONS FROM PRELIMINARY BATCH LEACH TESTS USING AQUIFER ROCKS

XIAOPING ZHOU[1,2], KLAUS J. STETZENBACH[1], ZHONGBO YU[2], & KAREN H. JOHANNESSON[3]

[1] *Groundwater Chemistry Group, Harry Reid Center for Environmental Studies, University of Nevada, Las Vegas, Las Vegas, Nevada 89154-4009, USA*

[2] *Department of Geoscience, University of Nevada, Las Vegas, Las Vegas, Nevada 89154-4010, USA*

[3] *Department of Earth and Environmental Sciences, The University of Texas at Arlington, Arlington, Texas 76019-0049, USA*

1. Introduction

Rare earth elements (REE) have been used extensively to study petrologic and mineralogic processes, as well as trace element cycling in the oceans since the late 1970's and early 1980's (e.g., Hanson, 1980; Elderfield and Greaves, 1982; DeBaar et al., 1983; Cullers and Graf, 1984). More recently, with the improvement of analytical instrumentation, REEs can be readily measured at the low parts-per-trillion level in natural waters using, for example, inductively coupled plasma mass spectrometry (ICP-MS; Stetzenbach et al., 1994; Graham et al., 1996; Halicz et al., 1999). One consequence of the improvements in analytical instrumentation is the increase in the numbers of investigations of the REEs in groundwaters (Smedley, 1991; Fee et al., 1992; Gosselin et al., 1992; Halicz et al., 1999). Many of these studies indicate that groundwaters commonly have REE patterns that closely mimic the REE patterns of the rocks through which they flow. The similarities between groundwater and aquifer rock REE patterns suggest that the REEs may be useful tracers of groundwater-aquifer rock interactions. Many investigations have already demonstrated the utility of the REEs as chemical tracers of numerous geochemical weathering processes and possibly biogeochemical redox processes (Hanson, 1980; Moffett, 1990, 1994). Although previous studies have demonstrated that in many cases the REE concentrations observed in groundwaters mimic those of rocks through which they flowed, quantitative, controlled investigations of the REE signatures aqueous solutions acquire by reacting with different types of rocks at low temperatures have not been completed. In order to gain a better understanding of how different rock types can affect the concentrations of REEs in aqueous solutions that react with the rocks and to obtain additional information concerning REE behavior during rock-water reactions, we conducted a series of batch reactor experiments involving compositionally different rock types, and two aqueous solutions (i.e., distilled deionized water and an acidic solution). It should be noted that the current study represents the results of a preliminary investigation within a larger study that includes the PhD dissertation research of the lead author.

K.H. Johannesson,(ed), Rare Earth Elements in Groundwater Flow Systems , 141-160.

2. Analytical Methods

2.1. ROCK SAMPLE COLLECTION

We chose the Paleozoic stratigraphic section at Frenchman Mountain near Las Vegas, Nevada as our major sampling location because: (1) the stratigraphic sequence in this section is very similar to the lower part of the regional carbonate aquifer of southern Nevada (e.g., Winograd and Thordarson, 1975); and (2) information gained by studying the REE signatures of these rocks and their resulting leachate solutions can be applied to our studies of REEs in groundwaters from this aquifer (e.g., Stetzenbach et al., 1994; Johannesson et al., 1997). In addition, the stratigraphic section is exposed on the eastern edge of the city of Las Vegas, which allows for easy access and detailed sampling. The Paleozoic strata exposed at the site consist of shales, sandstones, and especially abundant carbonate rocks (e.g., dolomite, dolomitic limestone, and limestone) that range in age from Cambrian up to Permian (Rowland, 1987; Rowland et al., 1990). Ordovician and Silurian rocks are, however, missing from this section. Sixteen out of 19 rock samples used for the batch tests in this study were collected from the lower portion of the Frenchman Mountain section, including samples from the Tapeats Sandstone (Tapeats1), Bright Angel Formation (shale or sandstone) [BAS(SS)1, BAS(SL)2, BAS3, and BAS6], and Bonanza King Formation (dolomite) (BKD1, BKD2, BKD3, BKD4, BKD5, BKD6, BKD7, BKD8, BKD9, BKD10, and BKD11). In addition, one sample (FR8Oaal) was collected from the Aysees Member of the Lower Ordovician Antelope Valley Limestone at Fossil Ridge; 20 miles northwest of Las Vegas, Nevada (see Johannesson et al., 2000a). One rhyolitic rock sample (Surprise1) was also examined in the batch tests. Therefore, the samples used in the batch tests represent five different rock types, including shale [BAS(SL)2, BAS(SL)3, BAS6], fine- and medium-grained sandstone (BAS(SS)1, Tapeats1], limestone (FR8Oaal), dolomite (BKD1 through BKD11), and felsic volcanic rock (Surprise1).

2.2. BATCH TESTS

2.2.1. Preparation and leachates

Approximately 5 - 10 kg of rock was collected from each sample location. Rock samples were subsequently broken into 0.5-1.0 cm size fragments using a sledge hammer. Fragments without weathering rinds, calcite veins, and obvious signs of alteration were carefully selected until approximately 500 to 1000 grams of fragmental material were obtained for each rock sample. The fragmental rock samples were further crushed to ~70 mesh in size, which were subsequently used in three batch tests employing different solutions and variable reaction times (Table 1). In addition, an aliquot of each crushed rock sample was preserved for REE analysis of the bulk rock (see below).

For Batch Tests 1 and 2, 60 to 220 g of distilled-deionized water (pH=7; 18 MΩ-cm) was added to between 30 g to 80 g of each of the crushed rock samples within acid-washed, high-density linear polyethylene bottles (mass ratio of solution to rock sample ranged from 2:1 to 3:1; Table 1). The crushed rock samples and distilled-deionized water was allowed to react at room temperature ($\sim 25^{\circ}C$) for 40 days (Batch Test 1) and 67 days (Batch Test

2), respectively. In Batch Test 3, a dilute nitric acid solution (pH=4) was prepared from ultrapure HNO_3 (Seastar, Inc. double sub-boiling, distilled in quartz) and distilled-deionized water, and subsequently reacted with the crushed rock samples for 42 days at room temperature ($\sim 25^\circ C$; Table 1). During each batch test period, all sample bottles containing the crushed rock - leaching solution mixtures were agitated for 5 minutes each day to allow a complete reaction between rock sample and leaching solution. Following completion of the reaction period, the leachate solutions were separated from the crushed rock using a centrifuge, and then filtered through 0.45 μm Nuclepore® filters (Johannesson and Zhou, 1999).

2.2.2. *REE analysis of leachates*

All leachate samples were acidified, subsequent to filtration, using ultrapure nitric acid (Seastar, Inc. double sub-boiling, distilled in Teflon®), and analyzed by inductively coupled plasma mass spectrometry (ICP-MS; Perkin-Elmer® Elan 5000) with ultrasonic nebulization. The ultrasonic nebulization increased the analytical sensitivity and decreased the potential interferences from oxide formation in the plasma stream (Hodge et al., 1998). Due to the small amount of leachate from each batch test, pre-concentration by cation-exchange was not employed. The following REE isotopes were selected for analysis since they have no or less elemental isobaric interferences: ^{139}La, ^{140}Ce, ^{141}Pr, ^{146}Nd, ^{149}Sm, ^{151}Eu and ^{153}Eu (mean value), ^{157}Gd, ^{159}Tb, ^{163}Dy, ^{165}Ho, ^{166}Er, ^{169}Tm, ^{172}Yb, and ^{175}Lu. During ICP-MS analysis, a series of 5 standards of known concentrations (i.e., 0.1 μg/kg, 0.5 μg/kg, 1.0 μg/kg, 5.0 μg/kg, and 10.0 μg/kg) were prepared and routinely monitored in order to calibrate the instrument, check the calibration, and calculate the sample REE concentrations. The REE values determined in the Method Blanks were subtracted from the leachate solutions to obtain the leachate concentrations.

2.3. ANALYSIS OF REE IN ROCK SAMPLES

The REEs were measured in 17 rock samples using ICP-MS (Perkin-Elmer® Elan 5000) as discussed in Johannesson and Zhou (1999). Briefly, approximately 0.25 grams of each rock (i.e., powder) sample were placed in precleaned Teflon®-lined microwave digestion bombs, followed by 5 mL of ultrapure HF (Seastar, Inc. double sub-boiling, distilled in Teflon®) and 5 mL of ultrapure HNO_3 (Seastar, Inc. double sub-boiling, distilled in quartz). The Teflon®-lined microwave digestion bombs were then sealed and placed in a microwave oven (CEM Corporation MDS-2100) and heated to 189°C and pressurized to 8.62×10^5 Pa (125 p.s.i.) for 25 minutes. After 25 minutes, the samples were allowed to cool before 30 mL of a saturated boric acid solution was added to each sample. The samples were subsequently heated again in the microwave for 5 minutes at 100°C and 6.9×10^4 Pa - 1.38×10^5 Pa (10 - 20 p.s.i.). The dissolved rock samples were then decanted into clean polyethylene bottles and diluted by a factor of 180 before analysis by ICP-MS. The U.S. Geological Survey rock standard W-2 (diabase) was included as a check standard during the analysis from which our measurements never deviated by more than 10% (Guo, 1996, Guo et al., this volume).

Table 1. Parameters used for three batch tests.

Parameters	Batch Test 1	Batch Test 2	Batch Test 3
Rock Types	shale, sandstone, limestone, dolomite, rhyolite	shale, sandstone, dolomite	shale, sandstone, limestone, dolomite, rhyolite
Solutions	distilled-deionized water	distilled-deionized water	acidic water solution
Sample Weight (g)	30	75 - 82	50
Solution Weight (g)	60	200 - 220	100
Solution/Rock Weight Ratio	2 : 1	3 : 1	2 : 1
pH (starting)	7	7	4
Temperature (^0C)	~ 25	~ 25	~ 25
Reaction Time (days)	40	67	42

3. Analytical Results

3.1. REE CONCENTRATIONS OF SOUTHERN NEVADA ROCKS

The REE concentrations (in ppm) for the 16 different sedimentary rock samples and one rhyolite (Surprise 1) examined in this study are presented in Table 2. Again, these same rock samples were used in the different batch leach tests of this study. The chondrite-normalized REE patterns for these rock (both silicate and carbonate rocks) samples are plotted in Fig. 1. The chondrite-normalizing factor that was employed for the normalization calculations is that tabulated by Hanson (1980).

The silicate rocks (Fig. 1a) have REE concentrations that are, on average, about a factor of 10 greater than the carbonate rocks examined in the study (Fig. 1b; Table 2). Moreover, every rock analyzed exhibits enrichment in the light REEs (LREE) over the heavy REEs (HREE) compared to chondrite (Fig. 1). The chondrite-normalized Nd/Yb ratios [(Nd/Yb)$_{CN}$, where CN = chondrite-normalized) range from 1.7 for one of the Bonanza

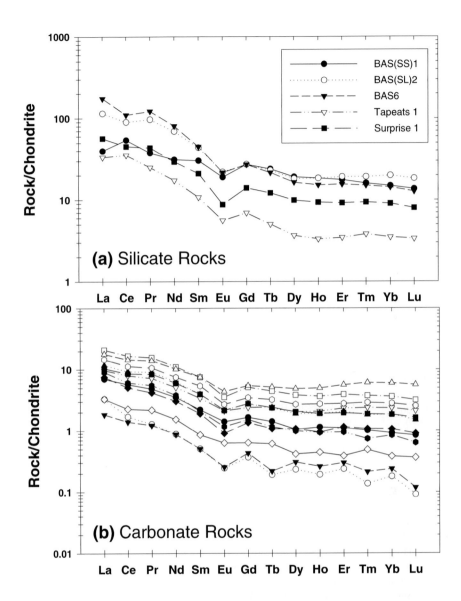

Figure 1. Chrondrite-normalized REE plots of (a) silicate rocks from Frenchman Mountain, Nevada and Death Valley, California, and (b) carbonate rocks (i.e., Bonanza King Dolomite) from Frenchman Mountain and Fossil Ridge (filled circle), Nevada. BAS = Bright Angel Shale. "Surprise" sample is from Death Valley. These rock samples were prepared and used in the batch tests.

King dolostone samples, up to 5.6 for sample BAS6 from the Bright Angel Formation (Table 2). All of the carbonate rocks have chondrite-normalized negative Ce anomalies, whereas the silicate rocks exhibit both negative and positive Ce anomalies (Fig.1). When the rock samples are normalized to Average Shale, they exhibit relatively flat patterns (not shown) as demonstrated by their $(Yb/Nd)_{SN}$ ratios (where SN = shale-normalized), which are generally close to 1 (Table 2). The Average Shale values used to calculate these shale-normalized ratios is the composite "Average Shale" previously used by oceanographic researchers (e.g., Elderfield and Greaves, 1982; DeBaar et al., 1983; Sholkovitz, 1988), and in our previous investigations (Johannesson et al., 1997, 2000a).

The chondrite-normalized REE patterns for the silicate and carbonate rocks of southern Nevada demonstrate that the carbonate rocks contain lower abundances of all of the REEs than the silicate rocks (Fig. 1). The spiky HREE patterns exhibited by some of the carbonate rocks (i.e., BKD1 and BKD2) reflect the difficulty encountered in quantifying the HREEs in these two rock samples using the digestion and analytical methods employed in this study. Many of the HREE for BKD1 and BKD2 approach or are below the method detection limits for our ICP-MS using the above described digestion techniques (e.g., Guo, 1996; Guo et al, this volume).

3.2 REES OF THE LEACH SOLUTIONS

3.2.1. *Leachable fraction using distilled water*
The REE data for each batch test are presented in Table 3. The data show that individual REE concentrations in the leachate solutions from Batch Test 1 are generally low, ranging from 1.4 pmol/kg to 2,570 pmol/kg (Table 3). In Batch Test 1, the solution that reacted with rhyolitic pumice (Surprise1) had the highest REE concentrations, whereas the solutions that reacted with Ordovician limestone (FR8Oaal), and with Cambrian dolomite (BKD3) exhibit the lowest REE concentrations (Table 3).

Rare earth element concentrations for each of the leachate solutions from Batch Test 1 are plotted in Fig. 2, where each leachate has been normalized to the REE concentrations determined in the respective rock with which each reacted. All of the leach solutions from Batch Test 1 have substantially lower REE concentrations than the respective rocks (e.g., between a factor of 10^5 to 10^6 lower). Interestingly, except for Eu in each case, and La for the leachates that reacted with BKD3 and FR8Oaal, the rock-normalized REE patterns for the leachate are relatively flat (Fig. 2). These flat rock-normalized pattern is best developed for the leachates that reacted with the rhyolitic rock sample, Surprise 1, with shale BAS(SL)2, and with fine sandstone BAS (SS)1 [(Yb/Nd)RN = 1.21, 1.04, 1.10, respectively; Table 3]. These flat patterns indicate that REEs were leached from these rocks in essentially the same relative proportions in which they occur in their respective rock samples. There is evidence for a slight enrichment in the HREEs in the Surprise 1 leach solutions, as the rock-normalized pattern for this leach solution exhibits a shallow positive slope between Tb and Lu (Fig. 2). There may also be some enrichment in the HREEs of the leachate solution that reacted with FR8Oaal, however insufficient data exists for the HREEs in this leachate (e.g., only Dy and Yb were quantified by our method). On the other hand, the leachates that reacted with dolomite (BKD3) and the

Table 2. Concentrations of rare earth elements (ppm) in aquifer materials (rocks) used for batch tests. CN = chondrite-normalized, SN = shale-normalized.

	La	Ce	Pr	Nd	Sm	Eu	Gd	Tb	Dy	Ho	Er	Tm	Yb	Lu	(Nd/Yb)$_{CN}$	(Yb/Nd)$_{SN}$	Ce/Ce*	Eu/Eu*
BAS(SS)1	13.1	47.6	4.2	18.7	5.5	1.3	6.8	1.1	6.2	1.3	3.5	0.5	3.0	0.5	2.1	1.7	0.2	0.0
BAS(SL)2	37.8	79.4	10.9	41.6	8.0	1.5	6.8	1.1	5.9	1.3	3.8	0.6	4.0	0.6	3.5	1.0	0.0	-0.1
BAS6	57.7	97.0	13.7	48.5	8.2	1.5	6.8	1.0	5.4	1.1	3.1	0.5	2.9	0.4	5.6	0.6	-0.1	-0.1
Tapeats 1	11.0	31.3	2.8	10.4	2.0	0.4	1.7	0.2	1.2	0.2	0.7	0.1	0.7	0.1	5.0	0.7	0.1	0.0
Surprise 1	18.7	39.8	4.9	17.7	3.8	0.6	3.5	0.6	3.2	0.7	1.8	0.3	1.8	0.3	3.3	1.1	0.0	-0.1
FR8Oaal	2.3	5.1	0.5	2.1	0.4	0.1	0.4	0.1	0.3	0.1	0.2	0.0	0.2	0.0	3.8	0.9	0.0	0.0
BKD1	1.1	1.5	0.1	0.5	0.1	0.0	0.1	0.0	0.1	0.0	0.0	0.0	0.0	0.0	5.0	0.7	-0.1	-0.1
BKD2	0.6	1.2	0.1	0.5	0.1	0.0	0.1	0.0	0.1	0.0	0.1	0.1	0.0	0.0	3.7	1.0	0.0	-0.1
BKD3	3.2	7.0	0.8	3.1	0.6	0.1	0.6	0.1	0.7	0.1	0.5	0.1	0.5	0.1	2.2	1.7	0.0	0.0
BKD4	3.4	7.5	0.9	3.6	0.7	0.1	0.7	0.1	0.7	0.1	0.4	0.1	0.4	0.1	3.3	1.1	0.0	0.0
BKD5	6.9	14.7	1.8	6.7	1.4	0.2	1.3	0.2	1.2	0.3	0.8	0.1	0.7	0.1	3.1	1.2	0.0	-0.1
BKD6	2.5	4.5	0.5	1.8	0.3	0.1	0.3	0.1	0.3	0.1	0.2	0.0	0.2	0.0	2.9	1.3	0.0	-0.1
BKD7	1.1	2.0	0.2	0.9	0.2	0.0	0.2	0.0	0.1	0.0	0.1	0.0	0.1	0.0	4.0	0.9	0.0	0.1
BKD8	3.9	8.2	1.0	3.8	0.7	0.2	0.7	0.1	0.6	0.1	0.4	0.1	0.4	0.1	3.4	1.1	0.0	0.0
BKD9	5.9	12.8	1.6	6.1	1.4	0.3	1.4	0.2	1.6	0.3	1.1	0.2	1.2	0.2	1.7	2.1	0.0	0.0
BKD10	3.0	5.4	0.6	2.3	0.4	0.1	0.4	0.1	0.3	0.1	0.2	0.0	0.2	0.0	4.5	0.8	0.0	0.0
BKD11	4.9	9.9	1.2	4.5	1.0	0.2	0.9	0.2	0.9	0.2	0.6	0.1	0.6	0.1	2.9	1.3	0.0	-0.3

Ce/Ce* = log {2Ce$_{SN}$/[La$_{SN}$ + Pr$_{SN}$]}.
Eu/Eu* = log {2Eu$_{SN}$/[Sm$_{SN}$ + Gd$_{SN}$]}.

Table 3. Concentrations of rare earth elements (pmol/kg) in leachates from three batch tests. RN = rock-normalized. Ratios are shale-normalized.

	La	Ce	Pr	Nd	Sm	Eu	Gd	Tb	Dy	Ho	Er	Tm	Yb	Lu	(Yb/Nd)	Ce/Ce*	Eu/Eu*
Test 1																	
BKD3	179.3	135.6	18.1	78.7	19.9	43.9	19.0	1.4	11.5	1.5	4.5	0.0	5.3	0.0	0.5	-0.3	1.0
Surprise 1	1594.6	2574.2	320.8	1201.8	216.1	95.1	232.1	30.0	174.2	37.7	117.5	18.7	124.1	19.8	1.2	-0.1	0.4
BAS(SL)2	185.0	464.6	59.7	226.4	47.1	26.5	50.3	7.2	38.0	6.3	20.1	3.5	18.7	3.4	1.0	0.1	0.4
BAS(SS)1	199.4	498.1	56.3	270.0	98.7	50.0	145.3	21.1	116.9	21.0	50.9	7.3	39.5	5.0	1.1	-0.1	0.3
FR8Oaal	194.4	28.0	6.6	17.6	1.5	4.3	3.2	0.0	1.7	0.0	0.0	0.0	1.8	0.0	1.4	-0.9	0.9
Test 2																	
Tapeats 1	122.5	213.7	22.1	79.6	21.9	146.3	15.4	2.0	8.8	2.1	6.9	1.4	2.1	0.7	0.5	-0.1	1.6
BAS6	506.5	475.8	77.2	341.7	72.5	27.6	91.2	12.3	58.3	9.9	33.6	3.0	36.7	5.3	2.2	-0.2	0.2
BKD1	43.9	120.4	11.2	54.9	7.4	22.1	24.5	4.2	22.9	2.8	9.6	1.8	7.1	0.9	2.3	0.1	0.7
BKD2	41.4	51.1	5.2	21.8	7.4	56.0	8.0	0.6	2.4	0.5	2.4	0.4	1.5	0.2	0.9	-0.1	1.6
BKD4	15.5	19.7	3.1	12.6	3.3	7.9	4.8	0.4	4.2	0.6	1.7	0.5	1.0	0.2	0.9	-0.2	1.0
BKD5	22.3	34.0	5.3	15.1	3.9	23.7	4.9	0.5	2.1	0.9	2.3	0.8	3.7	0.4	2.7	-0.1	1.5
BKD6	28.0	20.7	3.2	8.9	2.8	15.3	1.7	0.0	2.7	0.6	0.7	0.5	0.5	0.1	0.6	-0.3	1.6
BKD7	17.4	26.1	4.1	12.5	6.2	8.5	2.4	0.2	3.3	0.6	1.6	0.5	1.3	0.3	1.5	-0.1	0.8
BKD8	20.0	16.1	1.6	7.0	2.8	18.2	2.7	1.0	0.9	0.5	1.1	0.5	0.9	0.4	1.5	-0.2	1.5
BKD9	28.5	39.2	4.9	15.7	6.2	19.5	4.4	0.3	5.7	0.2	1.7	0.3	1.6	0.2	0.6	-0.1	1.2
BKD10	20.4	29.7	3.6	12.1	4.3	12.7	2.8	1.3	2.9	0.7	0.9	1.3	3.1	0.9	4.2	-0.1	1.2
BKD11	399.2	821.6	104.3	392.0	77.2	27.4	67.9	10.5	60.1	12.2	29.8	4.3	29.3	4.4	0.7	0.0	0.3
Test 3																	
FR8Oaal	221.7	245.5	43.5	183.7	37.2	15.0	37.5	6.6	24.7	5.8	22.3	1.4	11.4	1.4	0.9	-0.3	0.2
BAS6	334.0	1829.1	67.1	242.0	80.5	60.6	76.9	10.3	42.4	14.5	20.7	4.7	29.9	5.7	2.5	0.5	0.6
BKD1	138.9	2635.6	56.8	153.2	13.9	27.8	6.2	3.8	16.8	1.8	16.2	0.0	12.8	1.4	1.5	0.8	0.8
Tapeats 1	63.7	1513.0	16.1	88.7	11.0	223.1	19.7	0.0	4.8	0.0	2.1	0.0	3.0	0.0	0.6	0.9	1.8

Ce/Ce* = log {2Ce$_{RN}$/[La$_{RN}$ + Pr$_{RN}$]}.

Eu/Eu* = log {2Eu$_{RN}$/[Sm$_{RN}$ + Gd$_{RN}$]}.

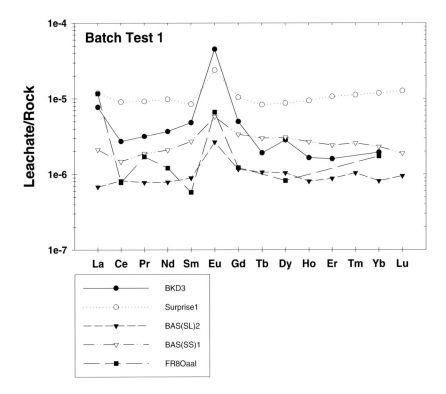

Figure 2. Rock-normalized REE patterns of the leachate solutions from Batch Test 1.

sandstone [BAS(SS)1] may have slight enrichments in the middle REEs (MREE) compared to their respective rocks. The Batch Test 1 leachates have variously enriched and depleted Ce concentrations. The leachate that reacted with BAS(SL)2 exhibits weak positive Ce anomaly (Ce/Ce* = 0.05; Table 3), whereas Surprise 1, BKD3, FR8Oaal, and BAS(SS)1 have negative Ce anomalies (Ce/Ce* = -0.07, -0.30, -0.93, -0.07, respectively; Table 3). On the other hand, the Eu concentration in each leachate is higher than expected based on the rock-normalized values for Sm and Gd ($0.28 \leq$ Eu/Eu* ≤ 0.97; Table 3).

Table 3 contains rare earth element concentrations for each of the leachate solutions from Batch Test 2. The solutions that reacted with one of the Cambrian dolomites (BKD11), and the shale (BAS6) had the highest REE concentrations, whereas the solutions that reacted with other Cambrian dolomites (e.g. BKD4 and BKD8) exhibit the lowest REE concentrations (Table 3). The chief difference in Batch Test 2 compared to Batch Test 1 is the greater solution/rock sample ratio (by weight, 3:1 vs. 2:1, respectively; Table 1), and the greater duration of Batch Test 2 (67 days vs. 40 days for Batch Test 1; Table 1).

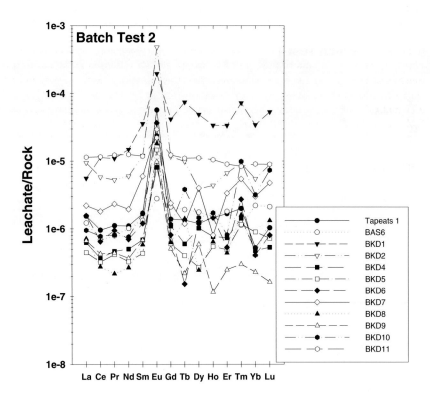

Figure 3. Rock-normalized REE patterns of leachate solutions from Batch Test 2.

Figure 3 is a rock-normalized REE plot of the leachate solutions from Batch Test 2 where each leach solution has been normalized to the respective rock with which it reacted. In general, the shapes of these rock-normalized leachate patterns resemble those from Batch Test 1 in that they are all relatively flat, and have large positive Eu anomalies ($0.21 \leq$ Eu/Eu* ≤ 1.61; Table 3). Only two leachate samples do not exhibit the same degree of positive enrichment in Eu compared to Sm and Gd; BAS6 and BKD11 have Eu/Eu* of 0.21 and 0.27, respectively (Table 3). These leachates also exhibit weak negative Ce anomalies (-$0.29 \leq$ Ce/Ce* ≤ -0.01), except for the leachate that reacted with BKD1 (Ce/Ce* = 0.13; Table 3). The chief differences between the results of these two batch tests are the smoother LREE patterns and more variability in the HREE patterns of the Batch Test 2 leachates compared to the Batch Test 1 leachates.

3.2.2. *Leachable fraction using acidic water solution*

Table 3 also lists the REE concentrations of leach solutions from Batch Test 3, in which crushed rock samples were reacted with a weak nitric acidic solution (pH = 4) for 42 days. Similar to the other batch tests, the REE concentrations of these leachates are also low, ranging from 1.1 pmol/kg to 2,630 pmol/kg (Table 3). The solution that reacted with

Cambrian dolomite (BKD1) had the highest REE concentrations, whereas the solution that reacted with Tapeats Sandstone (Tapeats1) had the lowest REE concentrations (Table 3).

Rock-normalized REE ratios of these weak nitric acid leach solutions are plotted in Fig. 4. These leachates have similar (i.e., order of magnitude) REE concentrations to those of Batch Test 1 (Table 3) and also exhibit similar rock-normalized ratios. Again, all of the leachates of Batch Test 3 exhibit positive Eu anomalies that range from a low of Eu/Eu* = 0.22 up to Eu/Eu* = 1.82 (Table 3). In addition, all but the leachate that reacted with the Ordovician Limestone (i.e., FR8Oaal) have large positive Ce anomalies (Fig. 4). The leachate that reacted with FR8Oaal instead exhibits a small negative Ce anomaly (Ce/Ce* = -0.26; Table 3). Neglecting the Ce and Eu anomalies, the leachates that reacted with FR8Oaal, BAS6, and Tapeats1 have relatively flat rock-normalized REE patterns. However, the rock-normalized REE pattern for the leachate that reacted with BKD1 exhibits an irregular shape (Fig. 4).

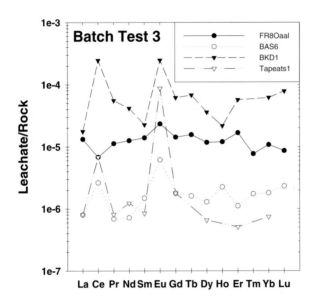

Figure 4. Rock-normalized REE patterns of leachate solutions from Batch Test 4.

4. Discussion

4.1. REE SIGNATURES OF LEACHATES

Many of the leachate solutions have REE signatures that are not substantially fractionated compared to the rock samples with which they reacted, and hence, do not differ from the source of the dissolved REEs in the leachates (Figs. 2 - 4). The similarities in the relative proportions of REEs between rock samples and leachates are especially apparent for the distilled-deionized water leachates (i.e., Batch Tests 1 and 2; Figs. 2 and 3). These data indicate that, in general, the REEs in these rocks must be in readily leachable sites. Although, overall the distilled-deionized water leachates exhibit relatively flat REE patterns when normalized to their respective rock sample, there are some subtle deviations that suggest minor fractionation of the REEs during the leaching process. For example, the leachate with perhaps the flattest rock-normalized pattern (i.e., Surprise 1, rhyolite) also exhibits a smooth increase in rock-normalized HREE values with increasing atomic number (Fig. 1). This smooth increase between Tb and Lu suggests that either (1) heavier REEs are located within progressively more readily leachable sites within the source rocks than the LREEs, or (2) solution complexation reactions in the distilled-deionized leachate become important as the water reacts with the rocks leading to preferential leaching and stabilization of HREEs within solution. Many investigators have argued that during the weathering process, LREEs released are subsequently captured by secondary clay minerals formed by weathering reactions, whereas the HREEs are preferentially liberated from the rock and removed from the site of weathering by solution complexation reactions involving inorganic and/or organic ligands (Nesbitt, 1979; Duddy, 1980; Schau and Henderson, 1983; Braun et al., 1993). It is well known that stability constants for REE complexes with carbonate ions increase with increasing atomic number (Cantrell and Byrne, 1987; Lee and Byrne, 1993). Moreover, production of bicarbonate and carbonate ions are a common product of weathering reactions involving silicate rocks and dissolution of carbonate rocks, respectively (e.g., Garrels and MacKenzie, 1967; Langmuir, 1971). Bicarbonate ions were likely produced in our Batch Test 1 and 2 as distilled-deionized water reacted with the crushed rock samples as vessels were initially open to the atmosphere.

Some of the leachates from the dolomites (Batch Test 2; Fig. 3) also exhibit enrichments in the HREEs relative to the LREEs, although the same, relatively smooth progressive increase in rock-normalized values reported for the Surprise 1 leachate is not observed. Instead, these dolostone leachates have a step-like rock-normalized REE pattern, with relatively lower LREE normalized values that step up to higher values between Nd and Sm, and are relatively constant from Tb through Lu (BKD1, Fig. 3).

Another subtle fractionation pattern of the rock-normalized leachate profiles is slight enrichments in the middle REEs. This subtle fractionation is only observed in leachates that reacted with the clastic sedimentary rock samples of the Bright Angel Formation [i.e., BAS(SS)1, BAS(SL)2; Fig. 2]. Johannesson and Zhou (1999) reported similar, albeit more pronounced, rock-normalized MREE enriched patterns for strong acid leach solutions that reacted with clastic sedimentary rocks containing Fe-Mn oxide/oxyhydroxide phases. Moreover, using a leach solution that specifically targets Fe-Mn oxide/oxyhydroxide phases (i.e., 0.04 M $NH_2OH \cdot HCl$ in 25% (v/v) CH_3COOH), these authors reported MREE enriched, rock-normalized REE patterns for those leachates that reacted with the clastic sedimentary rocks containing petrographically identifiable Fe-Mn oxide/oxyhydroxide phases. Johannesson and Zhou (1999) suggested that the MREE enriched patterns of the leachates reflected the distribution of the REEs in the Fe-Mn oxide/oxyhydroxide phases within these rocks. Others have shown that such oxide phases commonly exhibit MREE enrichments (Gosselin et al., 1992). Consequently, the slight MREE enrichments of the distilled-deionized water leachates for the two Bright Angel Formation samples compared to their respective rocks may reflect the presence of Fe-Mn oxide/oxyhydroxide phases within these clastic sedimentary rocks that supply MREEs to the leachate solutions. Further investigation may involve leaching these rocks with a hydroxylamine hydrochloride solution to specifically examine the possible contributions of Fe-Mn oxide/oxyhydroxide phases to their readily leachable REE fractions.

Interestingly, the distilled-deionized water leachate of Batch Test 1 that reacted with BKD3 exhibits higher rock-normalized LREE values than observed for the HREEs (Fig. 2). The REE pattern of this leach solution suggests that the LREEs occur in sites in BKD3 that are more readily leached by distilled-deionized water than the HREEs, or alternatively, the LREEs are preferentially leached from the rock owing to the formation of stronger LREE complexes with some ligand(s) than for the HREEs. It is difficult to identify a ligand that could occur in distilled-deionized water that more strongly complexes the LREEs in solution than the HREEs. Because BKD3 is a dolomite, carbonate ions are expected to increase in the distilled-deionized water leachate as a result of the dolomite reacting with the neutral pH solution. As a consequence, we would expect that the HREEs would be more stable in the leachate than LREEs owing to complexation of the REEs with carbonate ions. Therefore, the enrichment of the LREEs in this leachate likely reflects that the LREEs occur in more easily leached sites than the HREEs in BKD3.

The most interesting feature of Batch Test 3 is that all of the leachates except for the one that reacted with the Ordovician limestone, FR8Oaal, exhibit positive, rock-normalized Ce anomalies (Fig. 3). Batch Test 3 involved reacting each of the rock samples with a weak nitric acid solution (pH = 4), and hence was different than the other batch tests where distilled-deionized water (pH ≈ 7) was used. The Tapeats Sandstone sample is enriched in Ce relative to La and Pr when normalized to chondrite. Consequently, the positive Ce anomaly of the leachate sample that reacted with the Tapeats Sandstone indicates that the weak nitric acid leachate preferentially liberated Ce from the Tapeats Sandstone compared to both La and Pr. The same preferential leaching of Ce from the rocks can be argued for both BAS6 and BKD1 as a result of reacting these rocks with the weak acid solution (Fig. 4). Relatively more Ce must have been leached from the Tapeats Sandstone, however, for the leachate solution to be enriched in Ce with respect to La and Pr when normalized to the Ce enriched Tapeats Sandstone (compare Figs.1 and 4). The difference in the behavior of Ce between the distilled-deionized leachate solutions and the weak nitric acid solutions indicates that a Ce enriched mineral or amorphous phase, that is susceptible to the weak acid leach but not the distilled-deionized water solutions, is present in these southern Nevada sedimentary rocks.

As mentioned above, all of the leachates exhibit positive Eu anomalies when normalized to their respective rocks. These positive Eu anomalies may reflect either BaO^+ interference on the two naturally occurring Eu isotopes (i.e., ^{151}Eu and ^{153}Eu) during mass quantification by ICP-MS (i.e., Jarvis et al., 1989), or simply the fact that the rocks are depleted in Eu with respect to their chondrite-normalized Sm and Gd values (Fig. 1; Table 2). We suggest that for the vast majority of the leachates, the large positive Eu anomalies reflect Ba interference during quantification (i.e., false positive). For groundwater samples, the majority of the Ba is removed from the sample by extraction using diethylhexylphosphoric acid (Cerrai and Ghersini, 1966; Hodge et al., 1998). Typically, at least a 500 mL aliquot of each groundwater sample is required to quantify Eu after Ba is extracted from the sample. Unfortunately, insufficient leachate volumes were recovered during the batch tests in the current study (Table 1) to conduct the diethylhexylphosphoric acid extraction of Ba from these solutions, and subsequently correct the measured dissolved Eu concentrations.

Nonetheless, it is important to point out that for some of the leachate solutions; the Eu data suggests that false positives during mass quantification from Ba interferences may not be responsible for the positive Eu anomalies. These leachate solutions include those that reacted with BAS(SL)2, BAS(SS)1 of Batch Test 1, BAS6, BKD11 of Batch Test 2, and FR8Oaal and BAS6 of Batch Test 3. For the case of these leachate solutions, the positive Eu anomalies that are characteristic of the leachates are likely due, in part, to the large negative Eu anomalies of the rocks. In other words, because of the depletion in Eu (i.e., negative Eu anomalies) that is characteristic of the sedimentary (and felsic igneous) rocks from southern Nevada, normalization of leachate Eu values to the rock values leads, in part, to the positive rock-normalized Eu anomalies of the leachates. For example, the Eu concentration of the leachate that reacted with Surprise 1 rhyolitic pumice (i.e., 95.1 pmol/kg) is substantially lower than either the Sm or Gd concentrations (216 pmol/kg and 232 pmol/kg, respectively) in the leachate. Indeed, the Eu concentration of this leachate is 2.3 times lower than the corresponding Sm value, and 2.4 times lower than the Gd concentration. Moreover, the Eu concentration of the rhyolitic source rock is a factor of 6.3 lower than Sm, and 5.7 times lower than the Gd concentrations in this rock (Table 2). Hence, because the Eu concentration of the leachate in substantially lower than the corresponding Sm and Gd concentrations, the positive Eu anomaly or the rock-normalized leachate probably reflects the large Eu depletion of the rhyolitic rock, Surprise 1.

4.2. IMPLICATIONS OF BATCH TESTS

The results of our batch leaching tests are in general agreement, in terms of the similarities between the REE distributions of the leachates and the rock samples, with previous field investigations that report groundwaters with REE signatures that closely mimicked the REE signatures of the aquifer rocks through which they flowed (e.g., Smedley, 1991; Gosselin et al., 1992; Fee et al., 1992; Leybourne et al., 2000). In addition, the measured REE concentrations in the leachates from our batch tests are grossly similar, in magnitude, to those reported for groundwaters, including groundwaters from southern Nevada (Johannesson et al., 1997; 2000a). Consequently, the batch test data suggest that aquifer rocks are the primary source of REEs to groundwaters, and more importantly, REEs are leached from aquifer rocks by aqueous solutions in proportions similar to those found in the rock/aquifer materials. Consequently, the similarities between the results of our batch leach tests (i.e., REE concentrations and rock-normalized patterns) and reported observations of REE concentrations and patterns in actual groundwater-aquifer systems strongly suggests that groundwaters can inherit aquifer-rock like REE signatures via leaching reactions with the aquifer rocks without incurring significant fractionation of the REEs during the process.

It is generally well accepted that the acquisition of REEs by natural terrestrial waters from rock weathering reactions depends upon many factors including: (1) the distribution and abundance of REE-bearing minerals within the rocks; (2) the chemical composition of the natural water/weathering solution (e.g., major solute concentrations, pH, pe, concentrations of inorganic and organic complexing ligands, temperature); (3) the solubility of the REE-bearing minerals in the rocks with respect to the composition of the natural water/weathering solution; and (4) the ability of secondary minerals formed during water-rock reactions to accept REEs leached from the primary minerals of the unweathered rock (Humphris, 1984; Braun et al., 1990; Johannesson and Zhou, 1999). Moreover, previous investigators have examined REE fractionation as a function of chemical weathering of different types of parent rocks (e.g., Balashov et al., 1964; Ronov et al., 1967; Ludden and Thompson, 1978, 1979; Nesbitt, 1979; Duddy, 1980; Schau and Henderson, 1983; Humphris, 1984; Middelburg et al., 1988; Braun et al., 1990, 1993, 1998). Many of these studies indicate that weathered residual materials become enriched in LREEs and depleted in the HREEs compared to the parent rock during water-rock reactions owing, in part, to preferential mobilization of the HREEs (Balashov et al., 1964; Ronov et al., 1967; Nesbitt, 1979; Duddy, 1980). More specifically, LREEs are retained by secondary clay minerals formed by water-rock reactions, whereas HREEs are preferentially removed from the reaction site by solution complexation reactions involving inorganic and/or organic ligands (e.g., CO_3^{2-}, PO_4^{3-}, humic and/or fulvic acids) that are present in natural waters, including groundwaters (Nesbitt, 1979; Duddy, 1980; Schau and Henderson, 1983; Braun et al., 1993; Johannesson and Zhou, 1999). Therefore, solution and surface (i.e., adsorption) complexation can play important roles in fractionating REEs during water-rock reactions by enriching the aqueous solutions in the HREEs, and preferentially concentrating the LREEs in the residual phases (e.g., clay minerals). On the other hand, in the absence of important inorganic and organic complexing ligands, the REEs exhibit greater tendencies to sorb to surface sites as a function of increasing atomic number (e.g., Roaldset, 1974; Aagaard, 1974; Tang and Johannesson, in prep.).

The fact that the leachates resulting from the batch leach tests do not exhibit significant fractionation of the REEs compared to the aquifer rocks suggests that solution and surface complexation reactions involving these heavy metals were not important in the batch tests. More importantly, the observations reporting that many groundwaters have similar REE signatures to the aquifer rocks through which they flow implies that, at least in some cases, REEs are not fractionated during the weathering/leaching process whereby groundwater acquire their REE signatures. The results of our batch leaching tests and the reported REE patterns of actual groundwaters appear to contradict the observations and models of the behavior of REEs during the weathering process which argue for enrichment of the HREEs in the weathering solutions, and hence fractionation of the REEs. In other words, our batch tests suggest that, at least in some cases, preferential removal of HREEs compared to the LREEs from the site of active chemical weathering of aquifer rocks by groundwaters does not appear to be important. Instead, these leachates and real groundwaters appear to inherit REE distributions that closely resemble the source (i.e., aquifer) rocks.

However, more recent investigations have reported groundwaters with highly fractionated REE patterns compared to the aquifer/aquitard patterns (Johannesson et al., 1999, this volume; Johannesson and Hendry, 2000; Leybourne et al., 2000). For example, groundwaters from a shallow, basin-fill aquifer in southern Nevada, a till and clay-rich aquitard in Saskatchewan, and a rhyolitic aquifer in central México, all exhibit large enrichments in HREEs compared to LREEs when normalized to the REE concentration in their respective aquifer materials (Johannesson et al., 1999, 2000a, this volume; Johannesson and Hendry, 2000). The fractionation of REEs in these groundwaters is consistent with the formation of carbonato and dicarbonato complexes in solution. More specifically, the LREEs exhibit a greater affinity to sorb to aquifer/aquitard materials than the HREEs because they chiefly occur in these circumneutral pH groundwaters as positively charged, solution species (i.e., $LnCO_3^+$, Ln^{3+}, $LnSO_4^+$), whereas the HREEs occur as negatively charged dicarbonato species (Johannesson and Hendry, 2000). Consequently, solution and surface complexation reactions do exert controls on REEs in groundwater-aquifer systems. Therefore, it is conceivable that during the initial weathering of aquifer rocks by relatively acidic aqueous solutions (e.g., meteoric waters augmented by soil-zone CO_2), the REEs are leached from aquifer rocks in roughly the same proportion with which they occur in the rock. However, with flow within the aquifer and as more rock is weathered, solution complexation becomes progressively more important (e.g., production of HCO_3 during weathering raises the pH and increases the significance of carbonate complexes) leading to strong fractionation of the REEs (Johannesson et al., 2000b, this volume).

Although the current batch study shows some interesting results for REEs in the leachates during rock/water interactions, more leaching tests with systematic design parameters are needed to better constrain the leaching processes. The design parameters, such as pH values of solutions, particle sizes of crushed rock samples, water-rock batch contact time, and amount ratios of solution to rock sample, must be carefully controlled to learn more about these processes that occur as REEs transition from source (i.e., aquifer) rocks to natural waters. Such systematic leaching studies using different types of rocks and solutions will help us to understand the relationship between aquifer rock and groundwater REE signatures, the leaching mechanisms and kinetics of REEs during water/rock interaction, and the controlling factors on REE signatures and fractionation in natural waters.

5. Conclusions

Laboratory batch tests were conducted in order to study the affect that compositionally different rocks have on the behavior and concentration of rare earth elements (REEs) in low-temperature aqueous solutions. Different rock types (i.e., shale, sandstone, limestone, dolomite, and rhyolitic pumice) were reacted with distilled deionized water (pH=7; 18 MΩ-cm), and/or acidic aqueous solutions (pH=4). Rock samples were crushed to about 70 mesh in size. Different amounts of distilled deionized water or acidic aqueous solutions were mixed with 20 to 80 grams of individual crushed rock samples within acid-washed, pre-clean polyethylene bottles (mass ratio of aqueous solution to rock sample ranged from 2:1 to 3:1). The crushed rock samples and aqueous solutions were allowed to react for 40 to 65 days at room temperature (~25°C). The sample slurry was subsequently centrifuged and the supernatant filtered (0.45 μm poly-carbonate membrane) before quantification of dissolved REEs by ICP-MS. Concentrations of REEs in the leachate solutions are low, ranging from 0.05 pmol/kg to 2,570 pmol/kg. Solutions that reacted with rhyolitic pumice exhibit the highest REE concentrations, whereas the solutions that reacted with Ordovician limestone, and especially with Cambrian dolomite, have the lowest REE concentrations. When the leachates are normalized to the respective rock with which they reacted, they exhibit relatively flat REE patterns, except for large positive Eu anomalies. Interestingly, the solution that reacted with the pumice sample is slightly enriched in HREE, and the solution leached from Cambrian sandstone shows a weak MREE enrichment. These batch studies suggest that different types of rocks can play important roles in imparting REE signatures to natural waters, and in particular, groundwaters where rock/water ratios are high.

The results of our batch tests using aquifer materials and solutions allowed us to characterize REE signatures released from rock samples to leachates and to determine the easily exchangeable fractions of REEs during water/rock interaction under standard conditions. Moreover, this study provides information that may be applicable to REE behavior during low-temperature water/rock interaction in general. However, we should apply these batch test results to real groundwater systems with great caution because other factors, such as the filter size used, possible secondary minerals in rock samples and variations of laboratory conditions, could also affect REE patterns of the leachate solutions. More systematic batch tests are needed to better control the leaching processes during rock/water interactions.

Acknowledgments

We thank C. Guo and K. Lindley for assistance with the analytical work. We are grateful to Dr. E. H. DeCarlo for his helpful review of this contribution. The work was funded by the U.S. Department of Energy to KHJ and KJS.

References

Aagaard, P. 1974 Rare earth element adsorption on clay minerals. *Bull. Group. Franc. Argiles* **26**, 193-199.

Balashov, Y. A., Ronov, A. B., Migdisov, A. A., and Turanskaya, N. V. 1964 The effect of climate and facies environment on the fractionation of the rare earth elements during sedimentation. *Geochemistry International*, **5**, 951-969.

Braun, J. J., Pagel, M., Muller, J.P., Bilong, P., Michard, A., and Guillet, B. 1990 Cerium anomalies in lateritic profiles. *Geochimimica et Cosmochimica Acta*, **54**, 597-605.

Braun, J. J., Pagel, M., Herbillon, A., and Rosen, C. 1993 Mobilization and redistribution of REEs and thorium in a syenitic lateritic profile: A mass balance study. *Geochimimica et Cosmochimica Acta*, **57**, 4419-4434.

Braun, J.J., Viers, J., Dupré, B., Polve, M., Ndam, J., and Muller, J. P. 1998 Solid/liquid REE fractionation in the lateritic system of Goyum, East Cameroon: The implications for the present dynamics of the soil covers of the humid tropical regions. *Geochimimica et Cosmochimica Acta*, **62**, 273-299.

Cantrell, K.J., Byrne, R.H. 1987 Rare earth element complexation by carbonate and oxalate ions. *Geochimimica et Cosmochimica Acta*, **51**, 597-605.

Cerrai, E., Ghersini, G. 1966 Reversed-phase partition chromatography on paper treated with di-(2-ethylhexyl) orthophosphoric acid: a systematic study of 67cations in hydrochloric acid. *Journal of Chromatography*, **24**, 383-401.

Cullers, R.L., Graf, J.L. 1984 Rare earth elements in igneous rocks of the continental crust: intermediate and silicate rocks - ore petrogenesis. In: P. Henderson (ed.), Rare Earth Element Geochemistry, Elsevier (Amsterdam), pp. 257-316.

DeBaar, H.J.W., Bacon, M.P., Brewer, P.G., Bruland, K.W. 1983 Rare earth distributions with a positive Ce anomaly in the western Atlantic Ocean. *Nature*, **301**, 324-327.

Duddy, I.R. 1980 Redistribution and fractionation of the rare-earth and other elements in a weathering profile. *Chemical Geology*, **30**, 363-381.

Elderfield, H., Greaves, M. J. 1982 The rare earth elements in seawater. *Nature*, **296**, 214-219.

Fee, J. A., Gaudette, H. E., Lyons, W. B., Long, D. T. 1992 Rare earth element distribution in Lake Tyrrell groundwaters, Victoria, Australia. *Chemical Geology*, **96**, 67-93.

Garrels, R. M. and MacKenzie, F. T. 1967 Origin of chemical composition of some springs and lakes. In: W. Stumm (ed.) Equilibrium Concepts in Natural Water Systems, American Chemical Society, Advances in Chemistry Series, 67, (Washington, DC), pp. 222-242.

Gosselin, D.G., Smith, M.R., Lepel, E.A., Laul, J.C. 1992 Rare earth elements in chloride-rich groundwater, Palo Duro Basin, Texas, USA. *Geochimimica et Cosmochimica Acta*, **56**, 1495-1505.

Graham, E. Y., Ramsey, L. A., Lyons, W. B., and Welch, K. A. 1996 Determination of rare earth elements in Antarctic lakes and streams of varying ionic strengths. In: G. Holland and S. D. Tanner (eds.) Plasma Source Mass Spectrometry: Developments and Applications. The Royal Society of Chemistry (London), pp. 253-262.

Guo, C. 1996 Determination of fifty-six elements in three distinct types of geological materials by inductively coupled plasma-mass spectrometry. Unpublished M. S. thesis, University of Nevada, Las Vegas, 68 p.

Halicz, L., Segal, I., and Yoffe, O. 1999 Direct REE determination in fresh waters using ultrasonic nebulization ICP-MS. *Journal of Analytical Atomic Spectrom*etry, **14**, 1579-1581.

Hanson, G. N. 1980 Rare earth elements in petrogenetic studies of igneous systems. *Annual Review of Earth and Planetary Science*m **8**, 371-406.

Hodge, V.F., Stetzenbach, K.J., Johannesson, K.H. 1998 Similarities in the chemical composition of carbonate groundwaters and seawater. *Environmental Science and Technology*, **32**, 2481-2486.

Humphris, S.E. 1984 The mobility of the rare earth elements in the crust. In: P. Henderson (ed.), Rare Earth Element Geochemistry, Elsevier (Amsterdam), pp. 317-342.

Jarvis, K. E., Gray, A. L., and McMurdy, E. 1989 Avoidance of spectral interferences on europium in inductively coupled plasma mass spectrometry by sensitive measurement of the doubly charged ion. *Journal of Analytical Atomic Spectromemtry*, **4**, 743-747.

Johannesson, K.H., Stetzenbach, K.J., Hodge, V.F., Kreamer, D.K., Zhou, X. 1997 Delineation of ground-water flow systems in the southern Great Basin using aqueous rare earth element distributions. *Ground Water*, **35**, 807-819.

Johannesson, K. H. and Zhou, X. 1999 Origin of middle rare earth element enrichments in acid waters of a Canadian High Arctic lake. *Geochimimica et Cosmochimica Acta*, **63**, 153-165.

Johannesson, K.H. and Hendry, M. J. 2000 Rare earth element geochemistry of groundwaters from a thick till and clay-rich aquitard sequence, Saskatchewan, Canada. *Geochimimica et Cosmochimica Acta*, **64**, 1493-1509.

Johannesson, K.H., Farnham, I.M., Guo, C., and Stetzenbach, K.J. 1999 Rare earth element fractionation and concentration variations along a groundwater flow path within a shallow, basin-fill aquifer, southern Nevada, USA. *Geochimimica et Cosmochimica Acta*, **63**, 2697-2708.

Johannesson, K.H., Zhou, X., Guo, C., Stetzenbach, K.J., Hodge, V.F. 2000a Origin of rare earth element signatures in groundwaters of circumneutral pH from southern Nevada and eastern California, USA. *Chemical Geology*, **164**, 239-257.

Johannesson, K. H., Cortés, A., Ramos, L. J. A., Ramirez, A. G., and Durazo, J. 2000b "Rock-like" versus "seawater-like" REE signatures of groundwaters. *Geological Society of America Annual Meeting Abstracts with Programs*, **32**, A-188.

Langmuir, D. 1971 The geochemistry of some carbonate groundwaters in central Pennsylvania. *Geochimimica et Cosmochimica Acta*, **35**, 1023-1045.

Lee, J.H., Byrne, R.H. 1993 Complexation of trivalent rare earth elements (Ce, Eu, Gd, Tb, Yb) by carbonate ions. *Geochimimica et Cosmochimica Acta*, **57**, 295-302.

Leybourne, M. I., Goodfellow, W. D., Boyle, D. R., and Hall, G. M. 2000 Rapid development of negative Ce anomalies in surface waters and contrasting REE patterns in groundwaters associated with Zn-Pb massive sulfide deposits. *Applied Geochemisty*, **15**, 695-723.

Ludden, J. N. and Thompson, G. 1978 Behaviour of rare earth elements during submarine weathering of tholeiitic basalt. *Nature*, **274**, 147-149.

Ludden, J. N. and Thompson, G. 1979 An evaluation of the behavior of the rare earth elements during the weathering of sea-floor basalt. *Earth and Planetary Science Letters*, **43**, 85-92.

Middelburg, J. J., Van der Weijen, C. H., and Woittiez, J. R. W. 1988 Chemical processes affecting the mobility of major, minor and trace elements during weathering of granitic rocks. *Chemical Geology*, **68**, 253-273.

Moffett, J.W. 1990 Microbially mediated cerium oxidation in seawater. *Nature*, **345**, 421-423.

Moffett, J.W. 1994 The relationship between cerium and manganese oxidation in the marine environment. *Limnology and Oceanography*, **39**, 1309-1318.

Nesbitt, H.W. 1979 Mobility and fractionation of rare earth elements during weathering of a granodiorite. *Nature*, **279**, 206-210.

Roaldset, E. 1974 Lanthanide distributions in clays. *Bull. Group. Franc. Argiles*, **26**, 201-209.

Ronov, A. B., Balashov Y. A., and Migdisov A. A. 1967 Geochemistry of the rare earth elements in the sedimentary cycle. *Geochemistry International*, **4**, 1-17.

Rowland, S.M. 1987 Paleozoic stratigraphy of Frenchman Mountain, Clark County, Nevada. *Geological Society of America Centennial Field Guide - Cordilleran Section*, 52-56.

Rowland, S.M., Parolini, J.R., Eschner, E., McAllister, A.J., Rice, J.A. 1990 Sedimentologic and stratigraphic constraints on the Neogene translation and rotation of the Frenchman Mountain structural block, Clark County, Nevada. *Geological Society of America Memoir*, **176**, 99-122.

Schau, M. and Henderson, J. B. 1983 Archean chemical weathering at three locations on the Canadian Shield. *Precambrian Research*, **20**, 189-224.

Sholkovitz, E.R. 1988 Rare earth elements in the sediments of the North Atlantic Ocean, Amazon Delta, and East China Sea: reinterpretation of terrigenous input patterns to the oceans. *American Journal of Science*, **288**, 236-281.

Smedley, P.L. 1991 The geochemistry of rare earth elements in groundwater from the Carnmenellis area, southwest England. *Geochimimica et Cosmochimica Acta*, **55**,

2767-2779.

Stetzenbach, K.J., Amano, M., Kreamer, D.K., Hodge, V.F. 1994 Testing the limits of ICP-MS: determination of trace elements in ground water at the parts-per-trillion level. *Ground Water*, **32**, 976-985.

Winograd, I.J., Thordarson, W. 1975 Hydrogeologic and hydrochemical framework, south-central Great Basin, Nevada-California, with special reference to the Nevada Test Site. *USGS Profession Paper*, **712-C**.

Chapter 7

RARE EARTH ELEMENT CONTENTS OF HIGH pCO$_2$ GROUNDWATERS OF PRIMORYE, RUSSIA: MINERAL STABILITY AND COMPLEXATION CONTROLS

PAUL SHAND[1], KAREN H. JOHANNESSON[2], OLEG CHUDAEV[3], VALENTINA CHUDAEVA[4], & W. MIKE EDMUNDS[1]

[1] *British Geological Survey, Crowmarsh Gifford, Wallingford, UK*
[2] *Department of Earth and Environmental Sciences, The University of Texas at Arlington, Arlington, Texas, USA*
[3] *Far East Geological Institute, Russian Academy of Sciences, Vladivostok, Russia*
[4] *Pacific Institute of Geography, Russian Academy of Sciences, Vladivostok, Russia*

Abstract

The rare earth element (REE) geochemistry of cold, high pCO$_2$ groundwaters was studied in springs and boreholes in the Primorye region of the Russian Far East. The gas phase in these waters is dominated by mantle-derived CO$_2$ (up to 2.6 atm.), being introduced to shallow groundwaters along major fault systems. The aggressive nature of these moderately acidic groundwaters has led to unusual trace element characteristics with high concentrations of relatively immobile elements such as Al, Be, heavy REEs and Zr. They are also marked by extremely high concentrations of Fe and Mn, up to 80 mg l^{-1} and 4 mg l^{-1}, respectively. The REE patterns generally show enrichment in the middle to heavy REEs and low concentrations of the light REEs (La-Nd). Most groundwaters show relatively flat shale-normalised middle to heavy REE profiles, with the exception of Eu, which may form positive or negative anomalies depending on local mineralogy. A characteristic of many of the groundwaters is the presence of positive Sm-Eu anomalies. A range of potential ligands is present in the groundwaters and model calculations show that the dominant species are Ln^{3+}, carbonate complexes (i.e., predominantly LnCO$_3^+$), and LnF^{2+}. Concentrations of Cl$^-$ and SO$_4^{2-}$ are very low in most waters and nitrate and phosphate are below detection limit. The role of organic complexing is not known due to lack of data, but such complexes may be important since limited TOC data show that organic contents may be high. The middle to heavy REE enrichment found in the groundwaters is consistent with dissolution of Fe-Mn oxyhydroxides and release of adsorbed REE. Positive Eu anomalies in some groundwaters correlate with high Ca and Sr, pointing to control by plagioclase dissolution. High Y/Ho ratios and positive Y spikes on REY plots in high-F$^-$ waters suggest an important control by F$^-$ complexing, which is confirmed by speciation calculations. Extreme enrichments in the heavy REEs are found in two groups of mineral waters with Yb/La ratios up to 9.8. The extreme enrichments in heavy/light REEs present in these areas are too high to be simply controlled by speciation fractionation and it is suggested that a weathering phase with heavy REE enrichment is responsible. The source is suggested to be zircon, which typically displays such heavy REE enrichments. Although zircon is generally stable during low-temperature weathering, it is know to break down in acidic carbonated solutions. This is supported by high Zr concentrations (as well as high U, Be) in these groundwaters and a correlation between Zr and heavy REE enrichment.

K.H. Johannesson,(ed), Rare Earth Elements in Groundwater Flow Systems , 161-186.

1. Introduction

The behaviour of rare earth elements (REEs) in the aqueous environment has received considerable attention recently. Early studies of low temperature weathering processes focussed on the solid phase in weathering profiles (Nesbitt, 1979; Alderton et al., 1980). However, with the advent of sensitive multielement techniques such as ICP MS, the REEs (as well as a wide range of other ultra-trace elements) are routinely analysed in natural waters (e.g., Smedley, 1991; Verplanck et al., 2001). Such data have been used extensively in marine (Elderfield, 1988; Byrne & Kim, 1990; Sholkovitz et al, 1992) and fresh water environments (Smedley, 1991; Lee & Byrne, 1993; Johannesson et al., 1996) to study REE cycling and water-rock interaction in the aqueous environment.

The fractionation of REEs during high temperature and pressure processes is relatively well understood but it is only in the past decade that fractionation processes in the low temperature aqueous environment have been studied in detail (Lee & Byrne, 1993; Johannesson et al., 1996; Bau, 1999). The concentration of REEs in groundwaters is dependent on several factors:

(a)   release from weathering phases;
(b)   pH and redox status of the groundwater environment;
(c)   adsorption;
(d)   complexing ligands in groundwater;
(e)   hydrogeological factors (e.g. flow pathways, residence time).

In igneous and metamorphic rocks, most of the REEs are present in minor accessory minerals such as allanite, apatite, monazite, sphene and zircon (Hanson, 1980; Gromet & Silver, 1983). There is a general enrichment in light REEs from basic (basalt) to acidic (rhyolite) rocks, which is partly reflected in individual minerals, but basic alkaline igneous rocks may show extreme enrichments in the light REEs related to source characteristics. The dominant control on REEs in minerals is the mineral solid-liquid partition coefficient which gives individual mineral phases characteristic normalised REE profiles (Hanson, 1980). Most minerals do not selectively fractionate individual REEs relative to their neighbours but plagioclase feldspar may strongly fractionate Eu, which is redox sensitive at high temperature. Although the REEs generally form M^{3+} ions, under reducing conditions Eu^{2+} ions substitute for Sr^{2+} in plagioclase giving rise to a positive Eu anomaly. Likewise, Ce^{4+} may also be fractionated from its neighbours giving rise to mineral phases with Ce anomalies. Secondary minerals such as Fe-Mn oxides/oxy-hydroxides also strongly adsorb REEs and display characteristic REE profiles (Gosselin et al., 1992) related to adsorption controls and provide evidence of reacting mineral phases.

The acidity of natural waters exerts a significant control on REE solubility with concentrations commonly forming an inverse relationship to pH (Goldstein & Jacobsen, 1987; Smedley, 1991). The REEs also display strong sorption characteristics, particularly at high pH, onto mineral surfaces (Erel & Stolper, 1993; Sholkovitz et al., 1994) limiting their role as true conservative tracers. The light REEs are generally scavenged through sorption processes much more than the heavy REEs. The good correlation of REEs with flow in upland acidic catchments (Shand et al., 1999) and behaviour in acid groundwaters (Fee et al., 1991) may allow for their use as conservative tracers of flowpaths. Nevertheless, where strong REE-complexes are present in a groundwater system, the REEs may prove powerful tracers of groundwater flowpaths (Johannesson et al., 1997).

Several potential REE complexing agents are present in groundwater. The trivalent REEs are classed as hard ions and will complex preferentially with hard ligands containing highly electronegative donor atoms including F^-, SO_4, CO_3, PO_4 and OH (Brookins, 1989; Wood, 1990). Chloride and nitrate complexes are also known but are very weak and unimportant in most low-temperature natural waters. In general, simple ions (M^{3+}) and sulphate complexes will normally predominate at low pH with carbonate and dicarbanato species being predominant at circumneutral to basic pH (Wood, 1990; Johannessen et al., 1996). In addition, fluoride and phosphate complexes may be important where ligand concentrations are high (Wood, 1990; Gimeno et al., 2000). Complexing agents may, therefore, modify the concentrations and profiles of weathering-derived REEs. This can be further modified through adsorption/desorption reactions with Fe and Mn oxide/hydroxide phases and clay minerals. Clay and colloidal particles generally show light REE enrichments (Sholkovitz, 1995) in contrast to Fe and Mn oxide phases, which commonly display enrichments in the middle REEs displaying a "middle REE downward concavity" (Palmer & Elderfield, 1986; Johannesson & Zhou, 1999). Carbonates may also display such profiles (Gosselin et al., 1992). The dominant form of transport of the light REEs in river waters, especially at circumneutral pH is thought to be as sorbed species on colloidal particles, whereas the heavy REEs may be dominantly transported as complexes (Sholkovitz, 1985; Andersson et al., 2001).

The aim of this study is to characterise the REE concentrations and profiles in the pCO_2 groundwaters of Primorye, to model the speciation based on knowledge of potential complexing ligands present in the groundwater and to suggest sources for the unusual middle and heavy REE enrichment characteristic of these waters. Mineral waters with naturally high carbon dioxide contents are relatively restricted world-wide, in general only being found associated with major zones of seismicity (Barnes et al., 1978). Cold, high pCO_2 springs are generally found at destructive plate margins where subduction of oceanic crust or continental collision is taking place or has recently terminated. Previous work on such springs has focussed principally on the origin of the gas phase (Cornides, 1993; Wexsteen et al., 1988) with little published data (Frau, 1993; Michard et al., 1987; Shand et al., 1995) on the minor and trace element signatures.

2. Study Area

The geology of Primorye is highly complex (Figure 1). In general, the region can be sub-divided into two areas: Western Primorye where Proterozoic and Palaeozoic rocks are predominant and central-eastern Primorye (including the Sikhote-Alin mountains) where Mesozoic rocks are predominant. The structure of the area is dominated by large strike-slip and thrust boundaries, some of which are thought to represent terrane boundaries separating exotic terranes similar to those present in the eastern USA and the British Caledonides. Several periods of magmatic activity have also occurred from the Proterozoic to the Pliocene. The climate is monsoonal with high rainfall mainly in summer months, and cold winters. Much of the area is pristine wilderness covered with dense vegetation typified by deciduous, coniferous and mixed woodland (Taiga).

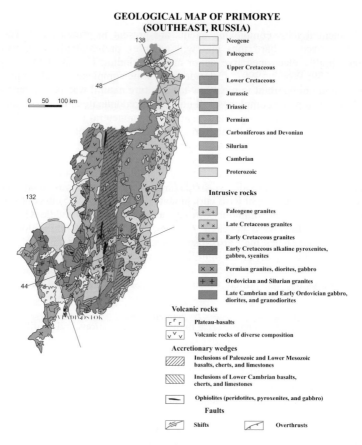

Figure 1. Geological map of the Primorye region of eastern Russia.

The CO_2-rich waters generally discharge as springs but several boreholes also source these waters. Some of the groundwaters (Schmakovka, Gornovodnoe) are used for spa treatment (Chudaeva et al., 1999), but in general they are not actively developed. The mineral waters have been subdivided on the basis of geography and vary significantly with bedrock geology (Figure 2) and a summary is given in Table 1. Unfortunately, there is little detailed information available on the mineralogy and petrology of rocks from the study areas.

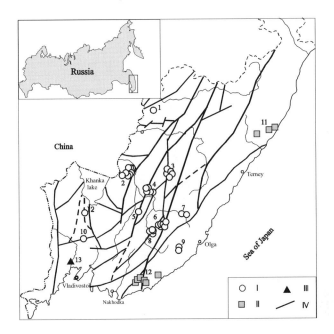

Figure 2. Location of high pCO$_2$ waters (I), thermal waters (II) and high TDS Na-HCO$_3$ waters (III) of Primorye. Faults are shown as lines (IV). Localities are 1. Lastochka, 2. Schmakovka, 3. Ariadnoy, 4. Shetukinskoya, 5. Pokrovka, 6. Leninskaya, 7. Lushki, 8. Chugooevskaya, 9. Gornovodnoe, 10. Rakavka.

3. Sampling and Analytical Techniques

A range of parameters were measured at each site including temperature (T), pH, specific electrical conductance (SEC), dissolved oxygen (DO), redox potential (Eh), and alkalinity (the latter determined by titration). Where possible the unstable parameters pH, DO and Eh were measured in an anaerobic flow-through cell.

Filtered (0.45 μm membrane filters) and acidified samples (1% v/v ultrapure HNO$_3$) were collected in high density polyethylene bottles for analyses of major cations, SO$_4^{2-}$ and a wide range of trace elements. Trace elements were measured by ICP-OES and ICP-MS (VG PQ1), with ICP-MS being used for determination of the REEs. Calibrations for cation analyses were performed using appropriately diluted standards, and both laboratory and international reference materials were used as checks of accuracy. The REEs were measured directly in solution without pre-concentration and

Location	lithology of borehole	mineralogy	age	surrounding geology	Depth of borehole
Gornovodnoe	Acid tuff, Ignimbrite, Felsite; dykes of granites	Ksp, qtz, volc. glass. 2^{γ}: Srct, ill, carb, alb, chl, Zeol, Py	K_2	Volcanic rocks of K_2 age. volcanic depression belonging to Eastern Sikhote-Alin volcanic belt	301m
Schmakovka	Granite, monzonite, syenite, skarn, marble, shl, sst, bslt	Qtz, Fsp, bt, phlog, ol, pyx, hbl, 2^{γ}: Ill, kaol, serp, talc, carb, alb	PR-N_2	Mainly granite and metamorphic limestones, covered by N_2 basalts. Junction of Khanka ancient massive with the Sikhote-Alin.	ca. 80m
Rakavka	Devonian acid volcanic-mafic dykes. Mes: sst, aleurolite. Pg-N terrigenous rocks with coal.	No data	D-Pg-N	Borehole located in graben which controls N-E faults	51 m
Shetykhinskaya	P_2-acid tuff, rare mafic tuff, aleurolite and sst. T: sst, aleurolite, gravels and cong. K_2 – plutonic rocks (acid - mafic)	No data	P_2- T-K_2	Western Sikhote-Alin and related to deep faults N-E orientation.	Only springs and wells.
Ariadnoy 1 Ariadnoy 2	P- sst, shl, porph and lens of limestone as well as silicate rocks. S-D silicate rocks and shale. K_1 -granite	No data	S-D-P-K_1	Cental Sikhote-Alin - Samarka accretionary wedge. Springs are controlled by deep N-E fault and subsidiary faults.	Only springs
Samarinskaya	P-T aleurolite, silic. rocks, sst, lst. J-K_1 – aleurolite, silicate rocks, gab, tuff, sst, tuff	No data	P-T-J-K_1	Springs located in the Central Sikhote-Alin in Samarka accretionary wedges	Only springs
Pakrovka	K_2 –volcanic rocks, sandstone, shl, with coal	No data	K_2	Shetukhinskaya group part of the Western Sikhote-Alin - related to N-E deep faults.	42.3m
Lushki	Acid tuff, volcanic glass, qtz, fsp, felsite; Sedimentary rocks: Sst and aleurolite	Qtz, fsp; volcanic glass. 2^{γ}: Ill and smec. minor carbonate , Ksp	K_2	K_2 Volcanics surrounding sandstones and aleurolites. Minor granites. Eastern Sikhote-Alin volcanic belt Volcanic depression.	60 m
Leninskaya	T-J aleurolite, sst, lst, bslt, spilite K-acid volcanics and tuff, sst K_2–plutonic rocks (diorite)	No data	T-J-K_2	Springs located in zone of deep faults N-E orientation so called theCentral Sikhote-Alin deep fault	Only springs
Chugooveskaya	sst with volcanics -K age; Camb silicate rocks and shl; Pg- andes and grn	No data	C-K-Pg	Close to deep N-E terrane faults : Samarka (accretionary wedge) and Zhuravlevka	Only springs

Table 1. Dominant geology of high pCO_2 water groups

precision was better than 2%. Instrumental drift during ICP-MS analysis was corrected using In and Pt internal standards. Filtered, unacidified samples were collected for anion analysis which was carried out by automated colorimetry.

4. Background Hydrochemistry

The CO_2-rich groundwaters have low temperatures (6-14°C) and pH varies from 4.07 to 6.18, although most are in the range pH 5 to 6 (Table 2). Most groundwater samples were moderately reducing but some springs contained dissolved oxygen concentrations up to 2.5 mg l^{-1}. A summary of the chemistry of groundwaters from these areas is given in Table 2. Total dissolved solids (TDS) varied widely from 23 to 2100 mg l^{-1}, with the lowest concentrations being associated with the most acidic groundwaters. The waters generally contain very high CO_2 contents (up to 2.6 atm), which effectively buffers the pH to values less than pH 6. Only the highest TDS samples reach saturation with respect to calcite (Figure 3). The concentrations of alkali metals, SO_4 and halogens are generally very low (e.g. Cl⁻ is generally less than 2 mg l^{-1}). However, F⁻ concentrations reach up to 2.1 mg l^{-1}. Silica concentrations are high for cold waters (up to 31 mg l^{-1} Si). Iron and Mn are also high in all groundwaters, reaching concentrations of 30 and 4 mg l^{-1}, respectively. Although trace element concentrations vary regionally in the different groups, the CO_2 waters are generally enriched in a range of trace elements (Li, Sr, Ba) compared with fresh groundwaters. Local enrichments also occur in relatively immobile elements such as Al, Be, Cs, Zr, and the heavy REEs (Shand et al, 1995; Chudaeva et al., 1999).

Shand et al. (1995) argued for a mantle origin of the CO_2 gas, on the basis of limited gas analyses, $\delta^{13}C$ data and the occurrence of similar springs across terranes of widely different age. They suggested a model whereby mantle degassing is occurring beneath the region of Primorye and the gases dissolve *en route* to the surface in the shallow groundwater system. More recent work using He isotopes has confirmed a mantle origin for the gas phase (Chudaev et al., 2001). Thermal waters (< 30°C) are not common in Primorye, occurring locally in eastern Primorye in association with granite intrusions of Cretaceous age (Chudaeva et al., 1995).

5. Rare Earth Element Characteristics and Profiles

Rare earth element concentrations vary considerably in the high pCO_2 groundwaters but show some consistency within geographical groups. Total REEs vary from 0.3 to 12 μg l^{-1} and selected data are shown in Table 3. Shale (NASC) normalised profiles for these groundwaters are shown in Figures 4. Samples collected a year apart showed similar profiles. Spatially associated fresh waters in the vicinity of the CO_2 waters and a number of rivers and fresh springs (Figure 4) from Primorye are also shown in order to assess the differences caused by weathering due to the addition of the gas phases. The acidic (pH 4.2 - 4.6) Rakovka springs show relatively flat shale-normalised profiles with a relative depletion in the light (La to Nd) REEs. These groundwaters have low TDS and are not considered to have reacted significantly with the host bedrock.

Parameter Locality	pH	SEC μS cm⁻¹	Na mg l⁻¹	Ca mg l⁻¹	HCO₃ mg l⁻¹	SO₄ mg l⁻¹	Cl mg l⁻¹	F mg l⁻¹	Si mg l⁻¹	Fe mg l⁻¹	Mn mg l⁻¹	Sr μg l⁻¹	Be μg l⁻¹	U μg l⁻¹
Rakavka N=2	4.19-4.56	67-129	3.2-129	1.2-6.0	<0.5-47	3.0-3.3	2.2-2.8	0.05-0.1	8.2-17	1.0-21	0.03-0.29	14-36	0.19-0.34	0.16-0.36
Gornovodnoe N=3	5.94-6.04	1384-1968	59-154	118-421	752-2019	0.3-27	3-7	1.0-2.1	11-30	12-88	1.8-4.2	1933-3156	8.5-14	2.5-5.3
Schmakovka N=6	5.19-6.01	60-1447	21-33	11-282	164-1219	<0.2-4	0.6-2.2	0.19-0.99	25-37	3-24	0.3-1.3	104-1101	0.4-8.6	<0.05-3.0
Shetinskoya N=3	5.17-6.18	460-2055	18-63	56-414	284-1807	7-9	0.9-1.5	0.04-4.5	18-26	0.1-25	0.5-1.2	233-4289	0.2-8.4	<0.05-0.27
Ariadnoy 1 N=2	5.04-5.34	252-395	11-15	22-30	184-242	2.4-2.5	0.5-0.9	0.42-0.60	14-17	19-27	1.2-1.4	172-213	0.7-1.3	<0.05
Ariadnoy2 N=4	5.15-5.42	379-521	39-51	14-23	231-374	3-4	1.6-2.7	0.82-1.75	8-11	9-22	0.7-1.5	140-190	0.8-1.6	<0.05
Samarinskaya N=2	5.25-5.70	608-672	21-22	76-80	406-435	9-10	0.6-0.9	0.09-0.11	17	7.3-7.7	1.2-1.3	902-954	0.08-0.19	<0.05
Pakovka N=1	5.52	1518	191	155	1158	<0.2	1.4	0.32	27	7.5	0.4	1662	3.6	<0.05
Lushki N=1	5.72	1156	85	161	853	<0.2	3.3	0.09	15	30	0.9	2217	0.7	0.23
Leninskaya N=3	4.07-5.83	114-1853	3-31	5-335	<0.5-1491	5-17	0.9-2.2	0.16-0.49	8-24	1.2-20	0.05-1.0	32-2248	0.2-2.9	<0.05-0.11
Chugooevskaya N=1	5.69	1847	103	271	1371	18	1.4	0.17	25	10	1.1	4138	0.8	0.12
Lastochka N=1	6.08		487	180	2038	4.6	3.4	<0.02	12	9	0.14	6260	0.5	0.05

Table 2. Summary of selected hydrochemical data for high pCO₂ groundwaters of Primorye.

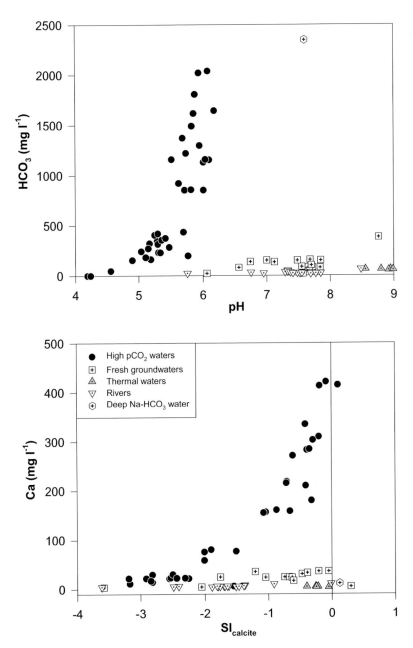

Figure 3. Plots of pH-HCO$_3$ and SI$_{calcite}$-Ca showing differences between high pCO$_2$ groundwaters and other waters of Primorye (from Shand et al., 1995).

Locality	Rakavka (20)	Gornovod-noe (84)	Schmako-vka (61)	Shetinskoy a (30)	Ariadnoy 1 (31)	Ariadnoy 2 (34)	Samarin-skaya (39)	Pakovka (40)	Lushki (48)	Leninska-ya (51)	Chugooev-skaya (54)	Lastochka (98)
La	0.74	0.26	0.27	1.34	0.13	0.34	0.09	0.08	0.54	1.20	0.31	0.17
Ce	2.07	0.63	0.48	2.65	0.33	0.62	0.12	0.15	1.27	1.95	0.46	0.27
Pr	0.28	0.10	0.08	0.31	0.04	0.12	<0.02	0.02	0.12	0.45	0.07	0.03
Nd	1.24	0.67	0.39	1.52	0.21	0.47	0.06	0.13	0.51	1.69	0.35	0.19
Sm	0.34	0.39	0.14	0.48	0.11	0.16	0.08	0.30	0.31	0.42	0.15	2.90
Eu	0.08	0.18	0.05	0.02	0.04	0.05	0.02	0.16	0.12	0.06	0.04	1.96
Gd	0.38	1.09	0.27	0.85	0.16	0.16	0.02	0.18	0.30	0.38	0.18	0.13
Tb	0.05	0.29	0.05	0.16	0.03	0.02	<0.01	0.04	0.03	0.06	0.03	0.02
Dy	0.28	2.45	0.33	1.17	0.17	0.12	0.03	0.25	0.32	0.34	0.19	0.20
Ho	0.06	0.64	0.12	0.27	0.04	0.02	<0.02	0.05	0.07	0.06	0.05	0.05
Er	0.17	2.23	0.36	0.82	0.11	0.05	0.03	0.18	0.23	0.13	0.14	0.18
Tm	0.03	0.33	0.06	0.10	0.01	<0.01	<0.01	0.02	0.03	0.02	0.02	0.02
Yb	0.20	2.38	0.30	0.60	0.07	0.05	0.02	0.12	0.24	0.11	0.10	0.13
Lu	0.03	0.39	0.05	0.08	<0.02	<0.02	<0.02	0.02	0.04	0.02	0.03	0.03
Y	1.51	20.68	4.10	10.61	1.61	0.69	0.40	2.57	2.97	1.39	2.09	3.0

Table 3. Selected REE and Y data (in $\mu g \ l^{-1}$) for high pCO_2 groundwaters of Primorye. (N) = number of samples in database.

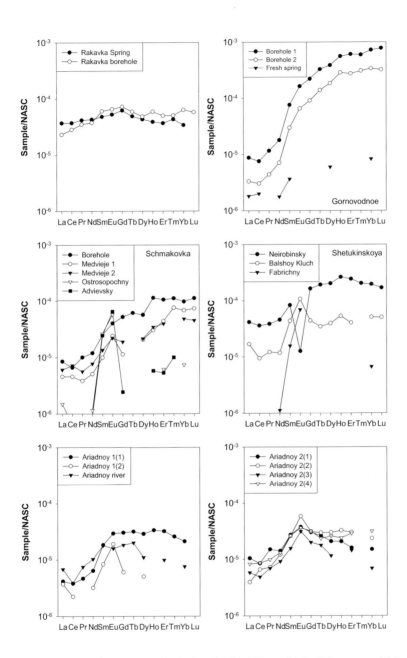

Figure 4. Rare earth element normalised plots for the different high pCO$_2$ waters of Primorye, fresh groundwaters and rivers. Sample points below the detection limit are not plotted.

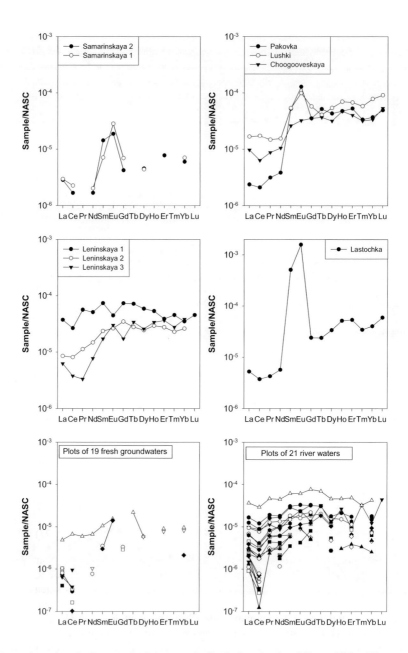

Figure 4 continued. Rare earth element normalised plots for the different high pCO$_2$ waters of Primorye, fresh groundwaters and rivers. Sample points below the detection limit are not plotted.

However, most other waters show a much more significant relative enrichment in the middle and heavy REEs. There is tendency towards relatively flat shale-normalised middle to heavy REE profiles in most groundwaters (e.g. Ariadnoy, Shetukinskoya, Samarinskaya), but commonly the trends are not particularly smooth. Several of the CO_2-rich springs also display a characteristic enrichment in both Sm and Eu. The groundwaters from Gornovodnoe and several from Schmakovka (Figure 4) show steep heavy REE enriched profiles with Yb/La ratios up to 9.4 (Yb_N/La_N up to 102). Such relative enrichments in natural waters are extremely unusual.

Europium anomalies are present in some samples and this is highlighted in the samples from Shetukinskoya (Figure 4) where both positive and negative anomalies are present in different springs. The Balshoy Kluch sample has a much higher TDS and pH and is particularly enriched in Sr (6200 $\mu g\ l^{-1}$) and Ca (415 mg l^{-1}), possibly implying a plagioclase feldspar source. Neirobinsky, on the other hand is enriched in Be, Al, Zn and As with a much lower pH. However, as noted earlier many of these springs display a characteristic positive Sm-Eu anomaly.

Most shallow groundwaters were found to have very low concentrations of REEs (Figure 4) with the majority being below the limits of detection. River waters on the other hand tended to have significantly higher concentrations and show relatively flat patterns or slight middle REE enrichment on shale normalised plots (Figure 4). Most also display negative Ce anomalies in contrast to most of the high pCO_2 waters.

6. Speciation of REEs

The speciation of the REEs in selected groundwater samples from the study region was determined using the approached outlined in Johannesson et al. (1996) based on a computer code originally developed by Millero (1992). The original model was updated by adding more recent stability constant data for REE complexes with carbonate ions (Lee & Byrne, 1993), hydroxyl species (Klungness & Byrne, 2000), phosphate ions (Lee & Byrne, 1992), and fluoride ions (Schijf & Byrne, 1999; Luo & Byrne, 2000). Free inorganic ligand concentrations used in the speciation modelling (e.g. $[CO_3^{2-}]_F$, $[SO_4^{2-}]_F$) were calculated from the major solute composition of the groundwater samples (i.e., Table 2) using the computer program PHREEQE (Parkhurst et al., 1980). It is important to point out that the model does not require the input of REE concentrations because trace element solution complexation is controlled by the free concentrations of complexing ligands in the natural water, along with the corresponding stability constants for the metal-ligand complexes (Millero, 1992; Bruno, 1997).

The results of our REE speciation calculations for the selected groundwater samples are shown in Figure 5 (a-p). We present the results for a light REE (Nd), middle REE (Gd), and a heavy REE (Yb) for each groundwater sample in Figure 5. In general, the speciation calculations indicate that REEs occur in the groundwater chiefly in the form of carbonate complexes (predominantly the carbonato complex, $LnCO_3^+$), as well as free metal ions (Ln^{3+}), and fluoride complexes (LnF^{2+}) (Figure 5). Carbonato complexes are predicted to be the dominant species for all REEs in the Rakavka-Borehole, both Gornovodnoe samples, the Schmakovka-Borehole, Medvieje, Balshoy Kluch, Ariadnoy Group 2, Samarinskaya, Pakovka, Lushki, Chugooveskaya, and Lastochka samples (Figure 5). Carbonate complexes are not generally expected to

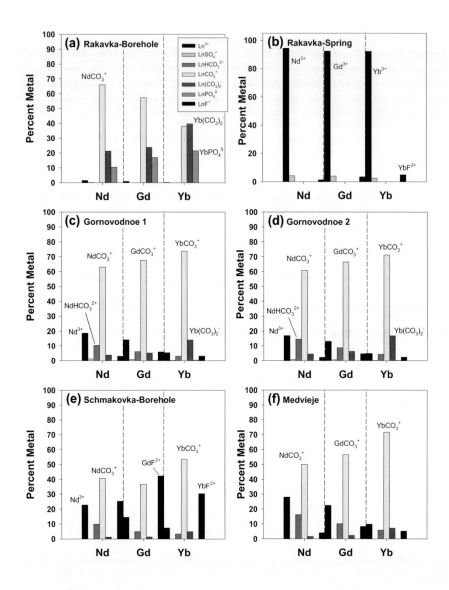

figure 5. Speciation of the light REE Nd, the middle REE Gd, and the heavy REE Yb in selected Primorye groundwaters. See text for discussion.

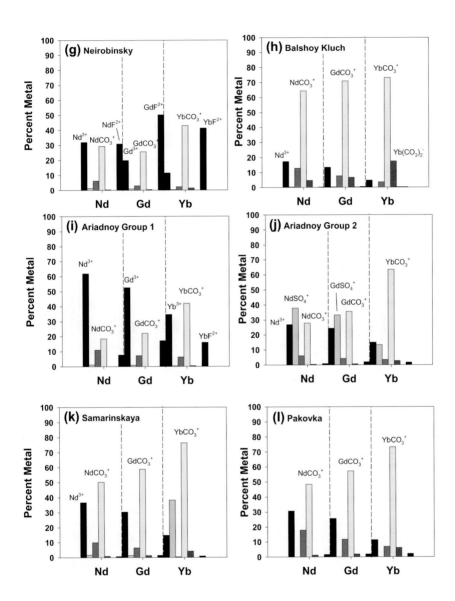

Figure 5 continued. Speciation of the light REE Nd, the middle REE Gd, and the heavy REE Yb in selected Primorye groundwaters. See text for discussion.

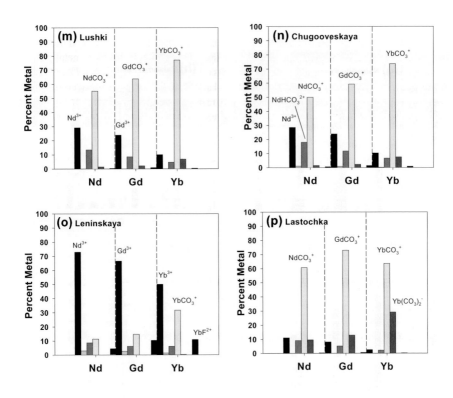

Figure 5 continued. Speciation of the light REE Nd, the middle REE Gd, and the heavy REE Yb in selected Primorye groundwaters. See text for discussion.

be important in acidic groundwaters, but the high DIC in many of the high pCO_2 samples means that significant CO_3 complexing is possible at pH < 6 (Wood, 1990). Furthermore, the model predicts that the percentage of each REE complexed as carbonato complexes tends to increase with increasing REE atomic number. Exceptions include the Rakavka-Borehole sample, for which carbonato complexes are predicted to account for progressively less of each REE with an increase in atomic number, and the Schmokovka-Borehole, Balshoy Kluch, Ariadnoy Group 2, and Lastochka samples for which the model predicts slightly lower amounts of middle REEs complexed as carbonato species (except Lastochka which is opposite) compared to light and heavy REEs (Figure 5). In general, carbonato complexes are predicted to account for between roughly 20% to 80% of the REEs in these groundwater samples. Dicarbonato complexes (i.e., $Ln(CO_3)_2^-$) are generally negligible in these natural waters, although this species is predicted to account for as much as 40% of Yb in the Rakavka-Borehole sample (Fig. 5a), and about 30% of Yb in the Lastochka sample (Fig. 5p).

The model predicts that REEs occur predominantly as free metal ion species (i.e., Ln^{3+}) in the Rakavka Spring, Neirobinsky, Ariadnoy Group 1, and Leninskaya groundwater samples (Figure 5). Indeed, for the Rakavka Spring water, our calculations indicate that 94%, 92%, and 92% of Nd, Gd, and Yb, respectively, occurs as free metal ions in solution (Fig. 5b). Alternatively, for the Leninskaya groundwater, 73%, 67%, and 50% of Nd, Gd, and Yb, respectively, are predicted to be in the form of the free metal ion species (Fig. 5o). Free metal ion species are also predicted to account for significant fractions of each REE in the Gornovodnoe, Schmokovka-Borehole, Medvieje, Neirobinsky, Balshoy Kluch, Ariadnoy Group 2, Smarinskoya, Pakovka, Lushki, and Chuhooveskaya samples, as well as light REEs in the Lastochka groundwater. On the other hand, fluoride complexes are predicted to account for substantial proportions of the REE species, and especially the heavy REEs, in the Schmokovka-Borehole, Neirobinsky, and Ariadnoy Group 1 samples (Figure 5). Fluoride complexes are expected to be at their most important in the pH range of these samples, and for the observed F^- concentrations.

7. Discussion

The unusual nature of these high pCO_2 waters makes them interesting from the point of view of the weathering environment in such groundwater systems. Samples of fresh groundwaters were collected from as close as possible to the high pCO_2 springs in order to understand the effect of CO_2-enrichment on the weathering process caused by addition of the gaseous phase. As discussed previously, most primary and secondary minerals show characteristic normalised REE profiles and combined with knowledge of the complexes available and their behaviour, it is possible to use solute REE concentrations to ascertain likely weathering mineral phases.

Several features of the REE signatures of these groundwaters require explanation including the depletion of the light REEs and the enrichment in the middle to heavy REEs. The most interesting waters are those from Gornovodnoe, both in terms of the extreme relative enrichments in the heavy REEs and the presence of unusually high immobile element concentrations, which may shed light on their origin from specific weathering mineral phases. These features are likely to be a combination of the source characteristics and modification due to the complexing agents present in the waters. The stability constants for SO_4 complexes do not vary significantly for the REEs and little or no fractionation would be expected due to aqueous sulphate complexing. However, the stability constants for REE $-$ F^- and $-$ CO_3 complexes show a steady increase with increasing atomic number (Lee & Byrne, 1993; Schijf & Byrne, 1999; Luo & Byrne, 2000) and could, therefore, fractionate the REEs and be responsible for the heavy REE enrichments found in the Primorye groundwaters. As mentioned previously, the REEs are predicted to be dominantly in solution in the Primorye groundwaters as carbonato complexes, free metal ions, and fluoride complexes.

Few minerals contain heavy REE profiles similar to those found in the Gornovodnoe and Schmakovka mineral waters. Similar REE profiles (although lower concentrations) were found in high pCO_2 waters by Michard et al. (1997) who interpreted such signatures as being due to solution-complex fractionation through bicarbonate ion-pairing. They noted that the S-shaped profiles produced are characteristic of waters with high contents of HCO_3 rather than high CO_3. Although bicarbonate does not form strong complexes (and probably does not fractionate the REEs) with the REEs (Wood, 1990; Lee & Byrne, 1993), carbonate complexes do and could be responsible for such heavy REE enrichments. However, even taking into account the role of complexing ligands such as F^- and CO_3, it is difficult to form such unique profiles with solution-complex fractionation alone and other processes must be responsible for such signatures. Sholkovitz (1995) showed that a large proportion of the dissolved REEs in river waters are transported in colloidal form. Colloidal particles are generally enriched in light REEs, and in a thorough study comparing different pores size filtration (including ultrafiltration), Sholkovitz (1995) demonstrated that the true dissolved concentrations show strong heavy REE enrichment. It is therefore possible that removal of colloidal material (and hence light REEs) could explain the heavy REE signatures of the Gornovodnoe waters. The true dissolved concentrations measured by Sholkovitz (1995) were found to be extremely low (pg l^{-1}), Yb/La ratios were much less extreme (< 10x), and the actual profiles were very different from those of the Gornovodnoe waters. Furthermore, the concentrations of heavy REEs in the rivers and fresh groundwaters are very low and, therefore, a source of these elements is required from a mineral phase.

It is, therefore, necessary to postulate a weathering phase that shows such enrichments. The two common mineral phases that display heavy REE enrichments are zircon and garnet (Hanson, 1982; Gromet & Silver, 1983). The Gornovodnoe and Shmakovka groundwaters are also particularly enriched in Be (up to 14 µg l^{-1}), Y (up to 22 µg l^{-1}), Cs (up to 13 µg l^{-1}), U (up to 5 µg l^{-1}) and Zr (up to 38 µg l^{-1}), which help to discriminate such sources. Zirconium, for example, is generally considered to be an extremely immobile element in most low-temperature environments. Frau (1993) also noted higher concentrations of Zr (up to 12 µg l^{-1}) in CO_2-rich groundwaters in Sardinia but suggested that the origin may be related to mafic Zr-bearing minerals. The groundwaters studied by Frau (1993) also show a heavy REE enrichment (and positive Sm-Eu anomalies), but not as extreme as the Gornovodnoe samples (authors' unpublished data). Such Zr-rich minerals, as well as garnet, are not present in the study area. The mobility of Zr may be enhanced due to the relatively high F^- concentrations in Primorye groundwater with which it forms a very stable complex. Moreover, Saxena (1966) and Strock (1941) showed that acidic, carbonated lime-rich waters readily attack zircon. Secondary alteration of zircon in pegmatites and some granites produces zircons enriched in several trace elements including Y, Th, U, Nb and Be (Goldschmit, 1958). The Gornovodnoe groundwaters also contain the highest concentrations of Be, Y and U and thus these secondary zircon phases may represent a potential source of the high Zr. Strock and Drexler (1941) considered that Zr in some acidic groundwaters was stabilised as calcium dicarbonato zirconylate, $Ca(ZrO)_2[CO_3]_2$ or as $ZrO[CO_3]_2^{2-}$, however Zr also forms very stable complexes with the fluoride ion (Baes & Mesmer, 1986), which is present at concentrations up to 2.1 mg l^{-1} in these waters.

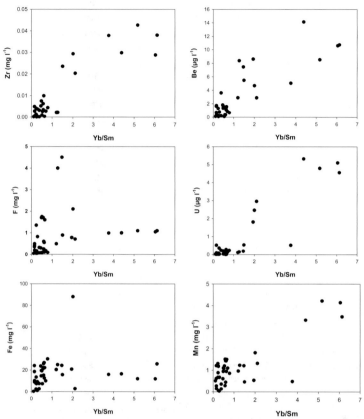

Figure 6. Plots of Yb/Sm (degree of heavy to light REE enrichment) plotted against a range of parameters for the high pCO₂ groundwaters.

Figure 6 shows Yb/Sm (which highlights the degree of enrichment of the heavy to middle REEs) plotted against a range of constituents. It can be seen that the groundwaters with high Yb/Sm are also high in Zr as well as other trace elements such as Be. There is no clear correlation with Fe and although the Gornovodnoe waters contain high dissolved Mn, this is not the case in all waters from this area or for the Schmakovka waters. It is therefore suggested that the unique patterns of heavy REE enrichment in these waters is due to zircon dissolution, the breakdown of which is caused by the F-rich carbonated groundwaters.

The relatively enriched middle to heavy REE profiles of most of the other Primorye waters may imply a common geochemical process as these waters are present in a range of contrasting geological terranes with different lithology and mineralogy. Sorption-desorption reactions associated with colloidal silicate phases are unlikely to impart such signatures. The affinity of REE adsorption onto colloidal clay particles is LREE >

MREE > HREE. With decreasing pH, the release of REEs occurs in the same order (Sholkovitz, 1995). It has been demonstrated with leachate experiments (Sholkovitz, 1995; Johannesson & Zhou, 1999) that Fe-Mn oxides/oxyhydroxides contain relative enrichments in the shale-normalised middle REEs. Bau (1999) calculated apparent REE and Y distribution coefficients (appD^{REY}) experimentally for iron oxyhydroxides using synthetic solutions and natural spring water. He showed that radius-independent fractionation is due to surface complexation and that REE scavenging is highly pH dependent. Negative anomalies for Y, La and Gd were also found to become more pronounced with increasing pH with profiles changing from flat at low pH to sigmoidal showing the M-type lanthanide "tetrad" effect at higher pH (Bau, 1999). The high pCO$_2$ waters are strongly enriched in Fe and Mn and the source of such high concentrations is most likely to be from dissolution of oxide/oxyhydroxide phases: the waters contain low SO$_4$ concentrations and relatively high transition element concentrations typical of oxyhydroxide minerals. Therefore, such profiles may be expected. The patterns of the apparent distribution coefficients in the pH range 4.6 to 6 show many similarities to the high pCO$_2$ waters. Selected distribution coefficients (Bau, 1999) are shown on Figure 7 along with some of the Primorye groundwaters. The flat REE pattern of the Rakavka springs is similar to appD^{REE} at low pH and the slightly more alkaline waters show similarities to the appD^{REE} at higher pH, including the relative enrichment in Sm and Eu and depletion of the light REEs. It is, therefore, considered that the dominant control on the REE signatures in most of these groundwaters is dissolution of Fe-Mn oxyhydroxides enhanced by the low pH buffering of the CO$_2$ gas. Such a control by these ubiquitous mineral phases explains the general similarities in groundwaters from such vastly different terrain in terms of rock type, mineralogy and age. Although silicate weathering has been important, it is likely that such inputs are overshadowed by sorption-desorption reactions. It is possible that the presence of Sm-Eu anomalies is an indicator of such reactions but further work is required on the solid phases present in the study areas.

The presence of Eu anomalies in some of the waters can be useful for discriminating weathering phases. Crustal rocks and a variety of minerals commonly show negative Eu anomalies as a consequence of a geochemical process with a previous history of plagioclase removal. Therefore, such profiles do not discriminate weathering sources well. However, the presence of positive Eu anomalies points to a plagioclase-rich source. The Balshoy Kluch samples, for example, display large positive Eu anomalies and taken together with very high Ca and Sr concentrations imply that plagioclase weathering has been important in their hydrochemical evolution. The presence of negative Eu anomalies in some of these shallow groundwaters is most likely inherited from the source rock minerals, rather than being a redox control.

The concentrations of the REEs in groundwaters are controlled by the interplay between source characteristics of the weathering phases, solution complexation and adsorption reactions. It has been shown that the most likely dominant source control on REEs in these high pCO$_2$ waters is dissolution of Fe-Mn oxyhydroxides. Colloidal transport in these pristine waters is unlikely to be important at the low pH's found in these groundwaters, although adsorption of the REEs onto clay minerals along flowpaths may be an important mechanism for lowering solute concentrations of the light REEs. Nevertheless, the role of speciation is considered to be important, particularly complexes with F$^-$ and CO$_3$, which account for significant percentages of the dissolved

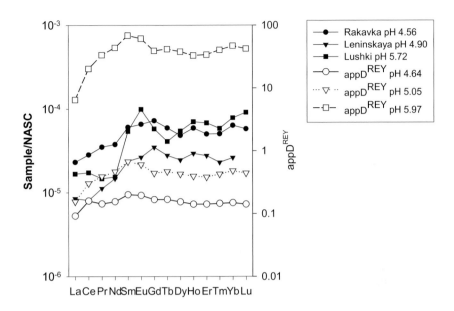

Figure 7. REE normalised plots for selected groundwaters with different pH and apparent distribution coefficients (Bau, 1990) at different pH. See text for details.

REEs. The presence of dissolved F^- may lead to modification and fractionation of the source REE characteristics (i.e., enhancing the heavy REE profiles on shale normalised plots). Although this is difficult to quantify in these CO_2-rich waters, a role for F^- is indicated by the relatively high Y concentrations, in addition to the significance of LnF^{2+} complexes in these waters as predicted by the speciation calculations (Figure 5). The Y/Ho ratio in most of these groundwaters (up to 56) is higher than crustal rocks (NASC = 26) and gives rise to spikes on shale normalised plots (Figure 8). Such features are typical of fluorine-rich fluids, due to much higher stability constants for Y than its neighbouring REEs (Bau & Dulski, 1995). In general, the pseudo-lanthanide Y behaves similarly to Ho due its similarity in ionic radius. This cannot be explained by dissolution of iron oxyhydroxide as the apparent distribution coefficients are much lower for Y than surrounding REEs and should give rise to negative, not positive, spikes. However, it is not clear at this stage whether such spikes may be due to dissolution of fluorite (with high Y/Ho) or to complexation fractionation related to differences in REE, Y-F stability constants. The role of P and organic complexes cannot be tested due to lack of data: P concentrations were below the limit of detection (0.2 mg l^{-1}) but may nevertheless be important at concentrations below this level. Of probable greater significance is the role of organic complexes as the only two analyses were very high (10 and 38 mg l^{-1}) in TOC and further study is required to assess if these are typical of high pCO_2 waters.

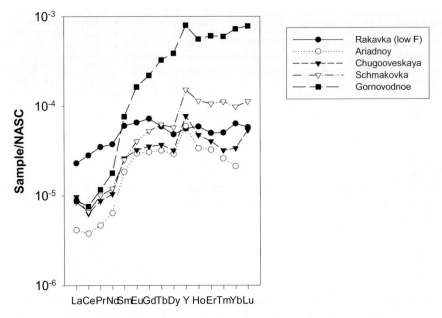

Figure 8. REE-Y shale normalised plots for selected high pCO$_2$ groundwaters showing positive Y anomalies except in the acidic low-F Rakavka groundwater.

A comparison of the high pCO$_2$ groundwaters with local fresh groundwaters and rivers, which generally contain low concentrations of the REEs (groundwater REEs are generally below detection limit), shows the significant effect that the addition of a CO$_2$-rich gas phase can have on weathering processes. The dominant effect is the creation of relatively acidic, aggressive groundwaters that will readily attack silicate minerals, Fe-Mn oxyhydroxide phases and, combined with complexing ligands such as F$^-$, accessory minerals such as zircon. However, in order to understand the roles of source and complexation controls, more detailed work is required on solid phases present in these groundwater systems. Nevertheless, this study of REEs has helped to highlight some of the processes and problems occurring in such complex geochemical systems.

8. Summary and Conclusions

High pCO$_2$ groundwaters in the Primorye region of the Russian Far East, associated with deep crustal fractures, are present across a series of geological terrains with contrasting lithology and age. They are moderately acidic owing to the buffering effect of CO$_2$ gas (up to 2.6 atm.) of mantle origin. These waters have some unusual geochemical characteristics including relative enrichments in the middle to heavy REEs. Most show relatively flat patterns from Sm to Lu on shale-normalised plots and there is a tendency to develop positive Sm and Eu anomalies. The dominant control on such profiles is considered to be dissolution of Fe-Mn oxyhydroxide phases as indicated by extremely high Fe and Mn concentrations. Samples from two locations (Gornovodnoe and Schmakovka) have heavy REE enriched profiles and the origin is interpreted to be from zircon. Although extremely stable in most low temperature environments, zircon can be broken down by such waters. This is substantiated by a correlation between heavy REE enrichment and solute Zr in these groundwaters. Plagioclase dissolution may also be indicated by the presence of high Ca and Sr in waters with a strong positive Eu anomaly. High Y/Ho ratios and positive Y spikes have also been noted in REE profiles of most samples and may be related to a control by F^- which has a high stability constant for Y (similar to Lu). Although the role of source characteristics is stressed in this paper, the role of complex-fractionation may be important in such waters in modifying the original REE characteristics of weathering minerals and during solute transport. Model calculations show that the dominant species of REEs in these groundwaters are Ln^{3+}, carbonate complexes (i.e., predominantly $LnCO_3^+$), and LnF^{2+}. There may be an important role for organic complexation but due to lack of data this cannot be tested at present. Furthermore, additional detailed work is also required on the mineralogy and geochemistry of the solid phases to test the above conclusions.

The effect of adding CO_2 gas to the groundwaters of Primorye is to enhance concentrations of the middle and heavy REEs as indicated by comparison with fresh surface and groundwaters in the area. Concentrations in fresh groundwaters are generally very low (mostly below detection limit). River water concentrations have similar concentrations of light REEs to the groundwaters but lower middle to heavy REEs and display flat profiles except for negative Ce anomalies on shale-normalised plots. This study has, therefore, highlighted the benefits of studying REEs in the unique weathering environment of CO_2-rich groundwaters.

Acknowledgements

The authors are grateful to Dr P Smedley for reviewing an early draft of this paper. Drs R Hannigan and A Michard also provided helpful and thought provoking reviews. P. Shand and W.M. Edmunds publish with permission of the Director of the British Geological Survey.

References

Alderton, D.H.M., Pearce, J.A. & Potts, P.J. 1980 Rare earth element mobility during granite alteration: evidence from southwest England. *Earth and Planetary Science Letters*, **49**, 149-165.

Andersson, P. S., Dahlqvist, R., Ingri, J., and Gustafsson, Ö. 2001 The isotopic composition of Nd in a boreal river: A reflection of selective weathering and colloidal transport. *Geochimica et Cosmochimica Acta*, **65**, 521-527.

Bau, M. Scavenging of dissolved yttrium and rare earths by precipitating iron oxyhydroxide: Experimental evidence for Ce oxidation, Y-Ho fractionation, and lanthanide tetrad effect. *Geochimica et Cosmochimica Acta*, **63**, No. 1, 67-77.

Bau, M. & Dulski, P. 1995 Comparative study of yttrium and rare-earth element behaviours in fluorine-rich hydrothermal fluids. *Contributions to Mineralogy and Petrology*, **119**, 213-223.

Baes, C.F. & Mesmer, R.E. 1986   The Hydrolysis of Cations, Krieger Publishing Company, Florida, 490 pp.

Barnes, I., Irwin, W.P. & White, D.E. 1978 Global distribution of carbon dioxide discharges, and major zones of seismicity. *United States Geological Survey, Water Resources Investigations*, Open File Report, **78-39**, 12p.

Brookins, D. G. 1989. Aqueous geochemistry of rare earth elements. In: B. R. Lipin and G. A. McKay (eds.) Geochemistry and Mineralogy of Rare Earth Elements. Reviews in Mineralogy, Vol. 21. Mineralogical Society of America (Washington, DC), pp. 201-225.

Bruno, J. 1997 Trace element modelling. In: I. Grenthe and I Puidomenech (eds.) *Modelling in Aquatic Chemistry*, OECD Nuclear Energy Agency (Paris), pp. 593-621.

Byrne, R.H. & Kim, K.H. 1990 Rare earth element scavenging in seawater. *Geochimica et Cosmochimica Acta*, **54**, 2645-2656.

Chudaeva, V.A., Chudaev, O.V., Chelnokov, A.N., Edmunds, W.M. & Shand, P. 1999 Mineral waters of Primorye (chemical aspects). In Russian with English summary. Dalnauka, Vladivostok.

Chudaeva, V.A., Lutsenko, T.N., Chudaev, O.V., Chelnokov, A.N., Edmunds, W.M. & Shand, P. 1995 Thermal waters of the Primorye region, eastern Russia. In, Y. Kharaka & O.V. Chudaev (eds.): Water Rock Interaction 8. A.A. Balkema, Rotterdam, 375-378.

Cornides, I. 1993 Magmatic carbon dioxide at the crust's surface in the Carpathian Basin. *Geochemical Journal*, **27**, 241-249.

Elderfield, H. 1988 The oceanic chemistry of the rare-earth elements. *Philosophical Transactions of the Royal Society of London*, A325, **105-126.**

Erel, Y. & Stolper, E.M. 1993 Modelling of rare-earth element partitioning between particles and solutions in aquatic environments. *Geochimica et Cosmochimica Acta*, **57**, 513-518.

Faure, G. 1977 Principles of isotope geology. John Wiley & Sons, New York, 589p.

Fee, J.A., Gaudette, H.E., Lyons, W.B. & Long, D.T. 1992 Rare earth element distribution in the Lake Tyrrell groundwaters, Victoria, Australia. *Chemical Geology*, **96**, 67-93.

Frau, F. 1993 Selected trace elements in groundwaters from the main hydrothermal areas of Sardinia (Italy) as a tool in reconstructing water-rock interaction. *Mineralogica Petrographica Acta*, **34**, 281-296.

Gimeno S., M. J., Auqué S., L. F., and Nordstrom, D. K. 2000 REE speciation in low-temperature acidic waters and the competitive effects of aluminium. *Chemical Geology*, **165**, 167-180.

Goldschmit, V.M. 1958 Geochemistry, Oxford University Press, 730 pp.

Goldstein, S.J. & Jacobsen, S.B. 1987 The Nd and Sr isotopic systematics of river-water dissolved material: implications for the sources of Nd and Sr in seawater. *Chemical Geology* (isotope Geosciences Section), **66**, 245-272.

Gosselin, D.G., Smith, M.R., Lepel, E.A. & Laul, J.C. 1992 Rare earth elements in chloride-rich groundwater, Palo Duro Basin, Texas, USA. *Geochimica et Cosmochimica Acta*, **56**, 1495-1505.

Gromet, L.P. & Silver, L.T. 1983 Rare earth element distributions among minerals in a granodiorite and their petrogenetic implications. *Geochimica et Cosmochimica Acta*, **47** 925-939.

Hanson, G.N. 1980 Rare earth elements in petrogenetic studies of igneous systems. *Annual Reviews of the Earth and Planetary Sciences*, **8**, 371-406.

Johannesson, K.H. & Lyons, W.B. 1994 The rare earth element geochemistry of Mono Lake water and the importance of carbonate complexing. *Limnology and Oceanography*, **39**, 1141-1154.

Johannesson, K.H., Stetzenbach, K.J., Hodge, V.F. & Lyons, W.B. 1996 Rare earth elements complexation behaviour in circumneutral pH groundwaters: assessing the role of carbonate and phosphate ions. *Earth and Planetary Science Letters*, **139**, 305-319.

Johannesson, K.H, Stetzenbach, K.J. & Hodge 1997 Rare earth elements as geochemical tracers of regional groundwater mixing. *Geochimica et Cosmochimica Acta*, **61**, 3605-3618.

Johannesson, K.H & Zhou, X. 1999 Origin of middle rare earth element enrichments in acid waters of a Canadian High Arctic Lake. *Geochimica et Cosmochimica Acta*, **63**, No. 1, 153-165.

Klungness, G. D. & Byrne, R. H. 2000 Comparative hydrolysis behaviour of the rare earths and yttrium: the influence of temperature and ionic strength. *Polyhedron*, **19**, 99-107.

Lee, J. H. & Byrne, R. H. 1992 Examination of comparative rare earth element complexation behaviour using linear free-energy relationships. *Geochimica et Cosmochimica Acta*, **56**, 1127-1137.

Lee, J.H. & Byrne, R.H. 1993 Complexation of trivalent rare earth element (Ce, Eu, Gd, Tb, Yb) by carbonate ions. *Geochimica et Cosmochimica Acta*, **57**, 295-302.

Luo, Y. –R. & Byrne, R. H. 2000 The ionic strength dependence of rare earth and yttrium fluoride complexation at $25^{\circ}C$. *Journal of Solution Chemistry*, **29**, 1089-1099.

Michard, A., Beaucaire, C. & Michard, G. 1987 Uranium and rare earth elements in CO_2-rich waters from Vals-les-Bains (France). *Geochimica et Cosmochimica Acta*, **51**, 901-909.

Millero, F. J. 1992 Stability constants for the formation of rare earth inorganic complexes as a function of ionic strength. *Geochimica et Cosmochimica Acta*, **56**, 3123-3132.

Nesbitt, H.W. 1979 Mobility and fractionation of rare earth elements during weathering of a granodiorite. *Nature*, **279**, 206-210.

Palmer, N.R. & Elderfield, H. 1986 Rare earth elements and neodymium isotopes in ferromanganese oxide coatings on Cenozoic foraminifera from the Atlantic Ocean. *Geochimica et Cosmochimica Acta*, **50**, 409-417.

Parkhurst, D. L., Thorstenson, D. C., and Plummer,L. N. 1980 PHREEQE – A computer program for geochemical calculations. *U. S. Geological Survey Water Resources Investigation Report* 80-96.

Saxena, S.K. (1966) Evolution of zircons in sedimentary and metamorphic rocks. *Sedimentology*, **6**, 1.

Schijf, J. & Byrne, R. H. 1999 Determination of stability constants for the mono- and difluoro-complexes of Y and the REE, using a cation-exchange resin and ICP-MS. *Polyhedron*, **18**, 2839-2844.

Shand, P., Edmunds, W.M., Chudaeva, V.A., Chudaev, O.V. & Chelnokov, A.N. 1995 High pCO_2 cold springs of the Primorye region, eastern Russia. In, Y. Kharaka & O.V. Chudaev (eds.): Water Rock Interaction 8. A.A. Balkema, Rotterdam, 393-396.

Shand, P., Edmunds, W.M., Wagstaff, S., Flavin, R. & Jones, H.K. 1999 Hydrogeochemical processes determining water quality in upland Britain. Hydrogeology Report Series of the British Geological Survey.

Sholkovitz, E.R. The aquatic chemistry of rare earth elements in rivers and estuaries. *Aquatic Geochemistry*, **1**, 1-34.

Sholkovitz, E.R., Landing, W.M. & Lewis, B.L. 1994 Ocean particle chemistry: the fractionation of rare earth elements between suspended particles and seawater. *Geochimica et Cosmochimica Acta*, **58**, 1567-1579.

Sholkovitz. E.R., Shaw, T.J. & Schneider, D.L. 1992 The geochemistry of rare earth elements in the seasonally anoxic water column and porewaters of Chesapeake Bay. *Geochimica et Cosmochimica Acta*, **56**, 3389-3402.

Smedley, P.L. 1991 The geochemistry of rare earth elements in groundwater from the Carnmenellis area, southwest England. *Geochimica et Cosmochimica Acta*, **55**, 2767-2779.

Strock, L.W. 1941 Geochemical data on Saratoga mineral waters - applied in deducing a new theory of their origin. *American Journal of Science*, **239**, 857.

Strock, L.W. & Drexler, S. 1941 Geochemical study of Saratoga mineral waters by a spectro-chemical analysis of their trace elements. *Journal Optical Society of America*, **31**, 167-173.

Verplanck, P. L., Antweiler, R. C., Nordstrom, D. K., and Taylor, H. E. 2001. Standard reference water samples for rare earth element determinations. *Applied Geochemistry*, **15**, 231-244.

Wexsteen, P., Jaffé, F.C. & Mazor, E. 1988 Geochemistry of cold CO_2-rich springs of the Scuol-Tarasp region, Lower Engadine, Swiss Alps. *Journal of Hydrology*, **104**, 77-92.

Wood, S.A. 1990 The aqueous geochemistry of the rare-earth elements and yttrium. 1. Review of available low-temperature data for inorganic complexes and the inorganic REE speciation of natural waters. *Chemical Geology*, **82**, 159-186.

Chapter 8

GEOCHEMISTRY OF RARE EARTH ELEMENTS IN GROUNDWATERS FROM A RHYOLITE AQUIFER, CENTRAL MÉXICO

KAREN H. JOHANNESSON[1], ALEJANDRA CORTÉS[2], JOSE ALFREDO RAMOS LEAL[2], ALEJANDRO G. RAMÍREZ[2], & JAIME DURAZO[2]

[1]*Department of Earth and Environmental Sciences, The University of Texas at Arlington, Arlington, TX 76019-0049, USA*
[2]*Instituto de Geofísica, Universidad Nacional Autónoma de México, 04510 México, D. F., México*

Abstract

Rare earth element (REE) concentrations were measured in groundwaters collected from wells finished in a fractured, rhyolitic (Cuatralba Ignimbrite) aquifer from the La Muralla region of the central Mexican State of Guanajuato. The study site is located within the Faja Volcanica Transméxicano (i.e., Trans-Mexican Volcanic Belt), an extensive region of active volcanism within central México. La Muralla groundwaters are relatively warm ($32.2 \pm 2.7\,°C$), dilute Na-Ca-HCO$_3$ waters (5.1 mmol/kg $\leq I \leq 9.5$ mmol/kg) of circumneutral pH ($7.27 \leq$ pH ≤ 8.01). Concentrations of REEs in La Muralla groundwaters are exceedingly low, as demonstrated by Nd values, which range from ~ 10 pmol/kg to 34 pmol/kg. La Muralla groundwaters exhibit enrichments in the heavy REEs (HREE) over the light REEs (LREE) compared to Average Shale, as well as volcanic rocks from the Trans-Mexican Volcanic Belt, including rhyolitic volcanic rocks similar to those of the Cuatralba Ignimbrite aquifer. Shale-normalized Yb/Nd ratios of La Muralla groundwaters range from 1.85 to 6.55, with a mean (± standard deviation) of 4.2 ± 1.2. Rare earth element concentrations for La Muralla groundwaters are normalized to the average REE values of 27 different calc-alkaline rhyolites (from the literature) from the Trans-Mexican Volcanic Belt. The average Trans-Mexican Volcanic Belt rhyolite-normalized Yb/Nd ratios for La Muralla groundwaters range from 1.57 to 5.55, with a mean (± standard deviation) of 3.52 ± 1. Speciation calculations predict that REEs occur principally as carbonate complexes in La Muralla groundwaters, with LREEs predominantly in the form of positively charged, carbonato complexes (LnCO$_3^+$), and to a lesser extent, free metal ions (Ln^{3+}), and HREEs chiefly in solution as negatively charged, dicarbonato complexes (Ln(CO$_3$)$_2^-$). The speciation model predictions suggest that the HREE enrichment of La Muralla groundwaters originate from solution and surface complexation reactions within the system. Specifically, the preferential complexation of HREEs as negatively charged, dicarbonato complexes acts to stabilize HREE is solution owing to both the strength of these complexes and their low affinity for aquifer surface sites. Because La Muralla groundwaters are of circumneutral pH, surface complexation sites within the Cuatralba Ignimbrite are expected to predominantly be negatively charged. Therefore, because LREEs occur primarily as positively charged, carbonato complexes in La Muralla groundwaters, they are preferentially removed from solution owing to complexation to

K.H. Johannesson,(ed), Rare Earth Elements in Groundwater Flow Systems, 187-222.

aquifer surface sites.

1. Introduction

Interest in the REEs in groundwater flow systems stems, in part, from their potential to be sensitive tracers for studying groundwater - aquifer rock interactions, and possibly for tracing groundwater flow. The utility of the REEs as tracers reflects their uniform trivalent charge (Ce^{4+} and Eu^{2+} can also occur), and the gradual decrease in their ionic radii with increasing atomic number (i.e., the lanthanide contraction) that accompanies the progressive filling of the $4f$-electron shell across the lanthanide series. Therefore, not only do REEs exhibit strong fractionation as a group due to size and charge, but they also exhibit significant "within-group" fractionation resulting from the lanthanide contraction. These unique properties can thus facilitate investigations of both complex and subtle geochemical processes that other, single element or single compound (i.e., molecules) tracers cannot discern. Moreover, because of the similar valence and ionic radii of the REEs and trivalent actinides (e.g., Am^{3+}, Cm^{3+}, and Cf^{3+}), the REEs are naturally occurring analogues for the radioactive transuranics, and can in some cases be used to study and predict the behavior of actinides in groundwater systems (Choppin, 1983, Brookins, 1986; Krauskopf, 1986; Silva and Nitsche, 1995). Radioactive isotopes of the REEs are also produced during the fission of uranium and plutonium based fuels (Roxburgh, 1987; Wood, 1990).

As a consequence of the interest in REEs, their occurrence and distribution in the oceans, rivers, and some estuaries has been extensively studied (e.g., Elderfield and Greave, 1982; Goldstein and Jacobsen, 1988; Elderfield, 1988; Elderfield et al., 1990; Bertram and Elderfield, 1993; Sholkovitz, 1993, 1995; Sholkovitz and Szymczak, 2000). Some investigations have even used Sm-Nd isotopes to fingerprint and trace global oceanic water masses (Piepgras and Wasserburg, 1980, 1987; Stordal and Wasserburg, 1986). Although great strides have been made towards understanding the oceanic, estuarine, and riverine geochemistry of REEs, considerably less study has focused on REEs in groundwater systems, although the number of such investigations is on the rise (e.g., Banner et al., 1989; Smedley, 1991; Fee et al., 1992; Gosselin et al., 1992 Banks et al., 1999; Leybourne et al., 2000; Négel et al., 2000; Dia et al., 2000). Perhaps the most significant finding of many of the earlier studies is the reported similarity between groundwater (pH $\leq$ 7) and aquifer rock REE patterns. Indeed, the fact that many groundwater samples exhibit REE patterns that closely resemble the REE patterns of the rocks through which they flow has important consequences for the study of groundwater-rock interactions. Recently, however, groundwater samples with REE patterns that differ dramatically from those of the aquifer rocks have also been identified (e.g., Johannesson et al., 1999; Johannesson and Hendry, 2000; Leybourne et al., 2000). These groundwaters have highly fractionated REE patterns compared to the aquifer rocks, and are typically enriched in the heavy REEs (HREE) over the light REEs (LREE). In general, the highly fractionated REE patterns of these groundwaters more closely resemble the relative distributions of REEs reported in seawater and some alkaline, saline lakes (e.g., Elderfield and Greaves, 1982; Bertram and Elderfield, 1993; Möller and Bau, 1993) than actual aquifer materials (i.e., fractured rocks/unconsolidated sediments).

In the present study we report REE concentration data and rock-normalized REE patterns for groundwater samples collected from a series of wells located within the central Mexican State of Guanajuato. The groundwater samples are principally from a fractured rhyolitic aquifer, and to a lesser extent, an overlying conglomeratic deposit composed chiefly of fragments of these rhyolitic rocks (CEASG, 2000). We develop a conceptual model involving solution complexation of the REEs with inorganic ligands to explain the origins of the highly fractionated REE patterns of these and other groundwaters.

2. Hydrogeologic Setting

The study area, known locally as La Muralla owing to its proximity to a small village by the same name, is located in the Mexican State of Guanajuato, approximately 35 km directly south of the city of León (Fig. 1). The study area is situated within the Faja Volcanica Transméxicano (i.e., Trans-Mexican Volcanic Belt), which is an extensive (~105,000 km^2) region of active volcanism within central México that extends from approximately Guadalajara in the west to near Veracruz in the east (Verma et al., 1985). The Trans-Mexican Volcanic Belt is characterized by basaltic to rhyolitic calc-alkaline volcanic rocks, as well as volcanic rocks with alkaline affinities (Pal, 1972; Negendank et al., 1985; Verma et al., 1985; Talavera Mendoza et al., 1995). Rocks in the La Muralla study area consist of recent volcanic and sedimentary deposits that overly older, (Tertiary) volcanic and sedimentary rocks (Fig. 1). The oldest stratigraphic unit in the region is the Tertiary Guanajuato Conglomerate, which is overlain by Tertiary rhyolitic volcanic rocks of the Cuatralba Ignimbrite. Contemporaneous and interstratified with the Cualtralba Ignimbrite are minor flows of the Dos Aquas Basalts (Quintero, 1986). Disconformably overlying these volcanics are Tertiary sedimentary rocks, including coarse clastics (i.e., conglomeratic rocks) composed, in part, of fragments of the Cuatralba Ignimbrite. These older Tertiary sedimentary rocks are themselves overlain by Pliocene to Quaternary basaltic flows, lacustrine deposits, and unconsolidated alluvial materials consisting of gravels, sands, clays, and residual soils (Fig. 1; CEASG, 1999).

The highly fractured Cuatralba Ignimbrite is the chief aquifer in the La Muralla district, whereas Quaternary alluvial deposits are responsible for local, shallow aquifers that may be hydraulically interconnected to the underlying Cuatralba Ignimbrite aquifer. Within the La Muralla region, the depth to groundwater ranges from 50 m to 131 m (Fig. 1; CEASG, 1999). Two regionally extensive fault systems, one oriented north-south, and the other northeast-southwest, converge in the La Muralla region (CEASG, 1999; Ramírez et al., 1999, 2000). La Muralla groundwaters are likely recharged 40 - 50 km to the northeast, and within the Sierra de Guanajuato, whereas they discharge, in part, to the Rio Turbio, ~10 km southeast of the study site (CEASG, 1999; Cortés et al., 2000). The warmest waters from the La Muralla region are distributed along a linear, northeasterly trend, suggesting control by regional faults or other linear geologic structures. Geothermal springs also occur within the La Muralla -Guanajuato-León region (e.g., Comanjilla and Palenque hot springs discharge waters between 97°C and 100°C; CEASG, 1999). It is possible, although not yet demonstrated, that these geo-

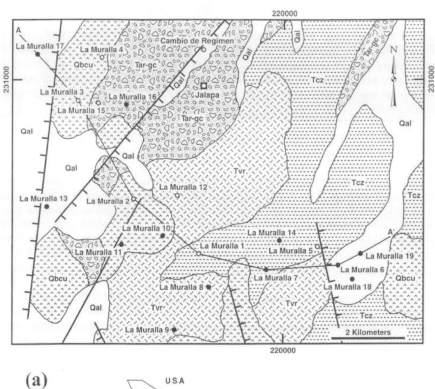

(a)

Figure 1. Location map of study site in the La Muralla region of Guanajuato, México. Geologic map and sample locations (filled circles) shown in (a), and schematic cross section along A - A' shown in panel (b) on following page. Map location shown in UTM and vertical exaggeration of cross section is 19x.

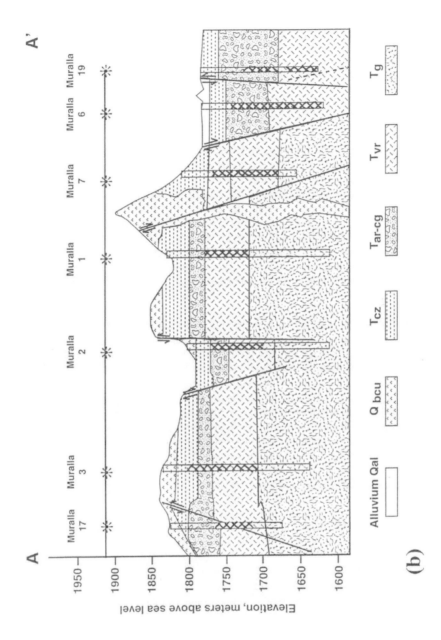

Figure 1 (continued). Symbols on map and cross section are Tg, Tertiary Guanajuato Conglomerate; Tvr, Tertiary Cuatralba Ignimbrite; Tar-gc, Tertiary coarse clastics including conglomeratic materials consisting of fragments of the Cuatralba Ignimbrite; Tcz, Tertiary lacustrine sediments, Qbcu, Quaternary basalts; and Qal, Quaternary alluvial deposits (CEASG, 1999). Screened portions of each well shown as x's in each well.

thermal waters communicate with the groundwaters of La Muralla via flow along faults in the regions or fractures within the ignimbrite flows (e.g., Ramírez et al., 1999, 2000).

3. Methods

3.1. SAMPLE COLLECTION

Prior to sample collection, all sample bottles were rigorously cleaned using trace element clean procedures. Briefly, high density linear polyethylene (HDPE) sample bottles were first triple washed with copious amounts of distilled-deionized water (18 MΩ-cm), and then placed in a 10-20% (v/v) reagent grade nitric acid bath for 7 to 10 days. The sample bottles were then removed from this acid bath, rinsed three times with distilled-deionized water, and subsequently immersed in another acid bath consiting of 10-20% (v/v) trace metal grade nitric acid (Fisher Scientific) for an additional 7 to 10 days. Following this second acid bath, the sample bottles were triple rinsed with distilled-deionized water, filled with distilled-deionized water, and doubled bagged in clean plastic bags. During sample bottle cleaning, laboratory personnel wore clean poly gloves. After sample bottles were cleaned and bagged, they were placed within a clean poly box for transport to and from the field collection sites.

All of the wells in the La Muralla region are production wells for the city of León, Guanajuato, and accordingly are pumped continuously. Consequently, waters from these wells are characteristic of the groundwater within the fractured rhyolitic (i.e., Cuatralba Ignimbrite) aquifer, and do not represent stagnant waters from the well bores. At each well-head, groundwater was collected with a large, collapsible polyethylene container (previously cleaned identically to the sample bottles). The collected groundwater sample was then immediately filtered through 0.45 μm Gelman Sciences in-line groundwater filter capsules (polyether sulfone membrane) by drawing the groundwater from the collapsible polyethylene container through Teflon® tubing (cleaned in the same fashion as the sample bottles) via a peristalic pump, and subsequently through the in-line filter capsule. Each individual sample bottle was rinsed three times with the filtered groundwater sample to condition the bottle before the bottle was filled with the actual filtered groundwater sample. The filtered groundwater samples for REE analysis were then immediately acidified to pH < 2 with ultra pure nitric acid (Seastar Chemicals, subboiling, distilled in quartz), doubled bagged within clean plastic bags, and returned to the large poly box for transport back to the laboratory. Samples for the major solutes were collected identically, except only a drop of nitric acid was used to preserve the cations, and the anion samples were not acidified. For each groundwater sample, pH, temperature (°C), and alkalinity were measured on site using separate sample aliquots, and following standard techniques.

3.2. ANALYTICAL METHODS

Major cations and anions (Table 1) were determined in each groundwater sample using ion chromatography (Dionex DX-500) following standard methods (e.g., Welch et al., 1996). Anions were determined using Ion Pac AS11 and AG11 columns, ASRS-ULTRA (4mm) self-regenerating anion suppressor, an EG-40 eluent generator, and

MilliQ water (18 MΩ-cm) as the reagent. The cations were measured with Ion Pac CS12A and CG12A columns, a CSRS-ULTRA (4mm) self-regenerating cation suppressor, and 20 mM methane sulfonic acid (MSA) as the reagent.

The REEs were measured in each groundwater sample (Table 2) via inductively coupled plasma mass spectrometry (ICP-MS). For each sample, the majority of the REEs were quantified directly using a Perkin-Elmer Elan 6000 ICP-MS as described by Graham et al. (1996). The approach and instrumentation was similar to the direct measurement technique used by Halicz et al. (1999) to measure REEs in groundwater samples from Israel, except instead of using an ultrasonic nebulizer, we employed a cross-flow nebulizer, with platinum sampler and skimmer cones. The REE isotopes ^{139}La, ^{140}Ce, ^{142}Ce, ^{141}Pr, ^{143}Nd, ^{146}Nd, ^{148}Sm, ^{149}Sm, ^{152}Sm, ^{151}Eu, ^{153}Eu, ^{157}Gd, ^{158}Gd, ^{159}Tb, ^{161}Dy, ^{163}Dy, ^{164}Dy, ^{165}Ho, ^{166}Er, ^{167}Er, ^{169}Tm, ^{172}Yb, ^{173}Yb, ^{174}Yb, and ^{175}Lu were used to quantify the REEs in La Muralla groundwater samples. Although, many of these REE isotopes are free of isobaric interferences (e.g., Smedley, 1991), it is advantageous, when possible, to monitor multiple isotopes of a particular element as an additional check for isobaric interferences (e.g., Hodge et al., 1998). Measured $REEO^+/REE^+$ ratios were ordinarily < 1%, and for those which were > 1%, appropriate corrections were made (Stetzenbach et al., 1994; Graham et al., 1996). Nevertheless, although BaO^+/Ba^+ ratios were low (i.e., < 0.1%), BaO^+ formation in the plasma stream was sufficient to cause substantial interferences on ^{151}Eu and ^{153}Eu during direct detection. Consequently, Eu was quantified on preconcentrated samples (see below).

Separate aliquots of each groundwater sample were preconcentrated in order to quantify Eu, which exhibits false positives from Ba interferences during direct analysis, and the mono-isotopic HREEs (Tb, Ho, Tm, and Lu), which occur at or below the direct method detection limits in La Muralla groundwaters (i.e., Graham et al., 1996). The preconcentration methods were similar to those described by Elderfield and Greaves (1983) and Greaves et al. (1989), and more recently by Klinkhammer et al. (1994), Schneider and Palmieri (1994), Stetzenbach et al. (1994), Johannesson and Lyons (1995), and Johannesson et al. (1997). The REE were first concentrated from the filtered and acidified sample aliquots by ferric hydroxide coprecipitation (e.g., Weisel et al., 1984; Welch et al., 1990; Johannesson and Lyons, 1995). The iron solution was prepared from high purity iron (III) nitrate (Alfa Aesar, Puratronic®, 99.999%) following the approach described by Welch et al. (1994). After addition of the iron (III) nitrate solution, sample pH was adjusted to between 8.0 and 9.0 with ultrapure ammonium hydroxide (Seastar Chemicals), and the resulting precipitate was allowed to form and settle for 24 hours. The precipitate was subsequently collected on acid-washed (10% v/v ultrapure HNO_3) Nuclepore® polycarbonate membrane filters (0.4 μm), and redissolved in 3 mL of a 10% v/v ultrapure HNO_3 (Seastar Chemicals, subboiling, distilled in quartz) solution. The ferric hydroxide coprecipitation separated the REEs and other trace elements in the groundwater samples from the major solutes, including a substantial fraction of the Ba (Welch et al., 1990).

The resulting solutions were subsequently loaded onto Poly-Prep columns (Bio-Rad Laboratories) packed with AG 50W-X8 (100-200 mesh, hydrogen form, Bio-Rad Laboratories) cation exchange resin at approximately 1 mL/min (Stetzenbach et al.,

1994). The iron and remaining Ba were eluted from the columns using 1.75 M ultrapure HCl (Seastar Chemicals, subboiling, distilled in quartz) and 2 M ultrapure HNO_3, respectively (Elderfield and Greaves, 1983; Greaves et al., 1989; Klinkhammer et al., 1994). Subsequently, the REEs were eluted from the columns with 8 M ultrapure HNO_3 (Stetzenbach et al., 1994). The remaining eluant was taken to dryness in Teflon® beakers and the residue was subsequently redissolved in 2 mL of a 1% v/v ultrapure HNO_3 solution (Klinkhammer et al., 1994; Stetzenbach et al., 1994). Final concentration factors ranged between 33 to 55 times the original sample values. The mono-isotopic HREEs and Eu were then quantified by ICP-MS as described above.

For both direct measurement and those involving the preconcentrated sample aliquots, the ICP-MS was calibrated and the sample concentrations verified using a series of REE calibration standards of known concentrations (1 ng/kg, 2 ng/kg, 10 ng/kg, 100 ng/kg, 500 ng/kg, and 1000 ng/kg). The calibration standards were prepared from NIST traceable High Purity Standards (Charleston, SC). Moreover, check standards prepared from Perkin Elmer multielement solutions were analyzed regularly during the analyses to certify accuracy (e.g., Graham et al., 1996; Hodge et al., 1998). In some cases, standard additions were used, where samples were spiked with 5 ng/kg additions of each REE, in order to better quantify the low concentrations of REEs in these samples (Graham et al., 1996). In addition, [115]In was added to each sample as an internal standard to monitor for matrix effects, differences in sample viscosity, solute build-up on the sampler and skimmer cones, and instrument drift during the analyses (Graham et al., 1996; Guo, 1996). Detection limits for both approaches were generally in the low pmol/kg level for the REEs (Table 3), and analytical precision, except for Eu, was typically better than 10% RSD (relative standard deviation), and as high as 2% RSD for Ce. Despite the preconcentration and attempts to separate Ba from sample aliquots, analytical precision for Eu ranged from 5.9% RSD to as high as 54.8% RSD, with a mean (± standard deviation) of 24.9 ± 15.6% RSD. Therefore, although Eu data are presented, they are not considered further.

3.3. SOLUTION COMPLEXATION MODELING

Complexation of the REEs in the La Muralla groundwaters was modeled using a combined specific ion interaction/ion pairing model initially designed for the REEs by Millero (1992). The model was updated by adding more recent experimental and linear free energy extrapolated complexation constant data for REE complexes with carbonate ions (Lee and Byrne, 1993), hydroxyl species (Byrne and Sholkovitz, 1996), and phosphate ions (Lee and Byrne, 1992). Free inorganic ligand concentrations used in the speciation modelling (e.g. $[CO_3^{2-}]_F$, $[SO_4^{2-}]_F$) were calculated from the major solute composition of the La Muralla groundwater samples (i.e., Table 1) using the computer program PHREEQE (Parkhurst et al., 1980). Although La Muralla groundwaters are dilute and do not require sophisticated techniques for calculating the variations in activity coefficients as a function of ionic strength, the solution complexation model was used because (1) it is valid at low and high ionic strengths, and (2) to allow for direct comparison of the speciation results for La Muralla groundwaters with other natural waters, including previously evaluated brines (e.g., Johannesson and Hendry, 2000). It is important to point out that the model does not require the input of REE concentrations because trace element solution complexation is controlled by the free concentrations of complexing ligands in the natural water, along with the corresponding stability constants for the metal-ligand complexes (Millero, 1992; Bruno, 1997).

Solution complexation of the REEs with naturally occurring organic ligands or anthropogenic organic acids may also be important in natural waters (Tipping, 1993; Wood, 1993; Lead et al., 1998; Tang and Johannesson, 2003). Although significant progress has been made in the development of metal-humic binding models (e.g., Tipping and Hurley, 1992; Tipping, 1998; Glaus et al., 2000; Hummel et al., 2000), it is still difficult to accurately model the solution complexation of REEs with naturally occurring organic matter owing to the complexity of humic substances (see Tang and Johannesson, 2003), and because of the absence of a precise technique to determine the activity coefficients and stoichiometric stability constants of REE complexes with natural organic substances. Additional attention is essential in order to adequately handle REE complexation with organic ligands in natural waters. Nevertheless, initial predictions suggest that organic complexation of REEs in groundwaters is likely to be substantially less significant than for surface waters (Tang and Johannesson, 2003).

4. Results

4.1. MAJOR SOLUTE COMPOSITION

Concentrations of the major cations and anions in La Muralla groundwaters are presented in Table 1, along with pH and water temperatures. La Muralla groundwaters can be classified as Na-Ca-HCO$_3$ waters based on the relative distribution of the major solutes (Fig. 2). The relatively high Na concentrations of La Muralla groundwaters likely reflect chemical reactions between the groundwaters and the rhyolitic rocks of the Cuatralba Ignimbrite. Furthermore, La Muralla groundwaters are of circumneutral pH, with values ranging from 7.3 to 8 (mean and standard deviation = 7.45 ± 0.2), and are relatively warm with temperatures ranging from 28°C to close to 38°C (mean and standard deviation = 32.2 ± 2.7°C; Table 1). La Muralla groundwaters are also relatively dilute, with ionic strength values ranging from 0.0051 moles/kg (La Muralla 13) up to 0.0095 moles/kg (La Muralla 16), and exhibit a mean ionic strength (± standard deviation) of 0.0077 ± 0.0012 moles/kg (Table 1).

4.2. RARE EARTH ELEMENT CONCENTRATIONS

The concentrations of the REEs in La Muralla groundwaters are presented in Table 2. The REE concentrations are similar to, or slightly lower than, REE concentrations reported for other groundwaters of circumneutral pH. For example, Nd concentrations of La Muralla groundwaters range from approximately 10 pmol/kg up to a high of 34 pmol/kg (mean and standard deviation = 17.1 ± 6.6 pmol/kg). By comparison, groundwater from carbonate aquifers of southern Nevada and eastern California, USA (pH = 7.16 ± 0.29; mean and standard deviation for n = 11 springs) have Nd values that range from 17 pmol/kg to 35 pmol/kg (mean and standard deviation = 27.3 ± 5.9 pmol/kg; Johannesson et al., 1997, 2000). Moreover, La Muralla groundwaters have Nd

Table 1. Major solute compositions (in mmol/kg) of groundwaters collected from wells from the La Muralla (e.g., LM 6 = La Muralla well # 6) region of the State of Guanajuato, México. Fluoride is in µmol/kg, and alkalinity is as HCO_3^-. Also tabulated are the pH, temperature (°C), and ionic strength (I; in moles/kg) of these groundwaters.

	pH	T°C	I moles/kg	Ca mmol/kg	Mg mmol/kg	Na mmol/kg	K mmol/kg	Cl mmol/kg	Alk.* mmol/kg	SO$_4$ mmol/kg	F µmol/kg	NO$_3$ mmol/kg
LM 6	7.41	35.1	0.0091	1.02	0.228	4.09	0.161	0.154	6.23	0.867	49.5	0.121
LM 7	7.58	31.2	0.0073	1.12	0.263	3.11	0.132	0.092	3.92	0.693	45.8	0.13
LM 8	7.32	30.3	0.0067	1.26	0.249	2.94	0.124	0.116	2.56	0.614	47.9	0.127
LM 9	7.31	30.7	0.0072	0.89	0.324	2.73	0.127	0.091	5.36	0.451	42.1	0.106
LM 10	7.37	28	0.0083	1.1	0.323	3.2	0.133	0.19	4.6	0.947	81.6	0.231
LM11	7.41	33.1	0.0084	1.03	0.2	3.86	0.123	0.228	4.08	1.15	86.3	0.142
LM 13	7.33	28.6	0.0051	0.87	0.192	1.52	0.108	0.024	4.21	0.081	26.3	0.125
LM 14	7.49	32.5	0.0079	0.98	0.226	3.74	0.14	0.166	4.0	0.958	57.4	0.125
LM 16	7.27	37.7	0.0095	1.24	0.186	4.18	0.111	0.273	4.68	1.38	84.2	0.125
LM 17	7.58	31.7	0.0086	1.15	0.23	3.89	0.148	0.254	3.96	1.19	95.3	0.124
LM 18	7.34	34.3	0.0073	0.9	0.253	3.33	0.147	0.126	4.32	0.685	60	0.143
LM 19	8.01	32.9	0.007	0.9	0.249	3.28	0.17	0.138	3.24	0.84	52.6	0.114

*As HCO_3^-

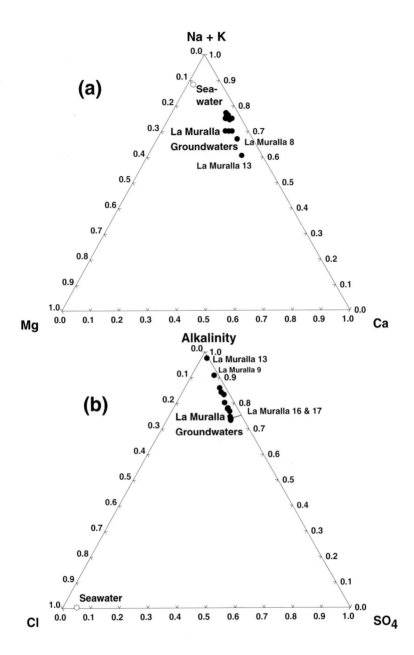

Figure 2. Ternary plots of (a) cations and (b) anions concentrations of La Muralla groundwaters. Seawater is shown for comparison.

Table 2. Rare earth element concentrations (in pmol/kg) for groundwaters collected from wells in the La Muralla region (e.g., LM 6 = La Muralla well # 6) of the State of Guanajuato, México. Shale-normalized, average rhyolite, and average calc-alkaline volcanic rock-normalized Yb/Nd ratios and Ce (Ce/Ce*) anomalies are also included (see text and Table 4).

	LM 6	LM 7	LM 8	LM 9	LM 10	LM 11	LM 13	LM 14	LM 16	LM 17	LM 18	LM 19
La	10.6	11.2	12.8	8.9	12.8	16	31.4	12.6	20.4	12.6	11	11.1
Ce	3.9	1.1	1.3	2.3	3.4	5	54.6	4.7	17.6	3.4	6.3	2
Pr	3.9	3.9	4.7	3.2	3.8	4.9	9.9	3.9	6.1	3.8	4.2	3.5
Nd	12.6	18.6	21.5	11.2	14.2	18.2	34	13.1	21.9	14.1	15.7	9.6
Sm	3.6	6.2	5	2.5	5.5	7	7.5	5.1	4.4	3.3	5.8	4.7
Eu	0.58	2	2	0.86	0.93	1	0.62		2.6	1.4		
Gd	7.2	8.8	12.8	7.2	8.3	9.2	9.1	8.7	10	7.2	6.8	6.3
Tb	2.1	2	2.7	1.8	2.3	1.9	2	2.4	2.2	1.9	1.8	1.6
Dy	7.4	7.7	12.2	7.8	6.5	7.2	7.5	7.4	8.8	6.6	6.7	4.2
Ho	1.4	1.7	3.5	1.9	2.1	1.9	2.1	1.8	2.2	1.7	2	1.6
Er	4.6	4.6	8.9	5.3	5.3	4.1	5.2	5.5	5.7	4	4.6	4.6
Tm	0.89	1.2	1.2	0.86	1	1.1	0.92	1.1	1.2	0.9	0.78	0.82
Yb	4.1	6	7.7	5.6	4.4	4.2	4.8	4.4	5.3	4.2	5.8	3.8
Lu	1.22	1.18	1.48	1.33	1.3	1.18	1.14	1.49	1.12	1.27	1.26	1.15
Yb/Nd_{SN}	4.27	4.22	4.67	6.55	4.03	3.01	1.85	4.34	3.12	3.9	4.79	5.2
Yb/Nd_{RN}	3.62	3.57	3.95	5.55	3.41	2.55	1.57	3.68	2.64	3.3	4.05	4.4
Yb/Nd_{VN}	3.88	3.83	4.24	5.96	3.66	2.73	1.68	3.95	2.83	3.55	4.35	4.73
$Ce/Ce*_{SN}$	-0.84	-1.4	-1.39	-0.98	-0.92	-0.87	-0.12	-0.79	-0.42	-0.92	-0.65	-1.11
$Ce/Ce*_{RN}$	-0.91	-1.47	-1.46	-1.06	-0.99	-0.93	-0.19	-0.86	-0.48	-0.99	-0.73	-1.18
$Ce/Ce*_{VN}$	-0.87	-1.43	-1.41	-1.01	-0.95	-0.89	-0.15	-0.82	-0.44	-0.95	-0.68	-1.14

SN = Shale-normalized

RN = Normalized to average of rhyolitic rocks from the Faja Volcanica Transméxicano (Table 4)

VN = Normalized to average of calc-alkaline volcanic rocks (rhyolites, andesites, basalts) from the Faja Volcanica Transméxicano (Table 4)

$Ce/Ce*_a = \log \{2Ce_a/[La_a + Pr_a]\}$, where a = SN, RN, or VN

Table 3. Limits of detection (pmol/kg) for direct REE determinations using a quadrupole ICP-MS (Perkin Elmer Elan 6000) with cross-flow nebulization (Graham et al., 1996). Concentrations in La Muralla field blank determined by standard additions.

	Detection Limits[§]	Field Blank*
La	2.4	< 2.4
Ce	2.8	1.6
Pr	1.3	1.3
Nd	14	2
Sm	4.4	1.7
Eu	1.2	1.3
Gd	2.4	2
Tb	0.6	1.3
Dy	2.1	1.5
Ho	0.7	1.3
Er	5.2	1.7
Tm	0.8	1.4
Yb	2	1.2
Lu	0.6	1

[§]See Graham et al. (1996).
*Determined by spiking samples with 5 ng/kg additions.

concentrations that are slightly lower than the mean Nd concentration of groundwater samples collected from the Battleford Till aquitard (pH = 7.54 ± 0.28; mean Nd = 22 pmol/kg) in Saskatchewan, Canada (Johannesson and Hendry, 2000). On the other hand, groundwater from the Carnmenellis region of England (pH = 5.82 ± 0.44; mean and standard deviation for n = 11 springs) have remarkably higher Nd concentrations that range from a low of 3,500 pmol/kg to as high as 478,000 pmol/kg (mean Nd = 79,600 pmol/kg; Smedley, 1991). Rare earth element concentrations in La Muralla groundwaters are also lower, on average, than those recently reported for groundwaters from a mining district in New Brunswick, Canada, and from mineral springs of the Massif Central in France (Leybourne et al., 2000; Négrel et al., 2000).

Although La Muralla groundwaters exhibit REE concentrations that are slightly lower than some groundwaters of circumneutral pH, their REE concentrations are typically slightly higher than those reported for groundwaters from alluvial aquifers in southern Nevada, USA. For example, Nd concentrations for groundwater samples collected from the shallow alluvial aquifers of the Amargosa Desert, Nevada range from 5 pmol/kg to 12 pmol/kg, with a mean Nd (± standard deviation) of 7.7 ± 3.3 pmol/kg (n = 4; Johannesson et al., 1997). Moreover, groundwaters from the shallow alluvial aquifer of the Oasis Valley of southern Nevada, which flow through basin-fill deposits chiefly composed of fragments of felsic volcanic rocks, have Nd concentrations between 4 pmol/kg and 27 pmol/kg (Nd = 14 ± 9 pmol/kg; mean and standard deviation for n= 8 samples; Johannesson et al., 1999). For Oasis Valley groundwater, pH ranges from 7.65 to 8.4 (pH 8.17 ± 0.23; mean and standard deviation).

4.3. NORMALIZED REE PATTERNS

Shale-normalized REE patterns for La Muralla groundwaters are presented in Fig. 3. The shale composite employed in Fig. 3 is the Average Shale used previously by Elderfield and Greaves (1982), Klinkhammer et al. (1983), De Baar et al. (1985), and Sholkovitz (1988) for marine systems, as well as by Johannesson et al. (1997, 1999) and Johannesson and Hendry (2000) in studies of REEs in groundwater systems. All La Muralla groundwater samples are enriched in HREEs compared to LREEs when normalized to Average Shale (Fig. 3). Shale-normalized Yb/Nd ratios, for example, range from 1.85 for La Muralla 13, up to 6.55 for La Muralla 9, with a mean (± standard deviation) shale-normalized Yb/Nd ratio of 4.2 ± 1.2 (Table 2). In addition, all of the La Muralla groundwater samples have negative, shale-normalized Ce anomalies (Fig. 3). The shale-normalized Ce anomalies for La Muralla groundwaters, which are calculated as $Ce/Ce^*_{SN} = \log\{2Ce_{SN}/[La_{SN} + Pr_{SN}]\}$, where SN = shale-normalized, range from -1.4 (La Muralla 7 and 8) to -0.12 (La Muralla 13), with a mean Ce anomaly (± standard deviation) of -0.87 ± 0.36 (Table 2).

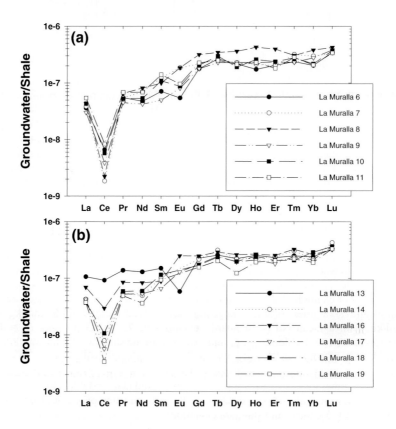

Figure 3. Shale-normalized REE plots of La Muralla groundwaters.

Table 4. Average concentrations (in ppm) of rare earth elements in rhyolitic rocks, and volcanic rocks (calc-alkaline basalts, andesites, and rhyolites), of the Faja Volcanica Transméxicano (i.e., Trans-Mexican Volcanic Belt), central México

	Average Rhyolite*	Average Volcanic Rock[†]
La	29.2	20
Ce	54.4	40.3
Pr	4.87	4.4
Nd	21.8	19.7
Sm	4.23	4.24
Eu	0.65	1.13
Gd	4.4	4.4
Tb	0.6	0.58
Dy	2.89	3.37
Ho	0.68	0.69
Er	2.4	2.16
Tm	0.33	0.29
Yb	2.38	2
Lu	0.39	0.3

*Mean of 27 analyses of different rhyolites (Ferriz and Mahood, 1987; Talavera et al., 1995; Verma, 1999, 2000; Gómez-Tuena and Carrasco-Núñez, 2000; Luhr, 2000).

[†]Mean of 27 rhyolites, 34 basalts, and 56 andesites (Luhr and Carmichael, 1980, 1981, 1985; Verma et al., 1985; Ferriz and Mahood, 1987; Hasenaka and Carmichael, 1987; Luhr et al., 1989; Lange and Carmichael, 1990; Wallace and Carmichael, 1992; Verma and Luhr, 1993; Ferrari et al., 1995; Talavera et al., 1995; Verma, 1985, 1999, 2000; Gómez-Tuena and Carrasco-Núñez, 2000; Luhr, 2000).

Although it is instructive to normalize REE concentrations of La Muralla groundwaters to Average Shale for purposes of comparison to earlier studies, it is more appropriate to normalize these groundwaters to the REE concentrations of Cuatralba Ignimbrite because aquifer rocks are thought to be the chief source of REEs to groundwaters (Banner et al., 1989; Smedley, 1991; Gosselin et al., 1992; Banks et al., 1999). To the best of our knowledge REE concentrations have not been determined for the rhyolitic rocks of the Cuatralba Ignimbrite. However, the REE concentrations of other calc-alkaline rhyolitic rocks from the Trans-Mexican Volcanic Belt have been measured and are reported in the literature (e.g., Ferriz and Mahood, 1987; Talavera et al., 1995; Verma, 1999, 2000; Gómez-Tuena and Carrasco-Núñez, 2000; Luhr, 2000). Consequently, we constructed an average rhyolite REE standard for the Trans-Mexican Volcanic Belt from existing published data to be used as a normalizing factor for La Muralla groundwaters. This approach assumes that: (1) because of the intimate contact between La Muralla groundwaters and the Cuatralba Ignimbite, these rhyolitic rocks are the chief source of REEs to the groundwaters; and (2) the Cuatralba Ignimbrite has REE concentrations and distributions that are similar to other calc-alkaline, rhyolitic volcanic rocks from the Trans-Mexican Volcanic Belt.   The constructed rhyolite standard consists of the average REE values from the analyses of 27 different calc-alkaline rhyolitic volcanic rocks from the Trans-Mexican Volcanic Belt (Table 4).

The chondrite-normalized REE patterns of each of the 27 rhyolitic volcanic rocks from the Trans-Mexican Volcanic Belt are shown in Fig. 4. These volcanic rocks exhibit chondrite-normalized REE patterns that are typical of rhyolitic volcanic rocks (e.g., Cameron, 1984; Kelleher and Cameron, 1990). Because other volcanic and plutonic rocks are common to the Trans-Mexican Volcanic Belt, as well as the León region of Guanajuato (e.g., the Dos Aquas Basalts, La Palma Diorite, El Cubilete Basalt; Quintero, 1986; Robles-Camacho and Armienta, 2000), we also constructed an average volcanic rock REE standard for the Trans-Mexican Volcanic Belt from existing published data (Table 4). This volcanic rock standard was calculated as the average of the same 27 rhyolitic volcanic rocks discussed above, along with REE analyses for 34 different calc-alkaline basalts, and 56 different calc-alkaline andesites from the Trans-Mexican Volcanic Belt (Table 4). The chondrite-normalized REE patterns for these basalts and andesites, which are typical of calc-alkaline andesites and basalts (Taylor and McLennan, 1985), are also presented in Fig. 4. Although alkaline volcanic rocks are reported from the Trans-Mexican Volcanic Belt, the Cualtralba Ignimbrite and other igneous rocks of the León region belong to the calc-alkalic series (Pal, 1972; Negendank et al., 1985; Verma et al., 1985; Quintero, 1986). Consequently, alkaline rocks were not used to construct the average rhyolite and "volcanic" rock normalizing standards for La Muralla groundwaters.

Figure 5 is a plot of REE concentrations for La Muralla groundwaters normalized to the average Trans-Mexican Volcanic Belt rhyolite standard from Table 4. The average rhyolite-normalized La Muralla groundwater patterns are similar to the shale-normalized patterns for these groundwater samples (compare Fig. 5 with Fig. 3). La Muralla groundwaters exhibit enrichments in the HREEs compared to the LREEs when normalized to the average, Trans-Mexican Volcanic Belt rhyolite standard. For example, the average rhyolite-normalized Yb/Nd ratios for La Muralla groundwaters range from 1.57 (La Muralla 13) to 5.55 (La Muralla 9), with a mean $(Yb/Nd)_{RN}$ ratio ($\pm$ standard deviation) of 3.52 ± 1 (Table 2; RN = rhyolite-normalized). In addition, La Muralla groundwaters exhibit substantial negative Ce anomalies when normalized to the Trans-Mexican Volcanic Belt rhyolite standard (Ce/Ce* = -0.94 $\pm$ 0.36; mean and standard deviation).

The REE patterns of La Muralla groundwaters normalized to the calculated, average Trans-Mexican Volcanic Belt volcanic rock (Table 4) are presented in Fig. 6. Again, the average volcanic rock-normalized REE patterns for La Muralla groundwaters are similar in shape to the shale-normalized, and average rhyolite-normalized REE patterns (Figs. 3, 5, and 6). That is, La Muralla groundwaters are strongly enriched in the HREEs compared to the LREEs, when normalized to average calc-alkaline volcanic rock from the Trans-Mexican Volcanic Belt, and exhibit significant negative Ce anomalies (Ce/Ce* = -0.9 $\pm$ 0.36; mean and standard deviation). The volcanic rock-normalized Yb/Nd ratios range from 1.68 (La Muralla 13) up to 5.96 (La Muralla 9), with a mean ($\pm$ standard deviation) of 3.78 ± 1.08.

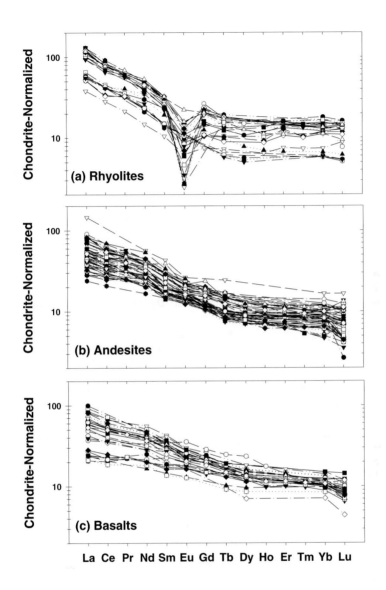

La Ce Pr Nd Sm Eu Gd Tb Dy Ho Er Tm Yb Lu

Figure 4. Chondrite-normalized REE plots of calc-alkaline (a) rhyolites, (b) andesites, and (c) basalts from the Trans-Mexican Volcanic Belt. The rhyolite, andesite, and basalt data represent 27, 56, and 34 different analyses of rhyolitic, andesitic, and basaltic volcanic rocks, respectively, from the Trans-Mexican Volcanic Belt obtained from the literature (see Table 4 and text).

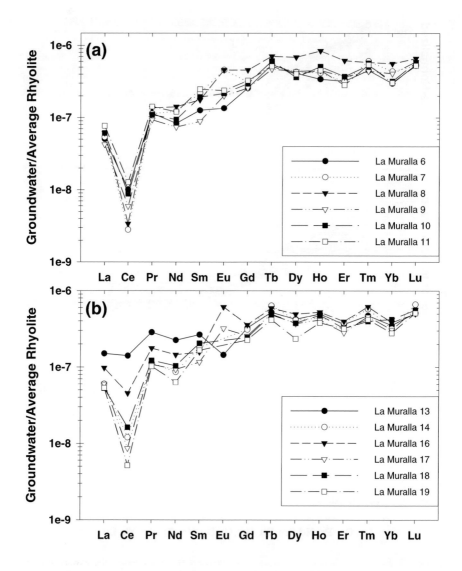

Figure 5. REE concentrations of La Muralla groundwaters normalized to the average of the 27 Trans-Mexican Volcanic Belt calc-alkaline rhyolites shown in Fig. 4, and presented in Table 4.

4.4. REE SOLUTION COMPLEXATION

The results of the inorganic solution complexation modeling for 7 of the 12 La Muralla groundwater samples (i.e., La Muralla wells 6, 7, 8, 10, 13, 16 and 19) are shown in Figs. 7 and 8. The speciation model results for the La Muralla groundwater samples are

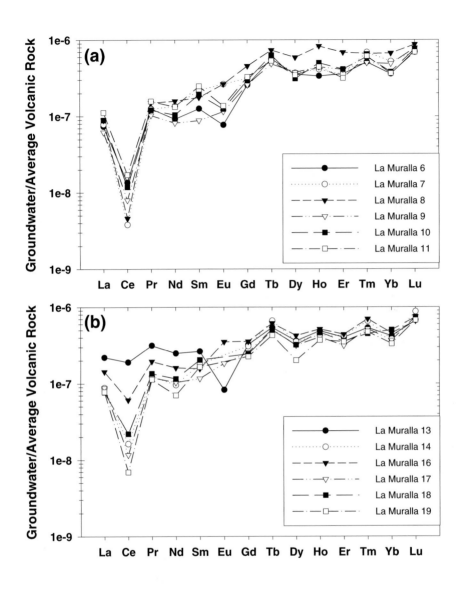

Figure 6. REE concentrations of La Muralla groundwaters normalized to the average the 27 rhyolites, 56 andesites, and 34 basalts from Trans-Mexican Volcanic Belt plotted in Fig. 4, and listed in Table 4.

relatively similar, except for La Muralla 6, and are representative of the speciation of REEs in La Muralla groundwaters as a whole. Other than La Muralla 6, which had a total phosphate concentration of 55.8 μmol/kg (not shown in Table 1), the other La Muralla groundwaters had total phosphate concentration below the method detection limit of 0.3 μmol/kg (ion chromatography). The relatively high total phosphate concen-

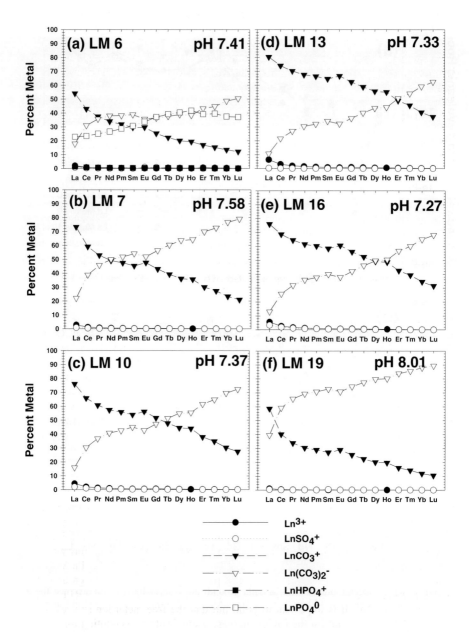

Figure 7. Results of the speciation model for groundwaters collected from La Muralla wells (a) La Muralla 6, (b) La Muralla 7, (c) La Muralla 10, (d) La Muralla 13, (e) La Muralla 16, and (f) La Muralla 19.

tration measured in La Muralla 6 groundwater significantly affects the distribution of the REE species in this groundwater (Fig. 7a) in that both $LnPO_4^0$, and to a lesser extent $LnHPO_4^+$, are predicted to account for important percentages of dissolved REEs. For example, our model calculations indicate that the $LnPO_4^0$ species ought to account for between 22% and 42% of each of the REEs, and for Gd, Tb, Dy, and Ho, the $LnPO_4^0$ species is predicted to be the dominant dissolved form of these REEs in La Muralla 6 groundwater (Fig. 7a). On the other hand, the $LnHPO_4^+$ species is predicted to account for ~ 1.5% of dissolved La, and about 0.8% of Ce in the La Muralla 6 groundwater, with the fraction of dissolved REEs occurring as $LnHPO_4^+$ decreasing with increasing atomic number to ~ 0.2 % for Lu (Fig. 7 a). The insignificance of REE phosphate complexation in the remaining La Muralla groundwater samples is underscored by calculating the speciation of the REEs in the least alkaline La Muralla groundwater sample (i.e., La Muralla 8; Table 1) assuming a total phosphate concentration equal to the ion chromatography method detection limit of 0.3 µmol/kg (Fig. 8). For these conditions, the model predicts that between 0.4% of La, and up to at most 1.8% of Lu, could occur in solution as the $LnPO_4^0$ complex (Fig. 8). It is unclear why La Muralla 6 has relatively high phosphate concentration, whereas phosphate in the other La Muralla groundwaters is the below detection limit. Possible sources of phosphate could include fast fracture flow from the surface, contamination from the well, or dissolution of localized phosphatic minerals.

Overall, the two inorganic REE species predicted to dominate in La Muralla groundwaters, including La Muralla 6 groundwater, are the carbonato ($LnCO_3^+$) and dicarbonato ($Ln(CO_3)_2^-$) complexes.   Specifically, for La Muralla 6, carbonato complexes are predicted to dominate for the LREEs up to Pr, after which dicarbonato complexes account for the majority of each dissolved REEs, except between Gd and Ho, where dicarbonato and $LnPO_4^0$ complexes account for roughly equal percentages of these dissolved REEs, with $LnPO_4^0$ species being slightly more abundant (Fig. 7a). For the remaining La Muralla groundwater samples, the model predicts that between 86% (La Muralla 8) and 98% (La Muralla 19) of La occurs in solution as carbonate complexes. Moreover, the model indicates that for REEs heavier than Dy, greater than 99% of each occurs as carbonate complexes in La Muralla groundwaters (Figs. 7 and 8).   In general, the LREEs are predicted to primarily occur as positively charged carbonato complexes, whereas the HREEs occur chiefly as negatively charged dicarbonato species. The transition along the REE series between where carbonato and dicarbonato complexes are predicted to dominate varies as a function of pH, with dicarbonato complexes dominating for progressively lighter REEs as pH increases (Table 1, Fig. 7). For example, carbonato complexes account for the majority of each REE, except the two heaviest REEs (Yb and Lu) in the lower pH (7.32) La Muralla 8 groundwater, whereas in the relatively high pH (8.01) La Muralla 19 groundwater, dicarbonato complexes are predicted to be the dominate form for all of the REEs except La (Figs. 7 and 8).   It is important to point out that the free metal ion species is also predicted to be important for the LREEs in these groundwaters, accounting for between 1% of La in solution (i.e., La Muralla 19) to as much as 9% of La (La Muralla 8).

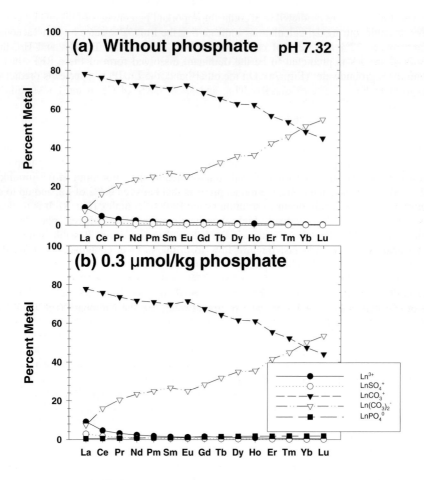

Figure 8. Results of speciation calculations for groundwater collected from La Muralla well 8 where (a) phosphate concentrations are below the detection limit as reported (Table 1), and (b) assuming the groundwater sample had a phosphate concentration equal to the method (ion chromatography) detection limit of 0.3 µmol/kg phosphate.

5. Discussion

5.1. REE FRACTIONATION PATTERNS OF LA MURALLA GROUNDWATERS

Figures 5 and 6 demonstrate that La Muralla groundwaters have different REE patterns than calc-alkaline volcanic rocks from the Trans-México Volcanic Belt, including local rhyolitic rocks similar to the Cuatralba Ignimbrite through which La Muralla groundwaters flow. In addition, Fig. 3 indicates that REE distributions in La Muralla groundwaters differ from composite shales, and hence, the bulk upper continental crust. The distribution of REEs in the upper continental crust is similar to that of composite shales such as Average Shale, PAAS, and NASC, although the crustal concentrations are slightly lower than in these composite shales (Taylor and McLennan, 1985). Inspection of Fig. 3 reveals that La Muralla groundwaters have shale-normalized REE patterns that are grossly similar to that of seawater in terms of their HREE enrichments and large, negative Ce anomalies (e.g., Elderfield and Greaves, 1982; Piepgras and Jacobsen, 1992; Bertram and Elderfield, 1993; Byrne and Sholkovitz, 1996).

Many previous investigators cite aquifer materials (i.e., rocks/sediments) as the primary source of REEs in groundwaters (e.g., Banner et al., 1989; Smedley, 1991; Fee et al., 1992; Gosselin et al., 1992). Moreover, in many cases, groundwaters have REE patterns that closely resemble those of the associated aquifer rocks (Smedley, 1991; Fee et al., 1992; Gosselin et al., 1992; Banks et al., 1999; Halicz et al., 1999; Leybourne et al., 2000; Möller, 2000). These observations suggest that groundwaters can inherit, via chemical weathering/solid-liquid exchange reactions, REE patterns that are similar to those of the aquifer rocks through which they flow. Because La Muralla groundwaters are in intimate contact with the Cuatralba Ignimbrite (Fig. 1), chemical weathering/solid-liquid exchange reactions between the groundwaters and these felsic volcanic rocks represents the most likely source of REEs to the groundwaters. However, the observed distributions of REEs in La Muralla groundwaters are strikingly different than the expected patterns for the Cuatralba Ignimbrite, which instead ought to closely resemble those of other rhyolitic volcanic rocks from the Trans-Méxican Volcanic Belt (Figs. 4, 5). Indeed, for La Muralla groundwaters to have inherited rock-like REE signatures from the Cuatralba Ignimbrite would require these rhyolitic rocks to be enriched in the HREEs compared to the upper continental crust (i.e., Average Shale) in general, and specifically, other rhyolitic volcanic rocks from the Trans-Méxican Volcanic Belt. Nevertheless, there is no evidence to suggest that rhyolitic rocks of the Cuatralba Ignimbrite have REE patterns that are strongly enriched in HREEs over the LREEs compared to shale, and especially other rhyolitic volcanic rocks (Fig. 4a). Moreover, it is difficult to conceive of a petrologic process that would produce rhyolitic rocks with "seawater-like" REE patterns similar to La Muralla groundwaters (Fig. 3). Consequently, although the Cuatralba Ignimbrite is the most likely source of REEs to La Muralla groundwaters, it is doubtful that these groundwaters inherited their "seawater-like" REE patterns directly from the Cuatralba Ignimbrite, or other rocks in the study region.

Alternatively, the HREE enriched patterns of La Muralla groundwaters may reflect chemical reactions between the groundwaters and HREE-enriched minerals or amorphous phases within the Cuatralba Ignimbrite, or secondary minerals/amorphous phases occurring along fractures in the aquifer. Primary minerals characteristic of felsic volcanic or plutonic rocks that have "seawater-like" REE patterns have not be identified, although the common accessory mineral zircon can exhibit substantial enrichments in HREEs relative to chondrite (e.g., Gromet and Silver, 1983; Fujimaki, 1986; Braun et al., 1993). However, zircon is not likely to be easily weathered by low-temperature aqueous solutions such as La Muralla groundwaters. On the other hand, if the secondary calcite within the Cuatralba Ignimbrite aquifer system (e.g., CEASG, 1999) have similar REE patterns to those from southern Nevada (e.g., Vaniman and Chipera, 1996), then it is possible that their dissolution by La Muralla groundwaters contributes to the unique REE patterns of these groundwaters, especially in terms of the Ce anomalies (Fig. 3). La Muralla groundwaters are all close to saturation with respect to calcite. Nonetheless, it is important to point out that the shale-normalized REE patterns for the secondary calcite studied by Vaniman and Chipera (1996; not shown) are substantially flatter than those of La Muralla groundwaters. For example, shale-normalized Yb/Tb ratios for secondary calcite studied by Vaniman and Chipera (1996) range from 0.37 to 2.39, with a mean (± standard deviation) of 1.05 ± 0.46 ($n = 59$). On the other hand, shale-normalized Yb/Tb ratios for La Muralla groundwaters range from 1.85 to 6.55, with a mean (± standard deviation) of 4.16 ± 1.19 (Table 2). Consequently, the greater HREE enrichments of La Muralla groundwater samples suggest that other processes besides quantitative dissolution of secondary calcite controls their REE fractionation patterns.

Instead, and as is the case for seawater, the HREE enriched patterns of La Muralla groundwaters are strongly suggestive of control by solution and surface complexation reactions within the aquifer system. The HREE enrichment characteristic of seawater, for example, are thought to reflect both solution complexation with carbonate ions, which acts to stabilize HREEs in solution relative to LREEs, and scavenging/adsorption by oxide- and/or organic material-coated detrital particles that preferentially remove LREEs from solution (e.g., Elderfield and Greaves, 1982; DeBaar et al., 1985; Elderfield, 1988; Byrne and Kim, 1990; Koeppenkastrop and DeCarlo, 1992, 1993; Sholkovitz et al., 1994; Byrne and Sholkovitz, 1996; Byrne and Liu, 1998). In the case of La Muralla groundwaters, the fractionating effect of solution and surface complexation reactions could conceivably occur at sites of active chemical weathering, as well as along groundwater flow paths. For example, incongruent silicate weathering involving feldspars and/or silicate glass within the Cuatralba Ignimbrite may contribute to the "seawater-like" REE patterns of La Muralla groundwaters by enriching the solutions in HREEs relative to the residual (i.e., clay) weathering products. Many investigators have shown that weathered residual materials are typically enriched in LREEs and depleted in HREEs compared to the parent rock owing to preferential capture of LREEs by secondary clay minerals formed during weathering, along with selected removal of HREEs from the site of active weathering by solution complexation with inorganic and/or organic ligands present in the weathering solutions (e.g., Ronov et al., 1967; Nesbitt, 1979; Duddy, 1980; Schau and Henderson, 1983; Braun et al., 1990, 1993). Furthermore, the formation of strong solution complexes could also inhibit the sorption of HREEs, relative to the LREEs, to aquifer surface sites along groundwater flow paths, and thus enhance the development of HREE enriched groundwaters (Johannesson et al., 1999).

The REEs are predicted to principally be complexed with carbonate ions in La Muralla groundwaters (Figs. 7, 8). Many other investigators have demonstrated that REE carbonate complexes dominate REE speciation in natural waters of neutral to high pH (e.g., Michard and Albarède, 1986; Michard et al., 1987; Wood, 1990; Lee and Byrne,

1992; Millero, 1992; Leybourne et al., 2000; Négrel et al., 2000). Moreover, the solution complexation model predicts that LREEs occur chiefly as positively charged, carbonato complexes ($LnCO_3^+$) in La Muralla groundwaters, whereas HREEs occur predominantly as negatively charged, dicarbonato complexes (i.e., $Ln(CO_3)_2^-$; Figs. 7, 8). A relatively small percentage of La, Ce, and Pr are also predicted to occur as positively charged free metal ions (i.e., Ln^{3+}; Fig. 7). Because LREEs primaily occur as positively charged, carbonato complexes and free metal ions in circumneutral pH La Muralla groundwaters, they ought to exhibit a greater affinity to sorb to the predominantly negatively charged, aquifer surface sites (e.g., Domenico and Schwartz, 1998) than the negatively charged, HREE - dicarbonato complexes. Benedict et al. (1997) measured solid-liquid partition coefficients (K_d = $[REE]_{solid\ phase}$ ($\mu g/g$)/ $[REE]_{liquid\ phase}$ ($\mu g/g$)) for an alluvial aquifer composed chiefly of fragments of rhyolitic volcanic rocks, and found that log K_d ranged from 3.26 for La to 1.9 for Tm (mean log K_d = 2.6), and that, in general, K_d decreased with increasing atomic number across the lanthanide series. The solid-liquid partition coefficients demonstrate that REEs are strongly complexed to surface sites within the alluvial rhyolitic aquifer, and that LREEs exhibit greater tendencies to adsorb to aquifer surface sites than HREEs (Benedict et al., 1997). Owing to the compositional similarities between the Cuatralba Ignimbrite and rhyolitic volcanic rock fragments of the alluvial aquifer studied by Benedict et al. (1997), it is reasonable to expect that REE solid-liquid partitioning coefficients specific to the Cuatralba Ignimbrite - La Muralla groundwater system should exhibit a similar decrease with increasing atomic number. In fact, a decrease in the solid-liquid partition coefficients with increasing atomic number across the lanthanide series is consistent with the predicted solution complexation of REEs in La Muralla groundwaters (Figs. 7, 8), because in the absence of inorganic and organic complexing ligands the tendency for REEs to adsorb to surfaces increases with increasing atomic number (Aargaard, 1974; Roaldset, 1974). Therefore, we suggest that the large HREE enrichments that characterize La Muralla groundwaters are principally the result of differential solution complexation across the lanthanide series and the effect that solution complexation has on REE adsorption across the series. Specifically, LREEs exhibit a greater tendency to sorb to aquifer surface sites owing to their occurrence as positively charged, carbonato complexes and free metal ions in solution, whereas HREEs exhibit greater affinities to remain in solution because of their occurrence as negatively charged, dicarbonato complexes.

5.2. CERIUM ANOMALIES

La Muralla groundwaters exhibit negative Ce anomalies when normalized to Average Shale as well as volcanic rocks of the Trans-Méxican Volcanic Belt (Table 2; Figs. 3 - 6). These negative Ce anomalies may reflect oxidative processes occurring within the aquifer, whereby Ce^{3+} is oxidized to less soluble Ce^{4+} (e.g., Braun et al., 1990; DeCarlo et al., 1998; Dia et al., 2000; Leybourne et al., 2000). Alternatively, they may reflect a signature inherited by reactions with the aquifer rocks (e.g., Smedley, 1991). For example, groundwaters from aquifers composed of Ce-depleted marine-derived carbonate rocks appear to obtain negative Ce-anomalies from the aquifer rocks (Halicz et al., 1999; Johannesson et al., 2000). Negative Ce anomalies, however, are not common to rhyolitic and other volcanic rocks of the Trans-Méxican Volcanic Belt (e.g.,

Fig. 4), or for that matter, upper continental crustal (i.e., silicate) rocks, in general (Taylor and McLennan, 1985). Therefore, it is difficult to ascribe a rock source to the shale-normalized and volcanic rock-normalized, negative Ce anomalies reported for La Muralla groundwaters (Figs. 3 - 6).

Another possible explanation is that the Ce anomalies of La Muralla groundwaters reflect solubility differences of Ce redox species related to pH conditions in the aquifer system. For example, because Ce^{3+} is more stable in low pH solutions (Brookins, 1988; Smedley, 1991), the Ce anomaly should become more negative (i.e., larger negative Ce anomalies) as pH increases. In fact, linear correlation coefficients suggest that a weak, inverse relationship exists between pH and the Ce anomalies of La Muralla groundwaters (e.g., $Ce/Ce*_{SN}$ vs. pH, r = -0.39; $Ce/Ce*_{RN}$ vs. pH, r = -0.39). However, these apparent correlations are not statistically significant (Davis, 1986). It is also possible that the Ce anomalies of La Muralla groundwaters reflect an increase in Ce oxidation with increasing pH (e.g., Goldstein and Jacobsen, 1988; Elderfield et al., 1990; Banks et al., 1999), although again, the lack of a statistically significant relationship between pH and Ce/Ce* does not support this explanation.

Elimination of rock sources and pH related solubility differences leaves oxidative processes as the most likely explanation for the negative Ce anomalies of La Muralla groundwaters. Although dissolved O_2 data for La Muralla groundwaters do not exist, the lack of evidence for active sulfide precipitation and general high quality (i.e., potability) of these waters lend qualitative support for oxic conditions in the aquifer. If La Muralla groundwaters are indeed oxic, then both biotic and abiotic oxidation of Ce^{3+} to Ce^{4+} could be responsible for the Ce depletions in these groundwaters. In the marine environment, Ce oxidation is known to occur by biologically mediated processes as well as abiotic scavenging (Moffett, 1990, 1994; Sholkovitz et al., 1994; DeCarlo et al., 1998). In the case of abiotic scavenging, because Ce exhibits greater sorption to some mineral surfaces (e.g., vernadite; δ- MnO_2) compared to other REEs, it is preferentially removed from seawater (Koeppenkastrop and DeCarlo, 1992; DeCarlo et al., 1998). Recent REE sorption studies indicate that Ce can also exhibit preferential adsorption to some aquifer substrates compared to other REEs (Benedict et al., 1997). Consequently, the negative Ce anomalies reported for La Muralla groundwaters may reflect greater adsorption affinity of oxidized Ce to aquifer surface sites or suspended particles relative to the other REEs. Finally, the anomalous behavior of Ce with respect to fluoride complexation recently reported by Schijf and Byrne (1999), led them to suggest that Ce anomalies in some natural waters may reflect complexation reactions as well as redox processes. It is conceivable that future measurements of REE stability constants with other ligands will also reveal anomalies for Ce that could not be resolved using linear free energy relationships (Lee and Byrne, 1992, 1993; Schijf and Byrne, 1999).

5.3. CONCEPTUAL MODEL OF GROUNDWATER REE PATTERN EVOLUTION

It is well known that the pH of meteoric (i.e., "pure") water, in equilibrium with atmospheric CO_2, is on the order of 5.67 (at $25°C$), whereas within the soil zone, groundwater pH may be as low as 4 to 5 (e.g., Nesbitt, 1979; Drever, 1997). Under these acidic conditions, which are likely to occur in recharge zones, the aquifer materials (i.e., rocks, minerals) will be susceptible to chemical weathering, resulting in release of REEs to the aqueous phase (Nesbitt, 1979; Duddy, 1980; Schau and Henderson, 1983; Braun et al., 1993). Inorganic and organic ligands will typically be protonated in these mildly acidic waters, as will many aquifer surface sites (e.g., Davis and Kent, 1990). Because REEs will be in the form of hydrous, free metal ions, or other positively charged solution (e.g., carbonato) complexes in these waters (e.g., Fig. 9a-c), they will exhibit relatively low sorption affinities for aquifer surface sites. Consequently, it is reasonable to expect that mildly acidic groundwaters near recharge zones may inherit and preserve rock-like REE patterns during initial contact with aquifer rocks. Indeed, dilute mildly acidic groundwaters (i.e., pH ≤ 7) from regions proximal to suspected recharge zones typically have REE patterns that closely resemble those of the aquifer rocks (e.g., Smedley, 1991; Halicz et al., 1999).

On the other hand, some groundwaters with pH greater than 7, but less than 9, including La Muralla groundwaters (Fig. 5), and those from Oasis Valley, Nevada ($7.65 \leq pH \leq 8.4$; Johannesson et al., 1999) and the Battleford Till, Saskatchewan, Canada ($7.04 \leq pH \leq 8.01$; Johannesson and Hendry, 2000), have REE patterns that are strongly fractioned (i.e., HREE enriched) with respect to the aquifer materials. These groundwaters can also be considered "mature" in terms of their distance from apparent recharge zones, and/or their age. For example, La Muralla groundwaters are thought to be recharged 40 to 50 km to the northeast, within the Sierra de Guanajuato (Fig. 1; CEASG, 1999; Cortés et al., 2000; Robles-Camacho and Armienta, 2000), whereas Oasis Valley groundwaters are recharged at least 40 km, and possibly as much as 250 km, north of the valley (Blankennagel and Weir, 1973; Davisson et al., 1999). Battleford Till groundwaters, however, are "mature" in terms of their age (e.g., 20 ka - 1.9 Ma BP; Hendry and Wassenaar, 1999; Hendry et al., 2000). Speciation modeling indicates that LREEs and HREEs occur as profoundly different species in these HREE enriched groundwaters, with LREEs predominantly in the form of positively charged, carbonato complexes, and to a lesser extent free metal ions and sulfate complexes, and HREEs occurring chiefly as negatively charged, dicarbonato complexes (Figs. 7, 9e, f; Johannesson et al., 1999; Johannesson and Hendry, 2000).

The differences in REE patterns and speciation between groundwaters collected near recharge zones and those further down gradient, suggest: (1) that REE concentrations and fractionation patterns evolve along groundwater flow paths; and (2) that the changes in REE concentrations and fractionation patterns are chiefly controlled by the effects of solution and surface complexation reactions along the flow path. Specifically, as mildly acidic recharge waters with aquifer-rock-like REE patterns react with the aquifer rocks, bicarbonate ions are produced, and pH rises (e.g., Garrels and MacKenzie, 1967). As groundwater pH rises, the relative distributions of inorganic ligands that can form solution complexes with REEs, tends to change (e.g., HCO_3^- increases relative to SO_4^{2-}). Some of the bicarbonate ions formed by rock weathering and dissociation of carbonic acid further dissociates to carbonate ions. Despite the low concentrations of carbonate ions in circumneutral pH groundwater, they form strong complexes with REEs owing to the large stability constants of REE carbonate complexes (Lee and Bryne, 1992, 1993). For groundwaters of pH ~ 7, positively charged, carbonato complexes (i.e., $LnCO_3^+$) dominate (Fig. 9 c, d). However, as groundwater pH rises above 7, negatively charged, dicarbonato complexes (i.e., $Ln(CO_3)_2^-$) account for progressively more of the HREEs than carbonato complexes (Fig. 9 d, e), and for groundwaters with pH > 8, the negatively charged, REE - dicarbonato complex is the dominant form for all REEs in solution (Fig. 6 f; Johannesson et al., 1999; Johannesson and Hendry, 2000). Moreover, cation exchange reactions involving, for example, Ca^{2+} for Na^+, that occur along groundwater flow paths, will lead to an increase in free inorganic ligand concentrations (e.g., HCO_3^- and CO_3^{2-}) as less Ca^{2+} and Mg^{2+} are available to form ion pairs with these ligands. The resulting higher $[CO_3^{2-}]_F$ concentrations will subsequently be available for complexation with REEs.

In addition, as groundwater pH rises, aquifer surface sites that were initially positively charged become negatively charged (e.g., Drever, 1997; Domenico and Schwartz, 1998; and references therein). The pH at which the charge on aquifer surface sites changes from positive to negative depends upon the isoelectric points of the aquifer solids. Isoelectric points for kaolinite and montmorillonite are estimated to be roughly 3.5 and < 2.5, respectively, whereas for Fe oxides/oxyhydroxides, the range is generally between 6 and 7 (Drever, 1997; Domenico and Schwartz, 1998). Thus, positively charged REE solution species (e.g., Ln^{3+}, $LnCO_3^+$, $LnSO_4^+$) that are relatively stable in mildly acidic groundwaters near recharge zones, will exhibit progressively greater adsorption affinities for aquifer surface sites as groundwater pH rises to values above 7 along flow paths (e.g., Fig. 9d-f). On the other hand, negatively charged, dicarbonato complexes should inhibit REE sorption to aquifer surface sites as groundwater pH rises above 7. Consequently, the HREE enriched patterns of "mature", down gradient groundwaters (i.e., La Muralla, Oasis Valley, Battleford Till) likely reflect the combined effects of solution complexation, which stabilizes HREEs in solution, and adsorptive processes that preferentially remove positively charged LREE species. Therefore, we suggest that geochemical reactions occurring along groundwater flow paths that affect the distribution and concentrations of complexing ligands, the characteristics of aquifer surface adsorption sites, and groundwater pH, play a critical role in controlling the concentrations and fractionation patterns REEs in aquifers. Future investigations are planned to quantify changes in REE concentrations and fractionation patterns along groundwater flow paths.

6. Conclusions

Rare earth element patterns of La Muralla groundwaters are strongly fractionated with respect to the Average Shale composite, as well as local calc-alkaline volcanic rocks of the Trans-Mexican Volcanic Belt, including rhyolitic rocks similar to the aquifer rocks (i.e., Cuatralba Ignimbrite). Solution complexation modeling predicts that the REEs chiefly occur in solution as carbonate complexes. Moreover, the model predicts that the relative percentage of each REE that occurs in solution as positively charged, carbonato complexes ($LnCO_3^+$), free metal ions (Ln^{3+}), and negatively charged, dicarbonato complexes ($Ln(CO_3)_2^-$), varies across the lanthanide series, and with pH. In general, LREEs are predicted to occur in La Muralla groundwaters as positively charged, carbonato complexes, whereas the HREEs are expected to dominantly be in solution as negatively charged, dicarbonato complexes. Furthermore, the point along the REE series where carbonato and dicarbonato complexes are predicted to predominate varies with pH such that dicarbonato complexes become increasingly important for progressively lighter REEs as pH rises. The solution complexation model suggests that the HREE enriched patterns of La Muralla groundwaters reflect a greater tendency for LREEs to sorb to aquifer surface sites owing to their occurrence as positively charged, carbonato complexes and free metal ions in solution. On the other hand, the model indicates that HREEs exhibit greater affinities to remain in solution in La Muralla groundwaters owing to their occurrence as negatively charged, dicarbonato complexes.

Acknowledgments

We are grateful to Sr. José Luis Cruz José, project director at the Comision Estatal de Agua y Saneamiento de Guanajuato, for his interest in this project, and for providing us with access to the La Muralla wells, field assistance, and financial support (CEAS-XXVI-OD-UNAM-99-088) for much of this work. We thank E. Y. Graham of the University of Alabama for the ICP-MS analyses which were performed under NSF grant INT-9912159 to KHJ. We thank Dr. L. Douglas James of the Hydrologic Sciences division of NSF for partial support for this work through grant EAR-0303761 to KHJ.

References

Aagaard, P. 1974 Rare earth element adsorption on clay minerals. *Bull. Group. Franc. Argiles,* **26**, 193-199.

Banks, D., Hall, G., Reimann, C., and Siewers, U. 1999 Distribution of rare earth elements in crystalline bedrock groundwaters: Oslo and Bergen regions, Norway. *Applied Geochemistry,* **14**, 27-39.

Banner, J. L., Wasserburg, G. J., Dobson, P. F., Carpenter, A. B., and Moore, C. H. 1989 Isotopic and trace element constraints on the origin and evolution of saline groundwaters from central Missouri. *Geochimica et Cosmochimica Acta,* **53**, 383-398.

Benedict, F. C., Jr., DeCarlo, E. H., and Roth, M. 1999 Kinetics and thermodynamics of dissolved rare earth element uptake by alluvial materials from the Nevada Test Site, southern Nevada, U.S.A. In: X. Xuejin (ed.) Geochemistry, Vol. 19, Proc. Int'l. Geol. Congr., VSP (Utrecht), pp. 173-188.

Bertram, C. J. and Elderfield, H. 1993 The geochemical balance of the rare earth elements and neodymium isotopes in the oceans. *Geochimica et Cosmochimica Acta,* **57**, 1957-1986.

Blankennagel, R. K. and Weir, J. E., Jr. 1973 Geohydrology of the eastern part of Pahute Mesa, Nevada Test Site, Nye County, Nevada. *USGS. Professional Paper,* **712-B**.

Braun, J. J., Pagel, M., Muller, J. P., Bilong, P., Michard, A., and Guillet, B. 1990 Cerium anomalies in lateritic profiles. *Geochimica et Cosmochimica Acta,* **54**, 597-605.

Braun, J. J., Pagel, M., Herbillon, A., and Rosen, C. 1993 Mobilization and redistribution of REEs and thorium in a syenitic lateritic profile: A mass balance study. *Geochimica et Cosmochimica Acta,* **57**, 4419-4434.

Brookins, D. G. 1986 Natural analogues for radwaste disposal: elemental migration in igneous contact zones. *Chemical Geology,* **55**, 337-344.

Bruno, J. 1997 Trace element modelling. In: I. Grenthe and I Puidomenech (eds.) Modelling in Aquatic Chemistry, OECD Nuclear Energy Agency (Paris), pp. 593-621.

Byrne, R. H. and Kim, K. -H. 1990 Rare earth element scavenging in seawater. *Geochimica et Cosmochimica Acta,* **54**, 2645-2656.

Byrne, R. H. and Sholkovitz, E. R. 1996 Marine chemistry and geochemistry of the lanthanides. In: K. A. Gschneider, Jr. and L. Eyring (eds.) Handbook on the Physics and Chemistry of the Rare Earths, Vol. 23, Elsevier (Amsterdam), pp. 497-593.

Byrne, R. H. and Liu, X. 1998 A coupled riverine-marine fractionation model for dissolved rare earths and yttrium. *Aquatic Geochem*istry, **4**, 103-121.

Cameron, K. L. 1984 Bishop Tuff revisited: New rare earth element data consistent with crystal fractionation. *Science,* **224**, 1338-1340.

Choppin, G. R. 1983 Comparison of solution chemistry of the actinides and the lanthanides. *Journal of the Less-Common Metals,* **93**, 232-330.

CEASG 1999 Estudio isotópico para la caracterización del aqua subterránea en la zona de La Muralla, Guanajuato. Comision Estatal de Agua y Saneamiento de Guanajuato, Contrato CEAS-XXVI-OD-UNAM-99-088, 74 pp.

Cortés, A., Ramos, A., Durazo, J., and Johannesson, K. 2000 Isotopic conceptual model of La Muralla, Guanajuato, Mexico. *EOS Transactions of the American Geophysical Union,* **81**, F476.

Davis, J. A. and Kent, D. B. 1990 Surface complexation modeling in aqueous geochemistry. In: M. F. Hochella, Jr. and A. F. White (eds.), Mineral-Water Interface Geochemistry, Reviews in Mineralogy, vol. 23. Mineralogical Society of America (Washington, DC), pp. 177-260.

Davis, J. C. 1986 Statistics and Data Analysis in Geology, 2nd Edition. John Wiley and Sons, New York, 646 pp.

Davisson, M. L., Smith, D. K., Keneally, J., and Rose, T. P. 1999 Isotope hydrology of southern Nevada groundwater: Stable isotopes and radiocarbon. *Water Resources Research,* **35**, 279-294.

DeBaar, H. J. W., Bacon, M. P., Brewer, P. G., and Bruland, K. W. 1985 Rare earth elements in the Pacific and Atlantic Oceans. *Geochimica et Cosmochimica Acta,* **49**, 1943-1959.

DeCarlo, E. H., Wen, X. -Y., and Irving, M. 1998 The influence of redox reactions on the uptake of dissolved Ce by suspended Fe and Mn oxide particles. *Aquatic Geochem*istry, **3**, 357-389.

Dia, A., Grua, G., Olivié-Lauquet, G., Riou, C., Molénat, J., and Curmi, P. 2000 The distribution of rare earth elements in groundwaters: Assessing the role of source-rock composition, redox changes, and colloidal particles. *Geochimica et Cosmochimica Acta,* **64**, 4131-4152.

Domenico, P. A. and Schwartz, F. W. 1998 Physical and Chemical Hydrogeology. 2nd ed. John Wiley and Sons, New York, 506 pp.

Drever, J. I. 1997 The Geochemistry of Natural Waters: Surface and Groundwater Environments. 3rd ed., Prentice Hall, Upper Saddle River, NJ, 436 pp.

Duddy, I. R. 1980 Redistribution and fractionation of the rare-earth and other elements in a weathering profile. *Chemical Geology*, **30**, 363-381.

Elderfield, H. 1988 The oceanic chemistry of the rare-earth elements. *Philosphical Transactions of the Royal Society of London,* **A325**, 105-126.

Elderfield, H. and Greaves, M. J. 1982 The rare earth elements in seawater. *Nature*, **296**, 214-219.

Elderfield, H. and Greaves, M. J. 1983 Determination of the rare earth elements in sea water. In: C. S. Wong, E. Boyle, K. W. Bruland, J. D. Burton, and E. D. Goldberg (eds.) Trace Metals in Sea Water. Plenum Press (New York), pp. 427-445.

Elderfield, H., Upstill-Goddard, R., and Sholkovitz E. R. 1990 The rare earth elements in rivers, estuaries, and coastal seas and their significance to the composition of ocean waters. *Geochimica et Cosmochimica Acta,* **54**, 971-991.

Fee, J. A., Gaudette, H. E., Lyons, W. B., and Long, D. T. 1992 Rare earth element distribution in the Lake Tyrrell groundwaters, Victoria, Australia. *Chemical Geology*, **96**, 67-93.

Ferrari, L., Garduño, V. H., Innocenti, F., Manetti, P., Pasquarè, G., and Vaggelli, G. 1994 A widespread mafic volcanic unit at the base of the Mexican Volcanic Belt between Guadalajara and Querétaro. *Geofísica Internacional,* **33**, 107-123.

Ferriz, H. and Mahood, G. A. 1987 Strong compositional zonation in a silicic magmatic system: Los Humeros, Mexican Neovolcanic Belt. *Journal of Petrology,* **28**, 171-209.

Fujimaki, H. 1986 Partition coefficients of Hf, Zr, and REE between zircon, apatite, and liquid. *Contributions to Mineralogy and Petrology,* **94**, 42-45.

Garrels, R. M. and MacKenzie, F. T. 1967 Origin of chemical composition of some springs and lakes. In: W. Stumm (ed.) Equilibrium Concepts in Natural Water Systems, American Chemical Society, Advances in Chemistry Series, 67, (Washington, DC), pp. 222-242.

Glaus, M. A., Hummel, W., and Van Loon, L. R. 2000 Trace metal-humate interactions. I. Experimental determination of conditional stability constants. *Applied Geochemistry*, **15**, 953-973.

Goldstein, S. J. and Jacobsen, S. B. 1988 Rare earth elements in river waters. *Earth and Planetary Science Letters,* **89**, 35-47.

Gómez-Tuena, A. and Carrasco-Núñez, G. 2000 Cerro Grande volcano: the evolution of a Miocene stratocone in the early Trans-Mexican Volcanic Belt. *Tectonophysics,* **318**, 249-280.

Gosselin, D. G., Smith, M. R., Lepel, E. A., and Laul, J. C. 1992 Rare earth elements in chloride-rich groundwaters, Palo Duro Basin, Texas, USA. *Geochimica et Cosmochimica Acta,* **56**, 1495-1505.

Graham, E. Y., Ramsey, L. A., Lyons, W. B., and Welch, K. A. 1996 Determination of rare earth elements in Antarctic lakes and streams of varying ionic strengths. In: G. Holland and S. D. Tanner (eds.) Plasma Source Mass Spectrometry: Developments and Applications. The Royal Society of Chemistry (London), pp. 253-262.

Greaves, M. J., Elderfield, H., and Klinkhammer, G. P. 1989 Determination of the rare earth elements in natural waters by isotope-dilution mass spectrometry. *Analytica Chimica Acta*, **218**, 265-280.

Gromet, L. P. and Silver, L. T. 1983 Rare earth element distributions among minerals in a granodiorite and the petrogenetic implications. *Geochimica et Cosmochimica Acta*, **47**, 925-939.

Guo, C. 1996 Determination of fifty-six elements in three distinct types of geological materials by inductively coupled plasma-mass spectrometry. Unpublished M. S. thesis, University of Nevada, Las Vegas, 68 p.

Halicz, L., Segal, I., and Yoffe, O. 1999 Direct REE determination in fresh waters using ultrasonic nebulization ICP-MS. *Journal of Analytical Atomic Spectrom*etry, **14**, 1579-1581.

Hasenaka, T. and Carmichael, I. S. E. 1987 The cinder cones of Michoacán - Guanajuato, central Mexico: Petrology and Chemistry. *Journal of Petrology*, **28**, 241-269.

Hendry, M. J. and Wassenaar, L. I. 1999 Implications of transport of ⊠D in pore waters for groundwater flow and the timing of geologic events in a thick aquitard system. *Water Resources Research*, **35**, 1751-1760.

Hendry, M. J., Wassenaar, L. I., and Kotzer, T. 2000 Chloride and chlorine isotopes (^{35}Cl, ^{36}Cl, ^{37}Cl) as tracers of solute migration in a clay-rich aquitard system. *Water Resources Research*, **36**, 285-296.

Hodge, V. F., Stetzenbach, K. J., and Johannesson, K. H. 1998 Similarities in the chemical composition of carbonate groundwaters and seawater. *Environmental Science and Technology*, **32**, 2481-2486.

Hummel, W., Glaus, M. A., and Van Loon, L. R. 2000 Trace metal-humate interactions. II. The "conservative roof" model and its applications. *Applied Geochemistry*, **15**, 975-1001.

Johannesson, K. H. and Lyons, W. B. 1995 Rare-earth element geochemistry of Colour Lake, an acidic freshwater lake on Axel Heiberg Island, Northwest Territories, Canada. *Chemical Geology*, **119**, 209-223.

Johannesson, K. H. and Hendry, M. J. 2000 Rare earth element geochemistry of groundwaters from a thick till and clay-rich aquitard sequence, Saskatchewan, Canada. *Geochimica et Cosmochimica Acta*, **64**, 1493-1509.

Johannesson, K. H., Stetzenbach, K. J., Hodg,e V. F., Kreamer, D. K., and Zhou, X. 1997 Delineation of ground-water flow systems in the southern Great Basin using aqueous rare earth element distributions. *Ground Water*, **35**, 807- 819.

Johannesson, K. H., Farnham, I. M., Guo, C., and Stetzenbach ,K. J. 1999)Rare earth element fractionation and concentration variations along a groundwater flow path within a shallow, basin-fill aquifer, southern Nevada, USA. *Geochimica et Cosmochimica Acta*, **63**, 2697-2708.

Johannesson, K. H., Zhou, X., Guo, C., Stetzenbach, K. J., and Hodge, V. F. 2000 Origin of rare earth element signatures of groundwaters of circumneutral pH from southern Nevada and eastern California, USA. *Chemical Geology*, **164**, 239-257.

Kelleher, P. C. and Cameron, K. L. 1990 The geochemistry of the Mono Craters-Mono Lake islands volcanic complex, eastern California. *Journal of Geophysical Research*, **95**, 17,643-17,659.

Klinkhammer, G., Elderfield, H., and Hudson, A. 1983 Rare earth elements in seawater near hydrothermal vents. *Nature*, **305**, 185-188.

Klinkhammer, G., German, C. R., Elderfield, H., Greaves, M. J., and Mitra, A. 1994

Rare earth elements in hydrothermal fluids and plume particulates by inductively coupled plasma mass spectrometry. *Marine Chemistr,y* **45**, 170-186.

Koeppenkastrop, D. and De Carlo, E. H. 1992 Sorption of rare-earth elements from seawater onto synthetic mineral particles: An experimental approach. *Chemical Geology,* **95**, 251-263.

Koeppenkastrop,D. and De Carlo,E. H. 1993 Uptake of rare earth elements from solution by metal oxides. *Environmental Science and Technology,* **27**, 1796-1802.

Krauskopf, K. B. 1986 Thorium and rare-earth metals as analogues for actinide elements. *Chemical Geology,* **55**, 323-35.

Lange, R. A. and Carmichael, I. S. E. 1990 Hydrous basaltic andesites associated with minette and related lavas in western Mexico. *Journal of Petrology,* **31**, 1225-1259.

Lead, J. R., Hamilton-Taylor, J., Peters, A., Reiner, S., and Tipping, E. 1998 Europium binding by fulvic acids. *Analytica Chimimica Acta,* **369**, 171-180.

Lee, J. H. and Byrne, R. H. 1992 Examination of comparative rare earth element complexation behavior using linear free-energy relationships. *Geochimica et Cosmochimica Acta,* **56**, 1127-1137.

Lee, J. H. and Byrne, R. H. 1993 Complexation of trivalent rare earth elements (Ce, Eu, Gd, Tb, Yb) with carbonate ions. *Geochimica et Cosmochimica Acta,* **57**, 295-302.

Leybourne, M. I., Goodfellow, W. D., Boyle, D. R., and Hall, G. M. 2000 Rapid development of negative Ce anomalies in surface waters and contrasting REE patterns in groundwaters associated with Zn-Pb massive sulfide deposits. *Applied Geochemistry,* **15**, 695-723.

Luhr, J. F. 2000 The geology and petrology of Volcán San Juan (Nayarit, México) and the compositionally zoned Tepic Pumice. *Journal of Volcanology and Geothermal Research,* **95**, 109-156.

Luhr, J. F. and Carmichael, I. S. E. 1980 The Colima Volcanic Complex, Mexico I. Post-caldera andesites from Volcán Colima. *Contributions to Mineralogy and Petrology,* **71**, 343-372.

Luhr, J. F. and Carmichael, I. S. E. 1981 The Colima Volcanic Complex, Mexico: Part II. Late-Quaternary cinder cones. *Contributions to Mineralogy and Petrology,* **76**, 127-147.

Luhr, J. F. and Carmichael, I. S. E. 1985 Jorullo Volcano, Michoacán, Mexico (1759-1774): The earliest stages of fractionation in calc-alkaline magmas. *Contributions to Mineralogy and Petrology,* **90**, 142-161.

Luhr, J. F., Allan, J. F., Carmichael, I. S. E., Nelson, S. A., and Hasenaka, T. 1989 Primitive calc-alkaline and alkaline rock types from the western Mexican Volcanic Belt. *Journal of Geophysical Research,* **94**, 4515-4530.

Michard, A. and Albarède, F. 1986 The REE content of some hydrothermal fluids. *Chemical Geology,* **55**, 51-60.

Michard, A., Beaucaire, C., and Michard, G. 1987 Uranium and rare earth elements in CO_2-rich waters from Vals-les-Bains (France). *Geochimica et Cosmochimica Acta,* **51**, 901-909.

Millero, F. J. 1992 Stability constants for the formation of rare earth inorganic complexes as a function of ionic strength. *Geochimica et Cosmochimica Acta*, **56**, 3123-3132.

Moffett, J. W. 1990 Microbially mediated cerium oxidation in seawater. *Nature*, **345**, 421- 423.

Moffett, J. W. 1994 The relationship between cerium and manganese oxidation in the marine environment. *Limnology and Oceanography*, **39**, 1309-1318.

Möller, P. 2000 Rare earth elements and yttrium as geochemical indicators of the source of mineral and thermal waters. In: I. Stober and K. Bucher (eds.) Hydrogeology of Crystalline Rocks, Kluwer Academic Publishers (Dordrecht), pp. 227-246.

Möller, P. and Bau, M. 1993 Rare-earth patterns with positive cerium anomalies in alkaline lake waters from Lake Van, Turkey. *Earth and Planetary Science Letters*, **117**, 671-676.

Negendank, J. F. W., Emmermann, R., Krawczyk, R., Mooser, F., Tobschall, H., and Werle, D. 1985 Geological and geochemical investigations on the eastern Trans-Mexican Volcanic Belt. *Geofísica Internacional*, **24**, 477-575.

Négrel, Ph., Guerrot, C., Cocherie, A., Azaroual, M., Brach, M., and Fouillac, Ch. 2000 Rare earth elements, neodymium and strontium isotopic systematics in mineral waters: evidence from the Massif Central, France. *Applied Geochemistry*, **15**, 1345-1367.

Nesbitt, H. W. 1979 Mobility and fractionation of rare earth elements during weathering of a granodiorite. *Nature*, **279**, 206-210.

Pal, S. 1972 Reconnaissance geochemistry of some rocks of the Guanajuato Mineral District, Mexico. *Geofísica Internacional*, **12**, 163- 199.

Parkhurst, D. L., Thorstenson, D. C., and Plummer, L. N. 1980 PHREEQE – A computer program for geochemical calculations. *USGS Water Resources Investigations Report* **80-96**.

Piepgras, D. J. and Jacobsen, S. B. 1992 The behavior of rare earth elements in seawater: Precise determination of variations in the North Pacific water column. *Geochimica et Cosmochimica Acta*, **56**, 1851-1862.

Piepgras, D. J. and Wasserburg, G. J. 1980 Neodymium isotopic variations in seawater. *Earth and Planetary Science Letters*, **50**, 128-138.

Piepgras, D. J. and Wasserburg, G. J. 1987 Rare earth element transport in the western North Atlantic inferred from Nd isotopic observations. *Geochimica et Cosmochimica Acta*, **51**, 1257-1271.

Quintera, O. 1986 Geología de los alrededores de Comalja de Corona, Edo de Jalisco. Resumen Primer Simposio de Geología Regional de México. Instituto de Geología, UNAM

Ramírez, A. G., Cortés, A. S., Ramos, J. A. L., Cruz, J. L. J., and Johannesson, K. H. 1999 Geologic and hydrodynamic conceptual model from northern Guanajuato, Mexico. *EOS Transactions of the American Geophysical Union*, **80**, F398.

Ramírez, A. G., Cortés, A., Ramos, A., Johannesson, K., and Cruz, J. L. 2000 Relationship between structural subsoil systems and the hydrodynamic functioning of La Muralla, Guanajuato State, Mexico. *Geological Society of America Annual Meeting Abstracts with Programs*, **32**, A-89.

Roaldset, E. 1974 Lanthanide distributions in clays. *Bull. Group. Franc. Argiles*, **26**,

201-209.

Robles-Camacho, J. and Armienta, M. A. 2000 Natural chromium contamination of groundwater at León Valley, México. *Journal of Geochemical Exploration,* **68**, 183-199.

Ronov, A. B., Balashov, Y. A., and Migdisov, A. A. 1967 Geochemistry of the rare earth elements in the sedimentary cycle. *Geochemistry International,* **4**, 1-17.

Roxburgh, I. S. 1987 Geology of High-Level Nuclear Waste Disposal. Chapman and Hall, London, 229 pp.

Schau, M. and Henderson, J. B. 1983 Archean chemical weathering at three locations on the Canadian Shield. *Precambrian Research,* **20**, 189-224.

Schijf, J. and Byrne, R. H. 1999 Determining stability constants for the mono- and difluoro- complexes of Y and the REE, using a cation-exchange resin and ICP-MS. *Polyhedron,* **18**, 2839-2844.

Schneider, D. L. and Palmieri, J. M. 1994 A method for the analysis of rare earth elements in natural waters by isotope dilution mass spectrometry. *Woods Hole Oceanographic Institution, Technical Report* WHOI-94-06, 39 pp.

Sholkovitz, E. R. 1988 Rare earth elements in the sediments of the North Atlantic Ocean, Amazon Delta, and East China Sea: Reinterpretation of terrigenous input patterns to the oceans. *American Journal of Science,* **288**, 236-281.

Sholkovitz, E. R. 1993 The geochemistry of the rare earth elements in the Amazon River estuary. *Geochimica et Cosmochimica Acta,* **57**, 2181-2190.

Sholkovitz, E. R. 1995 The aquatic chemistry of the rare earth elements in rivers and estuaries. *Aquatic Geochemistry,* **1**, 1-34.

Sholkovitz, E. R. and Szymczak, R. 2000 The estuarine chemistry of rare earth elements: comparison of the Amazon, Fly, Sepik and Gulf of Papua systems. *Earth and Planetary Science Letters,* **179**, 299-309.

Sholkovitz, E. R., Landing, W. M., and Lewis, B. L. 1994 Ocean particle chemistry: The fractionation of rare earth elements between suspended particles and seawater. *Geochimica et Cosmochimica Acta,* **58**, 1567-1579.

Silva, R. J. and Nitsche, H. 1995 Actinide Environmental Chemistry. *Radiochimica Acta,* **70/77**, 377-396.

Smedley, P. L. 1991 The geochemistry of rare earth elements in groundwaters from the Carnmenellis area, southwest England. *Geochimica et Cosmochimica Acta,* **55**, 2767-2779.

Stetzenbach, K. J., Amano, M., Kreamer, D. K., and Hodge, V. F. 1994 Testing the limits of ICP-MS: Determination of trace elements in ground water at the parts-per-trillion level. *Ground Water,* **32**, 976-985.

Stordal, M. C. and Wasserburg, G. J. 1986 Neodymium isotopic study of Baffin Bay water: sources of REE from very old terranes. *Earth and Planetary Science Letters*, **77**, 259-272.

Talavera, O., Ramírez, J., Guerrero, M. 1995 Petrology and geochemistry of the Teloloapan subterrane: a Lower Cretaceous evolved intra-oceanic island-arc. *Geofísica Internacional,* **34**, 3-22.

Tang, J. and Johannesson, K. H. 2003 Speciation of rare earth elements in natural terrestrial waters: Assessing the role of dissolved organic matter from the modeling approach. *Geochimica et Cosmochimica Acta,* **67**, 2321-2339.

Taylor, S. R. and McLennan, S. M. 1985 The Continental Crust: its Composition and

Evolution. Blackwell Scientific Publications, Oxford, 312 pp.

Tipping, E. 1993 Modelling the binding of europium and the actinides by humic substances. *Radiochimica Acta,* **62**, 141-152.

Tipping, E. 1998 Humic ion-binding model VI: An improved description of the interactions of protons and metal ions with humic substances. *Aquatic Geochemistry,* **4**, 3-48.

Tipping, E. and Hurley, M. A. 1992 A unifying model of cation binding by humic substances. *Geochimica et Cosmochimica Acta,* **56**, 3627-3641.

Vaniman, D. T. and Chipera, S. J. 1996 Paleotransport of lanthanides and strontium recorded in calcite compositions from tuffs at Yucca Mountain, Nevada, USA. *Geochimica et Cosmochimica Acta,* **60**, 4417-4433.

Verma, S. P. 1985 Mexican Volcanic Belt, Part 2. *Geofísica. Internacional,* **24**, 461-464.

Verma, S. P. 1999 Geochemistry of evolved magmas and their relationship to subduction-unrelated mafic volcanism at the volcanic front of the central Mexican Volcanic Belt. *Journal of Volcanology and Geothermal Research,* **93**, 151-171.

Verma, S. P. 2000 Geochemical evidence for a lithospheric source for magmas from Los Humeros caldera, Puebla, Mexico. *Chemical Geology,* **164**, 35-60.

Verma, S. P. and Luhr, J. F. 1993 Sr-Nd-Pb isotope and trace element geochemistry of calc-alkaline andesites from Volcán Colima, Mexico. *Geofísica. Internacional,* **32**, 617-631.

Verma, S. P., Lopez-Martinez, M., and Terrell, D. J. 1985 Geochemistry of Tertiary igneous rocks from the Arandas-Atotonilco area, northeast Jalisco, Mexico. *Geofísica. Internacional,* **24**, 31-45.

Wallace, P. and Carmichael, I. S. E. 1992 Alkaline and calc-alkaline lavas near Los Volcanes, Jalisco, Mexico: geochemical diversity and its significance in volcanic arcs. *Contributions to Mineralogy and Petrology,* **111**, 423-439.

Weisel, C. P., Duce, R. A., and Fasching, J. L. 1984 Determination of aluminum, lead, and vanadium in North Atlantic seawater after coprecipitation with ferric hydroxide. *Analytical Chemistry,* **56**, 1050-1052.

Welch, K. A., Lyons, W. B., Graham, E., Neumann, K., Thomas, J. M., and Mikesell, D. 1996 Determination of major element chemistry in terrestrial waters from Antarctica by ion chromatography. *Journal of Chromatography,* **A739**, 257-263.

Welch, S. A., Lyons, W. B., and Kling, C. A. 1990 A coprecipitation technique for determining trace metal concentrations in iron-rich saline solutions. *Environmental Technology Letters,* **11**, 141-144.

Wood, S. A. 1990 The aqueous geochemistry of the rare-earth elements and yttrium. 1. Review of the available low-temperature data for inorganic complexes and inorganic REE speciation in natural waters. *Chemical Geology,* **82**, 159-186.

Wood, S. A. 1993 The aqueous geochemistry of the rare earth elements: Critical stability constants for complexes with simple carboxylic acids at 25°C and 1 bar and their application to nuclear waste management. *Engineering Geology,* **34**, 229-259.

RARE EARTH ELEMENT CONCENTRATIONS, SPECIATION, AND FRACTIONATION ALONG GROUNDWATER FLOW PATHS: THE CARRIZO SAND (TEXAS) AND UPPER FLORIDAN AQUIFERS

JIANWU TANG[1] & KAREN H. JOHANNESSON[2]

[1]*Department of Ocean, Earth, and Atmospheric Sciences, Old Dominion University, Norfolk, Virginia 23529-0276, USA*
[2]*Department of Earth and Environmental Sciences, The University of Texas at Arlington, Arlington, TX 76019-0049, USA*

Abstract

Groundwater samples were collected in two different types of aquifers (i.e., Carrizo sand aquifer, Texas and Upper Floridan carbonate aquifer, west-central Florida) to study the concentration, speciation, and fractionation of rare earth elements (REE) along the groundwater flow path in each system. Major solutes and dissolved organic carbon (DOC) were also measured in these groundwaters. In the Carrizo aquifer, groundwaters in the recharge zone are chiefly Ca-Na-HCO_3-Cl type waters and shift into Na-HCO_3 waters with flow down-gradient. DOC is generally low (0.65 mg/L) along the flow path. In the Upper Floridan Aquifer, groundwaters are Ca (Mg)-HCO_3 waters at the recharge zone, shift into Ca-Mg-SO_4-HCO_3 waters at the mid-reaches of the flow path, and finally become Ca-Mg-SO_4 waters at the discharge zone. DOC is higher (0.64 – 2.29 mg/L) than in the Carrizo and initially increases along the flow path and then decreases down-gradient. Rare earth element concentrations generally decrease along the groundwater flow path in the Carrizo sand aquifer, whereas in the Upper Floridan aquifer, the Nd concentrations increase first and then slightly decrease, Gd concentrations are erratic, and Er tends to decrease along the groundwater flow path. Shale-normalized REE patterns are enriched in the HREEs for groundwaters from the recharge regions of both aquifers and tend to flatten with flow down-gradient. Speciation calculations predict that carbonato ($LnCO_3^+$) and/or dicarbonato complexes ($Ln(CO_3)_2^-$) are the main dissolved species for all REEs in groundwaters from the Carrizo sand aquifer, whereas organic complexes of the REEs (LnHM) are dominant in all but two furthest down-gradient groundwaters from the Upper Floridan aquifer. The variations of REE concentrations, speciation, and fractionation patterns along the flow paths reflect water-rock reactions (mainly adsorption and solution complexation) experienced by REEs in groundwaters from both aquifers. The solution complexation of REEs affects the REE concentrations and fractionation patterns in groundwaters along the flow path via the following two ways: (1) the complexation capacity of REEs with carbonate ions and organic ligands varies with increasing atomic number of REEs; and (2) different complexation forms (i.e., negative or positive) at different geochemical conditions of groundwaters (i.e., pH, alkalinity, and DOC concentrations).

K.H. Johannesson,(ed), Rare Earth Elements in Groundwater Flow Systems , 223-251.

1. Introduction

Rare earth elements (REE) are of interest to chemical hydrogeologists because of their potential as sensitive tracers for studying groundwater-aquifer rock interactions and, in some cases, for tracing groundwater flow (Johannesson et al., 1997a, b; Dia et al., 2000; Möller et al., 2000). Although great strides have been made towards understanding the oceanic, estuarine, and riverine geochemistry of REEs (Goldstein and Jacobsen, 1988a, b; Elderfield, 1988; Elderfield et al., 1990; Bertram and Elderfield, 1993; Sholkovitz, 1993, 1995), considerably less effort has been directed towards their study in groundwater systems (e.g., Banner et al., 1989; Smedley, 1991; Fee et al., 1992; Gosselin et al., 1992; Leybourne et al., 2000). It is generally agreed that one of the more significant discoveries resulting from previous investigations of REEs in groundwaters is the observed similarities between normalized groundwater and aquifer-rock REE patterns. Indeed, it is precisely this similarity between groundwater and associated aquifer-rock REE patterns that underscores the facility of these trace elements as tools for water-rock reaction studies (e.g., Banner et al., 1989). Remarkably, few studies have examined how REE concentrations and fractionation patterns vary along groundwater flow paths in real aquifers (e.g., Johannesson et al., 1999; Dia et al., 2000). Because pH, redox conditions, and the major solute composition of groundwater change and evolve along groundwater flow paths owing to chemical weathering reactions and microbial processes occurring in the aquifers (Plummer et al., 1983; Hamlin, 1988; Edmunds et al., 2003), it is expected that groundwater REE concentrations, and their relative distributions across the lanthanide series, will reflect these changes as the REEs are known to be strongly controlled directly, or indirectly, by pH, redox conditions, and aqueous complexation (Johannesson et al., 1999; Dia et al., 2000; Tang and Johannesson, 2003).

At least two previous published studies (Banner et al., 1989; Johannesson et al., 1999) have attempted to study or document changes in REE concentrations along groundwater flow paths. In the earlier study, Banner et al. (1989) argued that the REE concentrations as well as the Nd isotope signatures of groundwaters reflect the most recent stages of water-rock interaction experienced by the water parcel. More recently, Johannesson et al. (1999) suggested that groundwater REE concentrations reflect complex combinations of water-rock reactions occurring in recharge zones (i.e., low pH weathering solutions) and geochemical reactions occurring along groundwater flow paths. The latter study argues that these geochemical reactions affect the distribution and concentrations of complexing ligands, the characteristics of surface adsorption sites, and pH, all of which influence dissolved REE concentrations, and especially their normalized fractionation patterns (Johannesson et al., 1999). More recent investigations support the arguments initially put forth by Johannesson and colleagues that groundwater REE concentrations are likely more dynamic than originally envisioned, and essentially "evolve" along groundwater flow paths as weathering and microbial metabolism alter groundwater composition, surface site charges, and redox conditions (Dia et al., 2000; Johannesson et al., this volume). Understanding how REE concentrations change along groundwater flow paths as a function of changing groundwater and surface site composition is crucial to developing predictive computer models to study the fate and transport of these progressively more important industrial heavy metals in the environment, assessing their use as tracers of water-rock reactions in aquifers, and for the use of REEs as chemical analogues for radioactive actinide-series elements (e.g., McCarthy et al., 1988a, b). For example, under specific conditions of especially stable aqueous complexation of the REE with carbonate ions, natural organic matter, and synthetic organo-phosphate ligands, these commonly particle-reactive heavy metals are reported to behave relatively conservatively in groundwater flow systems (e.g., Johannesson et al., 1997b; McCarthy et al., 1988a; Möller et al., 2000), significantly increasing their mobilization in the environment. Moreover, by better quantifying the behavior of REEs in groundwaters within aquifer systems, substantial insight into the biogeochemical cycling of these and other trace elements will be gained for a portion of most global biogeochemical cycles that remains poorly understand, namely aquifers. Consequently, to elucidate the geochemical behavior of REEs along groundwater flow paths, we collected groundwater samples from two well-characterized aquifers, the Carrizo sand aquifer of Texas and a portion of the Upper Floridan (carbonate) aquifer in west-central Florida. Particular emphasis was focused on: 1) quantifying REE concentrations and fractionation patterns along groundwater flow paths; 2) modeling REE solution speciation along the groundwater flow paths; and 3) evaluating geochemical processes that affect REE concentrations and fractionation patterns along the flow paths in these different aquifers. In a future publication, we will present REE adsorption data for both aquifers, from which a combined solution and surface complexation model will be developed and presented (Tang and Johannesson, in prep).

2. Hydrogeology of the studied areas

2.1. CARRIZO SAND AQUIFER, TEXAS

The Carrizo sand aquifer is a confined aquifer of Eocene age, which is overlain by the Recklaw formation in the eastern and northeastern parts of Texas, the Bigford formation in western parts, and underlain by the middle Wilcox Group (Hamlin, 1988). The aquifer crops out in a band that nearly parallels the Gulf Coast in southeastern Texas (Fig. 1). The aquifer dips towards the southeast and contains fresh water at depths as great as 1500 meters (Pearson and White, 1967). The lithology of the Eocene Carrizo sand is remarkably uniform, consisting mostly of fine- to medium-grained quartz sand with minor amounts of clay, lignite, calcite, and pyrite (Pearson and White, 1967). We collected groundwater samples along a flow path within Atascosa County, Texas, that essentially followed the same flow path studied previously by Pearson and White (1967), Stute et al. (1992), and Castro et al. (2000) (Fig. 1). Groundwater ages estimated from previous ^{14}C measurement show that groundwaters are modern recharge (e.g., zero ^{14}C years) at the outcrop area of Carrizo sand aquifer and gradually increase to 27,000 ^{14}C years at 35 miles down-flow (Pearson and White, 1967). Thus, rainfall recharges the aquifer where it outcrops in the northern part of Atascosa County, and groundwater flows down-dip toward the southeast. The geochemical evolution of groundwater is mainly affected by: (1) dissolution of calcium carbonate; (2) dissolution of chlorides; (3) oxidation of iron sulfides and precipitation of ferric hydroxides within shallow portions of aquifer; and (4) chemical weathering of Al-silicate minerals further down flow (Hamlin, 1988). The hydraulic conductivity decreases by a factor of 10

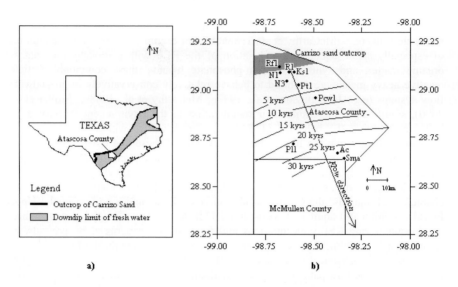

a)   b)

Figure 1. (a) Location of study area, outcrop and down dip limit of fresh water in the Carrizo sand aquifer; (b) Location of wells sampled, groundwater flow path, and carbon 14 ages (based on Pearson and White, 1967) of groundwater in the Carrizo sand aquifer

along the flow path ranging from about 20.4 m/day near the outcrop area to about 2.4 m/day in the down-dip (southeast) (Hamlin, 1988). The porosity of the Carrizo sand is estimated to range from 30 to 40% (Pearson and White, 1967).

2.2. UPPER FLORIDAN AQUIFER, WEST-CENTRAL FLORIDA

In the study area of west-central Florida (Fig. 2), the Upper Floridan aquifer (UFA) includes the highly permeable limestones of the Suwannee Limestone (Oligocene), Ocala Limestone (upper Eocene), and Avon Park (middle Eocene), and has a total thickness ranging from 60 to 640m (Miller, 1986). The UFA is overlain by the Hawthorn Formation (Miocene age), which is composed of sand, marl, clay, limestone, dolomite, and phosphatic deposits (Miller, 1986). The Hawthorn Formation is generally greater than 30 m thick in the study area and forms the upper confining unit of the UFA (Wicks and Herman, 1994). The UFA is separated from the Lower Floridan aquifer (LFA) by a less permeable confining unit, which consists of eight separate stratigraphic units and is commonly referred to as the middle confining unit (Miller, 1986). The middle confining unit in the study area is about 30-120 m thick with maximum of 200 m, and is chiefly composed of middle Eocene dolomite with intergranular anhydrite/gypsum.

The pre-development potentiometric surface of the UFA (Johnston et al, 1980) indicates that groundwater historically flowed radially outward from the potentiometric high at Polk City in the study area, which is also thought to be a major recharge area for

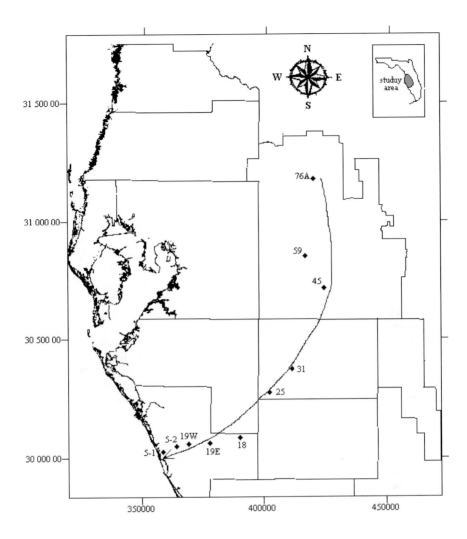

Figure 2. Location of study area, groundwater flow path investigated, and location of wells in the Upper Floridan aquifer

UFA. Groundwater samples were collected along a flow path from the potentiometric high at Polk City toward the southwest coast (i.e., near Venice, Florida; Fig. 2). The flow path chosen for study closely follows Path II of Plummer and Sprinkle (2001). Recharge to this flow path occurs at Polk City, where the confining Hawthorn Formation is thin. Groundwaters subsequently flow down gradient towards the coast and discharge into the Gulf of Mexico near Venice, Florida (Wicks and Herman, 1994). The adjusted ^{14}C ages from Plummer and Sprinkle (2001) show that groundwaters are roughly 1400 ^{14}C years in age at Polk City and then increase rapidly along the flow path. Most of groundwaters have adjusted ^{14}C ages of 20,000 to 30,000 years and do not show a simple progression in ^{14}C ages along the flow path. The adjusted ^{14}C ages of groundwaters indicate that much of recharge to the UFA occurred during the last glacial period and that little additional modern recharge has reached down-gradient parts of the UFA within the study area.

3. Methods

3.1. FIELD SAMPLING AND MEASUREMENT

All sample bottles were cleaned before being transported to the field site. Groundwater samples were collected from wells along the groundwater flow paths in the Carrizo sand aquifer in Texas (October 2002 and June 2003) and the Upper Floridan (carbonate) aquifer in west-central Florida (June 2003) for analysis of REEs, major solutes, and dissolved organic carbon (DOC). Alkalinity was titrated in the field using a digital titrator (HACH, Model 16900). pH, temperature, and conductivity of samples were measured on site using portable waterproof pH/conductivity meter (Fisher Accumet, AP85). All major solute samples were filtered through 0.45 μm in-line filters (Gelman Sciences groundwater filter capsules, polyether sulfone membrane). Cation samples (Ca^{2+}, Mg^{2+}, Na^+, K^+) were preserved with a drop of ultra-pure nitric acid (Seastar Chemicals), whereas anion samples (Cl^-, SO_4^{2-}, F^-, PO_4^{3-}) were not acidified. Groundwater samples for determination of total aqueous REE concentrations were filtered using identical Gelman Sciences in-line groundwater filter capsules and immediately acidified to pH < 2 with ultra-pure nitric acid. Cation, anion, and REE samples were stored cold until analysis. Dissolved organic carbon samples were filtered through 0.45μm in-line filters (polyether sulfone), immediately acidified to pH < 2 with ultra-pure HCl, and then quick frozen and stored frozen until analysis.

3.2. ANALYTICAL TECHNIQUES

Major cations of groundwaters were determined by inductively coupled plasma optical emission spectrometry (ICP-OES) and major anions of groundwaters were measured by ion chromatography (Dionex DX 600). Given the importance of the phosphate mineral control on REE solubility, we specifically estimated the detection limit for phosphate, which is about 0.4 μmol/kg. This value was used in our solution complexation calculations for those samples whose phosphate concentrations were below the detection limit. Dissolved organic carbon was determined on filtered samples by high temperature catalytic oxidation (HTCO) using a Shimadzu TOC-5000 total carbon analyzer.

The REEs were quantified directly, and without preconcentration, by high-resolution (magnetic sector) inductively coupled plasma mass spectrometry (ICP-MS; Finnigan MAT Element II) following techniques successfully employed by Johannesson and Lyons (1994, 1995), and Johannesson et al. (1996a, b, 1997a, 1999). In brief, each water sample was introduced to the ICP-MS directly from the sample bottle using a cross-flow nebulizer. The REE isotopes ^{139}La, ^{140}Ce, ^{141}Pr, ^{146}Nd, ^{147}Sm, ^{151}Eu, ^{157}Gd, ^{159}Tb, ^{163}Dy, ^{165}Ho, ^{166}Er, ^{169}Tm, ^{171}Yb, and ^{175}Lu were used to quantify the REEs in the groundwater samples. Although many of these REE isotopes are free of isobaric interference (Smedley, 1991), three REE isotopes [145]Nd, [161]Dy, and [167]Er were monitored as an additional check for isobaric interferences.

Analytical precision for La, Ce, and Eu was 5% relative standard deviation (RSD) or better. For Pr, Nd, Gd, Tb and Yb, the precision was 10% (RSD). For Sm, Dy, Ho, Er, and Lu, the precision was 12% (RSD), and the precision was 15% (RSD) for Tm. Replicate analyses of groundwater sample from Upper Floridan aquifer indicate that the measurement reproducibility was 10% (RSD) or better for La, Ce, Nd, Dy, Er, and Yb and was 13% (RSD) or better for Sm, Eu, Gd, and Lu. However, for Pr, Tb, Ho, and Tm, the reproducibility was 30%, 20%, 23.6%, and 20.6% (RSD), respectively.

3.3. SOLUTION COMPLEXATION MODELING

The solution complexation of the REEs in groundwaters of the Carrizo sand aquifer and Upper Floridan aquifer was modeled using a modified Humic Ion-Binding Model V. Humic Ion-Binding Model V, originally developed by Tipping and co-workers (Tipping and Hurley, 1992; Tipping, 1993, 1994), focuses on metal complexation with humic and fulvic acids. We modified the model to allow for prediction of organic and inorganic ligand complexation of all 14 naturally occurring REEs in natural terrestrial waters (Tang and Johannesson, 2003). The inorganic ligand complexation model followed that outlined by Millero (1992), and employed stability constants for REE inorganic ligand complexes tabulated in Millero (1992) and more recently determined values (Lee and Byrne, 1992, 1993; Klungness and Byrne, 2000; Luo and Byrne, 2000, 2004).

4. Results

4.1. STABILITY OF HYDROCHEMISTRY OF GROUNDWATERS AND AQUEOUS REES IN CARRIZO SAND AQUIFER

To evaluate the seasonal variation of groundwater hydrochemistry of the Carrizo aquifer, including the aqueous REEs, groundwater samples were collected from the same wells in the Carrizo in October 2002 and then again in June 2003. Additional sampling is planned to expand this aspect of the study. A comparison of the results for identical wells is illustrated in Figs. 3 and 4. Figure 3 shows that there are no substantial seasonal differences in hydrochemistry for the Carrizo groundwaters between October 2002 and June 2003. Absolute concentrations of the REEs (shown here as total REE concentrations), and their fractionation patterns as demonstrated by $(Er/Nd)_{SN}$ ratios, exhibit some seasonal variations in shallow groundwaters near the recharge area (Fig. 3). In general, however, the shale-normalized REE patterns for Carrizo groundwaters are remarkably similar for both sampling events (Fig. 4). During our sampling campaign in June 2003 we were unable to collect groundwaters from three of the wells we sampled in October 2002. Consequently, three other wells were substituted (F1, G1, and Pt2). However, owing to what we consider to be a probable contamination problem associated with the field blank for these samples, the REE data for groundwaters from these three wells may be spurious. Hence, we focus our discussion below on the October 2002 data only.

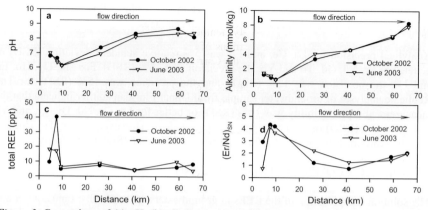

Figure 3. Comparison of (a) pH, (b) alkalinity, (c) total REE concentration, and (d) $(Er/Nd)_{SN}$ ratios of groundwaters from the Carrizo sand aquifer for October 2002 and June 2003

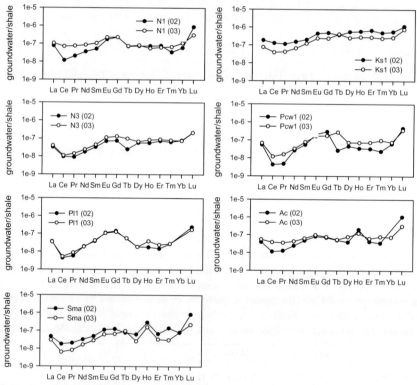

Figure 4. Comparison of shale-normalized REE patterns of groundwaters from the Carrizo sand aquifer for October 2002 and June 2003

4.2. ANALYTICAL RESULTS

Field measurements (i.e. pH, alkalinity, temperature, and conductivity), DOC, major solute compositions, and the concentrations of the REEs and some trace elements (i.e., Fe, Mn, and Al) of groundwaters from the Carrizo sand aquifer are presented in Table 1

Table 1. Concentrations of major solutes, REEs, and some other trace elements (Fe, Mn, Zn, Al, Cu) of groundwaters from the Carrizo sand aquifer, Texas. Also included are pH, temperature, conductivity, and DOC concentrations.

	Rf1	N1	R1	Ks1	N3	Pt1	Pcw1	Pl1	Ac	Sma
Distance (km)	0	4.29	6.9	7.4	9.37	14.51	26.05	41.05	59.02	65.77
Well depth (m)	61	158.5	167.6	122	182.9	289.6	487.7	655.3	1316.7	1420.4
pH	7.32	6.79	6.55	6.64	6.13	6.48	7.38	8.34	8.68	8.12
T (°C)	25.5	27	27.3	25.7	28.4	27.8	32.9	28.1	35.9	47.8
Conductivity (µS/cm)	-	586.5	-	318.5	11.5	361.8	554.9	506.9	918.6	1025.4
DOC (mg/l)	0.35	0.43	0.3	0.17	0.31	0.82	0.26	0.19	0.65	0.28
Ca (mmol/kg)	2.13	1.35	0.43	0.63	0.38	0.76	2.01	0.45	0.06	0.08
Mg	0.30	0.34	0.16	0.20	0.12	0.23	0.33	0.36	0.01	0.01
Na	1.98	1.42	1.14	1.12	0.71	1.16	1.15	3.96	8.74	10.24
K	0.25	0.19	0.14	0.13	0.13	0.20	0.14	0.15	0.05	0.06
Si	0.52	0.16	0.38	0.38	0.29	0.27	0.23	0.26	0.43	0.48
Alkalinity[a]	2.79	1.40	0.72	1.02	0.46	0.94	4.10	4.63	6.56	7.84
Cl	1.87	2.11	1.12	1.43	1.06	1.44	0.87	0.37	1.23	1.78
SO_4	0.92	0.72	0.18	0.27	0.22	0.36	0.37	0.35	0.79	0.71
PO_4	<0.4	0.93	0.82	<0.4	<0.4	1.85	<0.4	1.77	<0.4	<0.4
$[CO_3^{2-}]_F$[b] (µmol/kg)	3.12	0.46	0.13	0.23	0.03	0.15	6.17	57.65	202	105
total Fe	2.26	161.15	72.25	1.23	10.38	45.93	8.00	0.11	1.02	1.33
Mn	1.64	11.14	2.43	0.10	0.45	1.61	3.05	0.90	0.75	0.20
Zn	0.02	11.06	6.87	0.29	0.03	0.01	0.04	0.02	0.0	0.01
Al (nmol/kg)	5.61	7.60	24.94	23.60	14.38	785.51	22.77	24.52	648.03	899.71
Cu	7.58	6.81	3.73	2.94	2.52	3.73	5.06	4.96	3.55	5.23
La (pmol/kg)	61.86	22.63	13.65	54.84	9.75	8.44	16.47	10.69	11.66	14.12
Ce	23.11	7.01	11.67	76.37	5.61	1.84	2.56	2.57	6.80	10.67
Pr	6.50	1.49	1.71	8.20	0.64	0.29	0.34	0.41	0.91	1.47
Nd	32.77	9.51	8.38	40.54	4.65	4.06	7.16	5.08	6.38	8.60
Sm	6.95	2.58	2.43	10.09	1.65	2.04	2.79	1.89	2.39	2.46
Eu	3.76	1.84	0.86	4.78	0.75	1.31	2.28	1.19	0.87	1.20
Gd	16.75	9.26	2.77	19.71	3.03	6.30	11.97	5.70	2.90	5.06
Tb	2.07	0.55	0.27	2.89	0.20	0.20	0.21	-	0.40	0.61
Dy	16.30	2.52	1.85	18.80	1.96	1.18	1.57	0.63	1.34	2.13
Ho	4.69	0.60	0.46	4.79	0.46	0.22	0.29	0.14	1.62	2.37
Er	15.72	1.74	1.60	14.95	1.66	0.61	0.75	0.33	0.95	1.51
Tm	2.32	0.13	0.22	1.84	0.24	-	0.09	0.10	0.13	0.53
Yb	14.12	1.22	1.48	11.37	1.62	0.68	1.29	-	-	1.65
Lu	3.60	3.05	0.91	3.96	0.74	1.23	1.60	0.84	3.71	2.93
$(Er/Nd)_{SN}$	5.64	2.15	2.25	4.33	4.20	1.75	1.23	0.77	1.75	2.07
Ce/Ce*	-0.59	-0.61	-0.25	-0.07	-0.35	-0.72	-0.85	-0.68	-0.36	-0.28
Eu/Eu*	0.11	0.09	0.14	0.12	0.12	0.10	0.09	0.10	0.14	0.11

Distance (km): Distance from recharge area (km); [a]Alkalinity as HCO_3;
[b]Calculated from our measured alkalinity using PHREEQC (Parkhurst and Appelo, 1999);
SN = shale-normalized; Ce/Ce* = log { $2Ce_{SN}/[La_{SN}+Pr_{SN}]$ }; Eu/Eu* = log { $2Eu_{SN}/[Sm_{SN}+Gd_{SN}]$ };
- indicates concentration below detection limits;
All data except DOC and conductivity were measured in October, 2002;
DOC was measured in June, 2003 and conductivity was measured in June, 2004.

and for the Upper Floridan aquifer in Table 2. Our data indicate that groundwaters from the Carrizo sand aquifer are chiefly Ca-Na-HCO$_3$-Cl type waters at, or close to, the recharge zone where wells are relatively shallow (i.e., less than 168m, Table 1). With flow down-gradient, however, the groundwaters shift composition towards Na-HCO$_3$ type waters. Our observed change in Carrizo groundwater compositions is consistent with Hamlin's (1988) study results, which showed that groundwaters are mostly Ca-Na-Cl or Ca-Na-HCO$_3$-Cl type waters at depths less than 150m, shift to Ca-HCO$_3$ and Ca-Na-HCO$_3$ types at relatively shallow depths (less than 760m), and finally become Na-HCO$_3$ waters at well depths below 760m. The pH and alkalinity initially decrease with

Table 2. Concentrations of major solutes, REEs, and some other trace elements (Fe, Mn, Zn, Al, Cu) of groundwaters from the Upper Floridan aquifer, Florida. Also included are pH, temperature, conductivity, and DOC concentrations.

ROMP well number	76A	59	45	31	25	18	19E	19W	5-2	5-1
Distance (km)	0	32.2	46.3	78.8	90.8	112.1	122.5	132.8	137.7	144.1
Well depth (m)	80.5	61	100.6	140.2	91.4	153.9	125	125	155.4	150
pH	7.68	7.56	7.64	7.43	7.22	7.33	7.13	7.15	6.98	7.03
T (°C)	24.1	23.9	27.1	31.1	28.8	27.8	26.4	27.3	27.6	26.2
Conductivity (µS/cm)	297	340	372	532	1090	856	1236	1531	2645	2808
DOC (mg/l)	0.64	1.27	1.36	1.45	1.59	2.09	2.29	1.64	1.45	1.58
Ca (mmol/kg)	0.95	0.61	1.09	1.20	3.45	2.13	3.71	5.30	12.21	10.48
Mg	0.41	0.63	0.62	0.86	2.76	1.70	2.86	3.69	6.13	6.13
Na	0.24	0.39	0.37	0.42	0.62	0.93	1.21	1.37	1.05	2.26
K	0.03	0.02	0.04	0.06	0.09	0.07	0.08	0.14	0.11	0.12
Si	0.28	0.24	0.27	0.27	0.36	0.39	0.35	0.34	0.30	0.21
Alkalinity[a]	2.38	2.32	2.96	2.06	2.62	3.59	3.17	2.80	2.53	2.49
Cl	0.28	0.29	0.27	0.34	0.48	0.95	0.99	0.82	1.28	2.54
SO$_4$	0.12	0.14	0.31	1.12	5.64	2.27	5.10	7.89	17.17	15.39
PO$_4$ (µmol/kg)	1.68	0.48	0.51	<0.4	<0.4	<0.4	<0.4	<0.4	<0.4	<0.4
[CO$_3^{2-}$]$_F$[b]	5.84	4.31	7.14	3.23	2.47	4.36	2.35	2.19	1.34	1.44
total Fe	0.09	0.43	0.39	0.16	0.64	0.78	0.57	0.79	1.15	1.50
Zn	0.26	0.16	0.05	0.08	0.02	0.02	0.74	0.03	0.02	0.12
Al	0.11	0.04	0.03	0.01	0.06	0.01	0.42	0.01	0.01	0.04
Mn (nmol/kg)	9.65	8.58	2.95	0.12	7.09	0.88	2.04	4.77	0.93	4.74
Cu	3.07	1.41	0.26	1.33	14.54	17.87	11.70	22.72	10.02	14.44
La (pmol/kg)	1.58	1.63	5.17	6.93	9.96	7.66	4.29	3.51	6.59	
Ce	0.76	1.49	5.96	2.11	10.08	3.17	9.41	6.36	5.17	5.33
Pr	0.29	0.24	0.98	0.51	1.60	0.80	1.11	0.99	0.80	0.80
Nd	0.68	-	4.88	5.49	6.73	5.93	5.41	4.23	3.91	3.91
Sm	0.58	-	1.35	2.39	2.36	2.37	1.78	1.18	1.66	1.06
Eu	-	0.72	0.47	0.68	0.91	0.87	0.59	0.48	0.65	1.57
Gd	0.61	4.57	1.48	1.57	3.17	2.89	1.73	1.77	2.31	9.33
Tb	0.34	0.27	0.18	0.23	0.54	0.25	0.29	0.56	0.59	2.16
Dy	0.79	0.46	1.26	1.52	2.61	1.79	1.71	1.44	1.64	1.13
Ho	0.27	-	0.36	0.09	0.64	0.19	0.20	0.28	0.30	0.68
Er	1.19	0.20	1.22	0.31	1.30	0.48	0.43	0.69	0.53	0.50
Tm	0.26	-	0.20	-	0.25	0.11	-	0.13	0.14	0.19
Yb	1.77	-	1.16	-	0.87	-	-	-	0.65	0.78
Lu	0.30	0.34	0.20	-	0.33	0.18	-	0.10	0.18	0.42
(Er/Nd)$_{SN}$	20.60	n. c.	2.94	0.68	2.27	0.95	0.93	1.92	1.60	1.50
Ce/Ce*	-0.56	-0.25	-0.19	-0.63	-0.22	-0.54	-0.13	-0.12	-0.12	-0.27
Eu/Eu*	n. c.	n. c.	0.14	0.17	0.13	0.14	0.15	0.13	0.13	0.07

Distance (km): Distance from recharge area (km);
[a]Alkalinity as HCO$_3$;
[b]Calculated from our measured alkalinity using PHREEQC (Parkhurst and Appelo, 1999);
SN = shale-normalized; Ce/Ce* = log { 2Ce$_{SN}$/[La$_{SN}$+Pr$_{SN}$]}; Eu/Eu* = log
{ 2Eu$_{SN}$/[Sm$_{SN}$+Gd$_{SN}$]};
- indicates concentration below detection limits
n. c.: not calculated.

flow away from the recharge zone before increasing again down-gradient (Table 1, Figs. 3, 5). DOC is generally low (less than 0.82 mg/L), with no systematic variation along the groundwater flow path. REE concentrations are highest in groundwaters from the recharge zone (Nd 40.5 pmol/kg), decrease substantially with flow down-gradient, and reach relatively low and apparently stable values (Nd 6 ±1.7 pmol/kg) at roughly 10 km from the recharge zone (Table 1, Fig. 5). The decrease in REE concentrations with flow down-gradient is broadly consistent with our earlier investigation of REE in a shallow alluvial aquifer (Johannesson et al., 1999).

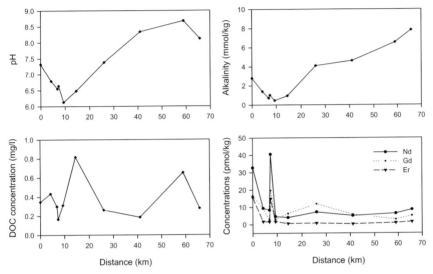

Figure 5. pH, alkalinity, DOC, and concentrations of Nd, Gd, and Er as functions of distance from the recharge zone in the Carrizo sand aquifer, Texas (October, 2002)

Groundwaters from the Upper Floridan Aquifer are Ca (Mg)-HCO$_3$ waters from the recharge zone to about 46 km down-gradient, shift to Ca-Mg-SO$_4$-HCO$_3$ waters along the mid-reaches of the flow path, and finally become Ca-Mg-SO$_4$ waters at the discharge area (Table 2). Ca, Mg, SO$_4$, and Cl concentrations generally increase along the groundwater flow path, whereas pH generally decreases (Table 2, Fig. 6). Alkalinity is relatively constant along the flow path (Fig. 6). Through a reaction model, Plummer et al. (1983) pointed out that the general pH decrease with relatively constant alkalinity was the result of dedolomitization reactions (gypsum and dolomite dissolution with calcite precipitation) along the groundwater flow path. Our chemical results are generally consistent with previous work (Back and Hanshaw, 1970; Sacks et al., 1995; Plummer and Sprinkle, 2001; J.D. Haber, SWFWMD, personal communication, 2003). High levels of sulfate, chloride, magnesium, and calcium in groundwaters at the southern portion of the study area are caused by saltwater intrusion as well as the upwelling of deeper water in the lower aquifer that has been in contact with the Evaporite layer (Sacks et al., 1995; J.D. Haber, SWFWMD, personal communication, 2003). For all but the Polk City well, the DOC is higher than in the Carrizo sand aquifer

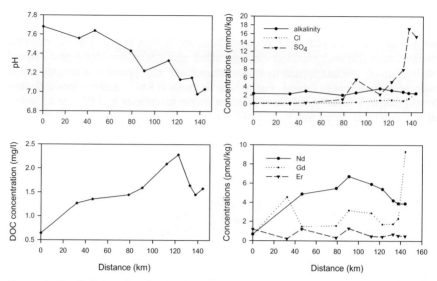

Figure 6. pH, alkalinity, Cl, SO₄, DOC, and concentrations of Nd, Gd, and Er as functions of distance from the recharge zone in the Upper Floridan aquifer, Florida

and initially increases along the flow path before decreasing abruptly in the last three groundwater samples (Fig. 6). Conductivity increases along the groundwater flow path and the groundwaters gradually become brackish in the discharge area (Table 2). Light REE (LREE, e.g., Nd) concentrations initially increase along the groundwater flow path, appear to "stabilize" towards the mid-reaches of the flow path, and decrease again further down flow. The middle REE (MREE, Gd) varies irregularly along the flow path, and the heavy REE (HREE, Er) concentrations generally tend to decrease along the flow path (Table 2, Fig. 6), although Er also exhibits fluctuations in concentrations along the flow path.

4.3. SHALE-NORMALIZED REE PATTERNS

Shale-normalized REE patterns are shown in Figs. 7 and 8 for groundwaters from the Carrizo sand aquifer and in Figs. 9 and 10 for Upper Floridan groundwaters. The composite shale used to normalize these groundwater REE concentrations is a mean of North American, European, and Russian shale, and has been employed extensively in the oceanographic literature, and more recently in groundwater studies (Elderfield and Greaves, 1982; Sholkovitz, 1988; Johannesson et al., 1997b, 1999; Johannesson and Hendry, 2000).

Carrizo groundwaters generally exhibit HREE-enriched shale-normalized patterns (Figs. 7 and 8). One interesting feature of these shale-normalized patterns is that, although the groundwaters show positive enrichment trends from the LREEs to the HREEs, each exhibits a change in the enrichment trend slope at Gd, such that the enrichment trend slope is generally steeper from Nd to Gd than from Gd to Lu (Fig. 7). All Carrizo

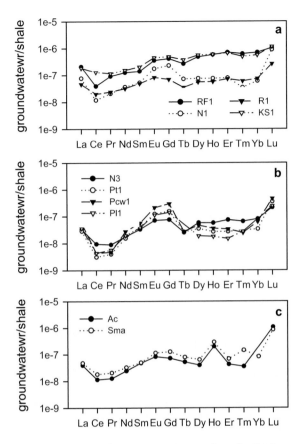

Figure 7. Shale-normalized REE patterns for groundwaters from the Carrizo sand aquifer, Texas (October, 2002)

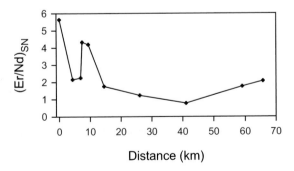

Figure 8. Shale-normalized Er/Nd ratios for groundwaters from the Carrizo sand aquifer as a function of distance from the recharge zone (October, 2002)

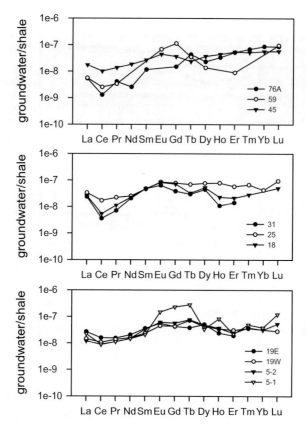

Figure 9. Shale-normalized REE patterns for groundwaters from the Upper Floridan aquifer, Florida

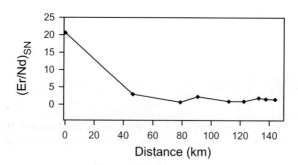

Figure 10. Shale-normalized Er/Nd ratios for groundwaters from the Upper Floridan aquifer as a function of distance from the recharge zone

groundwaters have slightly positive Eu anomalies (Eu/Eu* 0.09–0.14; Table 1). Because the Carrizo sand groundwaters possess anomalously elevated shale-normalized La concentrations compared to other LREEs, there are no actual negative Ce anomalies, although the calculated Ce/Ce* values are negative (-0.85 – -0.07; Table 1).

Although the general shale-normalized profiles for the Carrizo groundwaters are broadly similar, the shale-normalized REE patterns change gradually along the flow path (Figs. 7 and 8). At or close to the recharge area, where well depths are relatively shallow (i.e., 61-167.6m), REEs show relatively smooth HREE-enrichment patterns (Fig. 7a). At the mid-reaches of the flow path, the shale-normalized REE patterns are characterized by an overall "W-shape" (Fig. 7b). The two wells furthest down-gradient (i.e., Ac and Sma) show a flattened "W-shape" patterns, with Ho anomalies, which probably may represent analytical errors (Fig. 7c). The $(Er/Nd)_{SN}$ ratios, which are a measure of the fractionation between the HREEs and LREEs, are relatively high for groundwaters from shallow wells in and proximal to the recharge area (e.g., from 2.15 to 5.64). Moreover, these shallow groundwaters exhibit the greatest variation in HREE/LREE fractionation between samples N1 and Rf1 as indicated by their $(Er/Nd)_{SN}$ ratios (Table 1, Fig. 8). From a distance of ~10 km from the recharge zone, the $(Er/Nd)_{SN}$ ratios decrease for the next 40 km along the flow path before increasing again at the two furthest down-gradient wells (Fig. 8).

The Upper Floridan groundwaters also generally have HREE enriched shale-normalized patterns, although they differ from the shale-normalized patterns of Carrizo groundwaters (Fig. 9). For example, the shale-normalized REE patterns of UFA groundwater are commonly flatter than those for groundwaters from the Carrizo sand aquifer (cf. Figs. 7 and 9). All UFA groundwaters show positive Eu anomalies (0.06 – 0.17) and negative Ce anomalies (-0.12 – -0.63). The shale-normalized REE patterns of the Upper Floridan groundwaters also gradually change along the groundwater flow path. For example, near the recharge zone (i.e., Polk City well, Well ROMP 76A), UFA groundwater exhibits a smooth HREE-enrichment pattern (Fig. 9). However, with subsequent flow down-gradient, the REE patterns of the UFA groundwaters tend to flatten. The variation (i.e., flattening) in REE patterns along the flow path is clearly demonstrated in Fig. 10 by examination of the shale-normalized Er/Nd ratios of UFA groundwaters. The $(Er/Nd)_{SN}$ ratio decreases significantly from the recharge area and remains comparatively constant along the remaining flow path.

4.4. REE SOLUTION COMPLEXATION

Results of the REE speciation modeling are plotted in Fig. 11 for groundwaters from the Carrizo sand aquifer and in Fig. 12 for those from the Upper Floridan carbonate aquifer. Figure 11 indicates that the solution complexation of REEs is chiefly controlled by pH and alkalinity for groundwaters from the Carrizo sand aquifer. Because DOC is low (less than 0.82 mg/L) in groundwaters from the Carrizo sand aquifer, organic complexation of REEs is not expected to be significant (Fig. 11; see Tang and Johannesson, 2003). Carbonato ($LnCO_3^+$) and/or dicarbonato complexes ($Ln(CO_3)_2^-$) are the main dissolved species for all REEs in these groundwaters. Free REE ions (Ln^{3+}) are significant in those groundwaters with pH below 7.0. Generally, within the pH

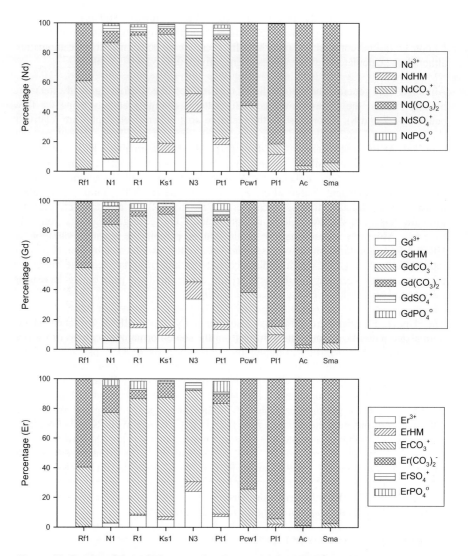

Figure 11. Results of the solution complexation model for Nd, Gd, and Er in percentage for groundwaters from the Carrizo sand aquifer, Texas (October, 2002)

range of these groundwaters (6.13-8.68), the proportion of Ln^{3+} decreases with increasing pH, whereas the importance of $Ln(CO_3)_2^-$ increases with pH and becomes the dominant species for all REEs when the pH is above 8.0 (see Johannesson et al., this volume). The percentage of $LnCO_3^+$ increases first and then decreases with increasing pH along the flow path. In each groundwater, from LREEs to HREEs (as shown by Nd, Gd, and Er in Fig. 11), the proportion of Ln^{3+} decreases and that of $LnCO_3^+$ and $Ln(CO_3)_2^-$ increases with increasing atomic number across the REE series where the pH

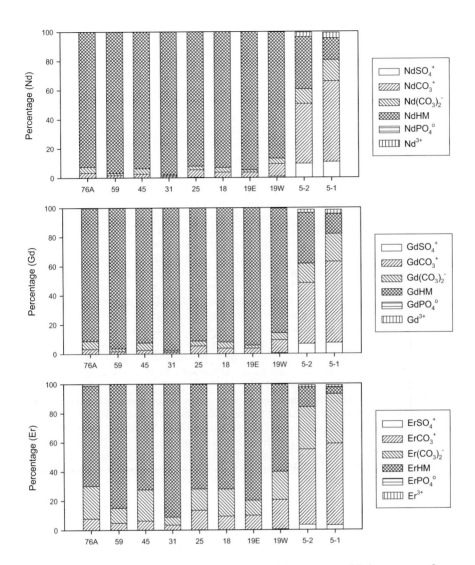

Figure 12. Results of the solution complexation model for Nd, Gd, and Er in percentage for groundwaters from the Upper Floridan aquifer, Florida

is below 7.0. Where the pH is above 7.0, the proportion of $LnCO_3^+$ decreases and that of $Ln(CO_3)_2^-$ increases with increasing atomic number. Where the pH is above 8.0, almost all REEs exist in solution as $Ln(CO_3)_2^-$. The results are in agreement with other studies of REE speciation in natural terrestrial waters (e.g., Johannesson and Lyons, 1994, 1995; Johannesson et al., 1996b; Leybourne et al, 2000; Johannesson and Hendry, 2000). Thus, for the Carrizo sand aquifer, groundwaters from the recharge area to the mid-reach of the flow path, compared to the LREEs, the HREEs tend to complex strongly with carbonate ions, and more of the HREEs exist as $LnCO_3^+$ where the pH is below 7.0 and as $Ln(CO_3)_2^-$ where the pH is below 8.0. In furthest down-gradient groundwaters, $Ln(CO_3)_2^-$ is dominant for all REEs.

Compared to groundwaters from the Carrizo sand aquifer, Upper Floridan aquifer groundwaters have relatively high DOC concentrations and low concentrations of competitive cations (Fe, Al and Mn) (Tables 1, 2). Because of the relatively high DOC concentrations of UFA groundwaters, our speciation calculations predict organic ligands complexes of the REEs (i.e., LnHM, where HM = humic matter) are dominant in all but two furthest down-gradient groundwaters (Fig. 12). This is a remarkable and entirely unexpected finding. It should be clearly pointed out, however, that we have yet to verify the predominance of organic ligand complexation by direct measurements for UFA groundwaters. Consequently, the predictions that all REEs are predominantly complexed with dissolved organic matter in UFA are currently tentative results, reflecting only the ability of the model (Humic Ion-Binding Model V; Tipping and Hurley, 1992; Tipping, 1993, 1994) to predict REE speciation (Tang and Johannesson, 2003). In these groundwaters from the discharge area to about 133 km down-flow, the model predicts that Nd (LREE) and Gd (MREE) will be almost completely complexed by dissolved organic matter, whereas $LnCO_3^+$ and $Ln(CO_3)_2^-$ account for 8-40% of dissolved Er (HREE, Fig. 12). The model suggests that DOC is a stronger complexing agent for the dissolved LREEs than for the dissolved HREEs, whereas free carbonate ions (CO_3^{2-}) are more important complexers to the dissolved HREEs. For the two furthest down-gradient groundwaters (i.e., ROMP 5-2 and 5-1), because of salt water intrusion and the up-coning of deeper groundwaters due to groundwater withdrawals (Sacks et al., 1995; J.D. Haber, SWFWMD, personal communication, 2003), the concentrations of Ca, Mg, SO_4, and Cl are abnormally high (Table 2). Our modeling results show that carbonato $(LnCO_3^+)$ and dicarbonato complexes $(Ln(CO_3)_2^-)$ are more important than LnHM in these blackish groundwaters. Sulfate complexes $(LnSO_4^+)$ and free ions (Ln^{3+}) of REEs also account for significant proportions of REEs in ROMP 5-2 and 5-1 well waters. Specifically, along flow from ROMP 5-2 to 5-1, the proportion of LnHM, $LnSO_4^+$ and Ln^{3+} decreases from Nd to Er and that of $LnCO_3^+$ and $Ln(CO_3)_2^-$ increases with increasing atomic number. These observations also suggest that DOC is more important in LREE complexation, whereas free carbonate ions (CO_3^{2-}) are more important for the HREEs.

5. Discussion

5.1. PROCESSES CONTROLLING REE CONCENTRATIONS

There are many processes that likely control REE concentrations along groundwater flow paths including: (1) release of REEs from host rocks; (2) surface complexation (adsorption/desorption) of REEs with surface sites on solid phases; (3) solution complexation of REEs with organic and inorganic ligands; (4) precipitation of REE phosphate minerals; and (5) mixing of groundwaters.

Previous studies (Nesbitt, 1979; Schau and Henderson, 1983; Braun et al., 1993) pointed out that during weathering REEs can be mobilized and released from the altered host rocks. Milliken et al. (1994) studied elemental mobility during burial in sandstones of the Frio Formation, south Texas. They found that REEs are mobilized during weathering of feldspars within the sandstones. These authors reported that whole-rock REE concentrations decrease from the shallowest to the deepest sandstones in the Frio Formation, and suggested, based on their observations, that as diagenesis progresses, REE are lost from the acid-insoluble fraction (i.e., silicate) of the rock and gained by the acid-soluble fraction (i.e., phosphates, carbonates, grain-surface coatings; Milliken et al., 1994). The Carrizo sand aquifer crops out in the northern part of Atascosa County, and thus north and west of the Frio Formation, and dips down towards the southeast to depths as great as 1500m. Furthermore, similar to the Frio Formation of south Texas, several of the same hydrochemical and hydrodynamic processes, including dissolution of calcium carbonate and feldspar leaching, are also occurring within the Carrizo sand aquifer (Hamlin, 1988; Milliken et al., 1994). Thus, although we did not study REE mobilization from the sand of the Carrizo aquifer, we expect that the release of REEs from Carrizo sand into groundwaters due to weathering and/or burial diagenesis probably affects the REE concentrations and variations along the groundwater flow path.

In the case of the Upper Floridan aquifer, previous studies (Plummer, 1977, Plummer et al., 1983; Jones et al., 1993) documented a series of water-rock reactions occurring in the Upper Floridan aquifer that we expect also likely affect the geochemical evolution of REE concentrations and fractionation along the groundwater flow path. Using a geochemical reaction model, Plummer et al. (1983) defined the main reactions occurring in the Upper Floridan aquifer as the dissolution of gypsum and dolomite and precipitation of calcite. The general process whereby gypsum dissolution drives calcite replacement of dolomite within the aquifer rocks is called dedolomitization (Plummer et al., 1983). A number of other reactions also occur within the UFA, including the oxidation of organic matter, sulfate reduction, ferric hydroxide dissolution, and pyrite precipitation (Plummer et al., 1983; Plummer and Sprinkle, 2001). We suggest that dedolomitization, along with microbially driven reductive dissolution of oxides/oxyhydroxides, release REEs into UFA groundwaters. Furthermore, these reactions, especially, dedolomitization, also result in a down-gradient increase of dissolved Ca, Mg, and SO_4, a down-gradient decrease in pH, and the low Fe concentration due to low ferric hydroxide and probably pyrite solubility in the upper Floridan aquifer. However, it is important to point out that the high major solute concentrations and total dissolved Fe of ROMP 5-2 and 5-1 can not be explained entirely by the geochemical evolution of UFA groundwater along the flow path via water-reactions between groundwater and aquifer rocks. Seawater intrusion and the up-coning of deeper groundwaters due to groundwater withdrawals for agricultural use have been invoked to account for these brackish groundwaters (Sacks et al., 1995; J.D. Haber, SWFWMD, personal communication, 2003). Upon mixing with seawater and deep brackish waters, we expect that dissolved organic matter in colloidal form may flocculate and be subsequently removed from solution (e.g., Sholkovitz, 1976). If this process does occur, we expect it would lead to the removal of some dissolved REEs from these groundwaters, especially the LREEs. Thus, we doubt that REE variations in the downstream groundwaters from the Upper Floridan aquifer might also reflect the mixing of deep groundwaters and seawaters.

Adsorption of REEs onto solid phases could also remove REEs from groundwaters and thus decrease their dissolved concentrations (Benedict et al., 1997; Johannesson et al., 1999). In contrast, solution complexation of REEs with inorganic and organic ligands can result in the formation of weakly- or non-adsorbing aqueous complexes of the REE, and thus may facilitate keeping the REE in solution (e.g., Johannesson et al., 1999; this volume; Johannesson and Hendry, 2000). In the Carrizo sand aquifer, due to the chemical weathering of Al-silicate minerals, the pH of groundwaters generally increases along the flow path (HCO_3^- is generated). Ferric hydroxide precipitation occurring in the shallow part of the aquifer and clay formation from feldspar weathering suggests that surface complexation (adsorption) is probably the most important process controlling REE concentrations in and proximal to the recharge area. Indeed, the variation of REE concentrations and fractionation patterns along the flow path can be explained by surface and solution complexation. In and proximal to the recharge area, where most of the Carrizo aquifer groundwaters have pH values less than 7.0, REEs mainly exist as positively charged species (i.e., $LnCO_3^+$ and Ln^{3+}; Fig. 11). Strong surface complexation (adsorption) of REEs may thus occur to negatively charged surface sites on solid aquifer phases (e.g., clay minerals), which will remove REEs from solution in these groundwaters. With further flow down-gradient, almost all of the REEs are predicted to exist as negatively charged dicarbonato complexes ($Ln(CO_3)_2^-$) in groundwaters of the Carrizo sand aquifer (Fig. 11). Consequently, within the down-gradient reaches of the Carrizo sand aquifer, surface complexation (adsorption) of REEs is unlikely to occur on the predominantly negatively charged surface sites of aquifer rocks. Thus, the concentrations of REEs tend to be stabilized in the down-gradient groundwaters. To quantitatively evaluate the role that surface and solution complexation play in controlling REE concentrations along the flow path in the Carrizo sand aquifer, we are currently conducting adsorption experiments in our laboratory, with the goal of developing a combined solution and surface complexation model.

In the Upper Floridan aquifer, the dissolution of gypsum and dolomite, microbial oxidation of organic matter, and subsequent ferric hydroxide dissolution may release REEs to solution and/or decrease the adsorption ability of solid phases. Previous studies of solid-liquid distribution coefficient of REEs (Koeppenkastrop and De Carlo, 1993; Sholkovitz, 1995; Benedict et al., 1997) indicate that, with increasing atomic number, the tendency of the REEs to be adsorbed onto aquifer solid phases decreases, i.e., LREE > MREE > HREE. Consequently, chemical weathering of minerals, and especially the congruent dissolution of minerals that make up the Upper Floridan aquifer, ought to result in the release of relatively more LREEs than HREEs into groundwaters. Thus, the aqueous concentrations of LREEs rather than HREEs are affected more by congruent dissolution of these phases. We suggest that the increase in Nd (LREE) concentrations from the recharge zone to mid-reach of the flow path in the UFA (Fig. 6) reflects the important role of the dedolomitization process in the aquifer.

The speciation modeling suggests that solution complexation is also an important process controlling the concentrations of REEs along the flow path in Upper Floridan aquifer. Our speciation modeling predicts that almost all of the REEs should be complexed with organic matter in UFA groundwaters except for waters from well

ROMP 5-2 and 5-1 (Fig. 12). The concentrations of Nd (LREE) show a general increase with increasing DOC along the flow path (the correlation coefficient, r, between the concentrations of Nd and DOC is 0.7; Fig. 6). However, the concentrations of the MREE Gd and the HREE Er do not increase along with DOC (Fig. 6), although their dominant species are predicted to be organic complexes (LnHM; Fig. 12). One possible explanation for this apparent disconnect between Gd and Er and DOC concentrations compared to Nd and DOC along the flow path is that our measured concentrations of Gd and Er are in error owing to the difficulty associated with measuring such low, sub-picomolal levels. Another possibility is that our model results over-predict the role that organic ligands play in complexing the MREEs and the HREEs in UFA groundwaters. If so, then the general decrease in the measured Er concentrations could be due to the decrease of free carbonate ions (i.e., $[CO_3^{2-}]_F$) along the flow path. The model actually predicts that carbonato and dicarbonato complexes $(ErCO_3^+ + Er(CO_3)_2^-)$ account for 10~40% of dissolved Er in all but the two furthest down-gradient groundwaters, where carbonate complexes of Er dominate (i.e., > 80% of Er; Fig. 12).

Precipitation of REE phosphate minerals could also exert control on REE concentrations along the groundwater flow path in the aquifers. The very low solubility products of REE phosphate salts (i.e., $10^{-26} < K_{sp}^{o} < 10^{-24}$; Jonasson et al., 1985; Byrne and Kim, 1993) suggest that the phosphate ions could set solubility limits on dissolved REEs in terrestrial waters (Johannesson et al., 1995). However, because pure REE phosphate phases do not form in natural solutions and REE phosphate phases are actual coprecipitates (Byrne and Kim, 1993; Byrne et al, 1996), it is unclear how to calculate the saturation states of REE phosphate coprecipitates. As a first approximation, we followed the summation method used by Byrne and Kim (1993) to calculate activity products for REE phosphate coprecipitates and compare these values with REE phosphate solubility products. The calculated activity products (AP $= \sum_{La}^{Gd} \{M^{3+}\}\{PO_4^{3-}\}$) of light rare earth phosphates in Carrizo and Upper Floridan groundwaters are given in Table 3. Table 3 shows that the calculated log(AP) values of light rare earth phosphates

Table 3. Activity products (AP $= \sum_{La}^{Gd} \{M^{3+}\}\{PO_4^{3-}\}$) of light rare earth phosphates in Carrizo and Upper Floridan groundwaters

Aquifer	Well number	Log (AP)	Aquifer	Well number	Log (AP)
Carrizo	Rfl	-23.64	Upper Floridan	ROMP 76A	-25.65
	N1	-23.66		ROMP 59	-26.34
	R1	-23.78		ROMP 45	-25.79
	Ks1	-23.45		ROMP 31	-26.12
	N3	-24.83		ROMP 25	-25.52
	Pt1	-23.92		ROMP 18	-25.89
	Pcw1	-24.45		ROMP 19E	-25.90
	Pl1	-24.42		ROMP 19W	-25.68
	Ac	-25.48		ROMP 5-2	-25.25
	Sma	-25.17		ROMP 5-1	-24.88

are more than -24.0 in the up-gradient Carrizo groundwaters, between -25.5 and -24.4 in the down-gradient Carrizo groundwaters and between -26.3 and -24.9 in the Upper Floridan groundwaters, which are slightly larger than or close to the estimated values of $LnPO_4 \cdot nH_2O$ solubility products at 25 °C and zero ionic strength (log K_{sp}°; see Table 6, Byrne and Kim, 1993). Although our calculated REE phosphate activity products for Carrizo sand and UFA groundwaters can only be viewed as estimates, the results (Table 3) indicate that REEs in groundwaters from both aquifers are saturated or close to saturation with respect to REE phosphate coprecipitates. Thus, REE phosphate coprecipitates may remove the dissolved REEs from the up-gradient Carrizo groundwaters and/or buffers the concentrations of REEs in the down-gradient Carrizo groundwaters and in the Upper Floridan groundwaters at the low values observed.

5.2. PROCESSES CONTROLLING REE PATTERNS

5.2.1. Carrizo groundwaters

During weathering of the host-rock, REEs are not only mobilized but also fractionated (Nesbitt, 1979; Schau and Henderson, 1983; Braun et al., 1993). The weathering process usually results in HREE-enrichment patterns in the weathering solutions (e.g., groundwaters), which is attributed to the selective alteration of HREE-enriched minerals and to the preferential formation of strong dissolved complexes of the HREE with carbonate ions, and perhaps some organic ligands, thereby facilitating the transport of HREE into the solution (Nesbitt, 1979). Our speciation modeling results indicate that for a given groundwater sample; a greater relative proportion of each REE is complexed with carbonate ions as a function of increasing atomic number across the REE series. The dominance of carbonate complexation for the HREE as compared to the LREE supports the release of HREE from host rocks and inhibition of HREE re-adsorption.

After REEs are released into groundwaters via chemical weathering and subsequently transported along the flow path, we submit that solution complexation of REEs will continue to affect the REE fractionation patterns. In and proximal to the recharge area, because of relatively low pH, the LREEs are more likely to be adsorbed than the HREEs onto the clay minerals and oxide/hydroxide minerals because a greater proportion of each LREE is predicted to occur as free metal ions and other positively charged species (Johannessonet al., 1999, this volume; Johannesson and Hendry, 2000). Moreover, LREEs typically form weaker solution complexes than HREEs, which is demonstrated by the general increase in stability constants for most REE aqueous complexes (i.e., carbonate) with increasing atomic number. Thus, the shallow groundwaters from near the recharge area have HREE-enrichment patterns and relatively high $(Er/Nd)_{SN}$ ratios (Figs 7 and 9). However, further down gradient, in Carrizo sand aquifer groundwaters, all REEs are predicted to chiefly exist as negatively charged dicarbonato complexes $(Ln(CO_3)_2^-)$ due to high pH values, which greatly diminishes the ability of REE to adsorb to clay minerals and oxide minerals. Thus, the deep Carrizo groundwaters tend to have relatively low $(Er/Nd)_{SN}$ ratios (Fig. 9).

Previous studies (Byrne and Kim, 1993; Byrne et al., 1996) indicate that for conditions where carbonate complexation dominates REE speciation as is the case in Carrizo sand groundwaters, REE phosphate coprecipitation tends to enrich the LREE in the solid phase and, hence, contributes to HREE-enrichments in solution. Table 3 shows that up-gradient groundwaters from the Carrizo sand aquifer have relatively high estimated LREE-phosphate activity products, and thus are expected to coprecipitate REE phosphate salts. Therefore, REE phosphate coprecipitation is another possible process that may contribute to the formation of the relatively HREE-enrichment patterns and relatively high $(Er/Nd)_{SN}$ ratios in these shallow, up-gradient groundwaters.

5.2.2. Upper Floridan groundwaters

The initial groundwater from the recharge area (i.e., ROMP 76A at Polk City) has a smooth HREE-enrichment pattern with a high $(Er/Nd)_{SN}$ ratio. However, along the groundwater flow path, REE fractionation patterns tend to flatten and the $(Er/Nd)_{SN}$ ratios generally decrease. These changes in REE fractional pattern along the flow path probably reflect the release of REEs in the Upper Floridan aquifer due to incongruent dissolution during dedolomitization. As discussed above, the dissolution of gypsum and dolomite, oxidization of organic matter, and ferric hydroxides dissolution occurring in the Upper Floridan aquifer could account for the release of REEs from solid aquifer phases. Due to different adsorption behavior, more LREEs are expected to be released into groundwaters and thus flatten the REE fractionation pattern.

Solution complexation of REEs can also affect the fractionation patterns of dissolved REEs (e.g., Sholkovitz et al., 1994; Johannesson and Lyons, 1994). As shown in Fig. 12, Nd (LREE) is predicted to mainly occur as organic complexes (NdHM) in UFA groundwaters, whereas a significant proportion (10-40%) of Er (HREE) exists as carbonato and dicarbonato complexes $(ErCO_3^+ + Er(CO_3)_2^-)$. The increase of DOC (indicated by our measured values) and decrease of free carbonate ion concentrations $([CO_3^{2-}]_F$, indicated by our calculation using PHREEQC, Parkhurst and Appelo, 1999) along the flow path (Table 2) could lead to an increase of LREE and decrease of HREE concentrations and thus, explain the REE fractionation along the flow path.

The high conductivity and increase in dissolved Ca, Mg, SO_4, and Cl concentrations for the furthest down-gradient groundwaters indicate that salt water intrusion and upconing of deep brackish waters occur in the discharge area (Sacks et al., 1995). The mixing of fresh groundwater with seawaters and deep brackish waters likely cause flocculation of organic colloids, subsequently inducing the removal of REEs in order of LREEs > MREEs > HREEs (Sholkovitz, 1995). Our measured DOC and Nd (LREE) concentrations support the possibility of this process. Thus, the mixing of groundwaters with seawaters and deep brackish waters in coastal regions also apparently fractionates the REE patterns.

6. Conclusions

Aqueous REE concentrations along groundwater flow paths exhibit regular but different variations in the Carrizo sand aquifer and Upper Floridan carbonate aquifer.

Rare earth element concentrations generally decrease along the groundwater flow path in the Carrizo sand aquifer, whereas in the Upper Floridan aquifer, the Nd concentrations increase first and then slightly decrease, Gd concentrations are erratic, and Er tends to decrease along the groundwater flow path. Shale-normalized REE patterns are enriched in the HREEs for groundwaters from the recharge regions of both aquifers and tend to flatten with flow down-gradient.

Speciation calculations predict that carbonato ($LnCO_3^+$) and/or dicarbonato complexes ($Ln(CO_3)_2^-$) are the main dissolved species for all REEs in groundwaters from the Carrizo sand aquifer, whereas organic complexes of the REEs (LnHM) are dominant in all but two furthest down-gradient groundwaters from the Upper Floridan aquifer. For the two furthest down-gradient groundwaters (i.e., ROMP 5-2 and 5-1) of UFA, carbonato ($LnCO_3^+$) and dicarbonato complexes ($Ln(CO_3)_2^-$) are more important than LnHM in these blackish groundwaters.

The variations of REE concentrations and fractionation patterns along the flow paths reflect water-rock reactions experienced by REEs in groundwaters from both aquifers. In the Carrizo sand aquifer, chemical weathering of Al-silicates and ferric hydroxide precipitation in the shallow, up-gradient portion of the aquifer, and adsorption appears to be important, removing dissolved REEs, especially LREEs, from the groundwaters as suggested by the decrease in REE concentrations and flattening in shale-normalized REE patterns along the flow path. In Upper Floridan groundwaters, due to dissolution of gypsum and dolomite, oxidation of organic matter and ferric hydroxide dissolution occurring along the flow path, release of REEs into the groundwaters is suggested to be an important process. The release of REEs, mainly LREEs, into groundwaters flattens the HREE-enrichment fractionation patterns.

The solution complexation of REEs affects the REE concentrations and fractionation patterns in groundwaters along the flow path through the following two main ways: (1) the complexation capability of REEs with carbonate ions and organic ligands varies with increasing atomic number of REEs; and (2) different complexation forms (i.e., negative or positive) at different geochemical conditions of groundwaters (i.e., pH, alkalinity, and DOC concentrations).

REE phosphate coprecipitates may set the upper limit of REE concentrations in groundwaters for both aquifers and tends to affect shale-normalized patterns of REEs by enriching the LREE in the solid phosphate phase.

Acknowledgements

The authors are especially thankful to Dr. L. Douglas James, Hydrologic Sciences program director for NSF hydrologic sciences for making the study possible, as part of NSF grant EAR-0303761 to Karen H. Johannesson. The authors would like to express gratitude to Ms. Shama Haque for drafting the maps, Dr. Matthew I. Leybourne at the University of Texas at Dallas for use of his ICP-OES and ICP-MS for some of these analyses, Dr. David J. Burdige at Old Dominion University for DOC analyses, Dr.

Andrew Kruzic at the University of Texas at Arlington for use the IC, and Drs. Zhonxing Chen and Cynthia Jones at Old Dominion University for the REE analyses.

References

Back W. and Hanshaw B. B. 1970 Comparison of chemical hydrogeology of the carbonate peninsulas of Florida and Yucatan. *Journal of Hydrology*, **10**, 330-368.

Banner J. L., Wasserburg G. J., Dobson P. F., Carpenter A. B., and Moore C. H. 1989 Isotopic and trace element constraints on the origin and evolution of saline groundwaters from central Missouri. *Geochimica et Cosmochimica Acta*, **53**, 383-398.

Benedict F.C., Jr., DeCarlo E.H., and Roth M. 1997 Kinetics and thermodynamics of dissolved rare earth uptake by alluvial materials from the Nevada Test site, southern Nevada, U.S.A. In Geochemistry (ed. X. Xuejin), Vol. 19, pp. 173-188. Proc. 30th Int'l. Geol. Congr., VSP, Utrecht.

Bertram C. J. and Elderfield H. 1993 The geochemical balance of the rare earth elements and neodymium isotopes in the oceans. *Geochimica et Cosmochimica Acta*, **57**, 1957-1986.

Braun J.-J., Pagel M., Herbillon A., and Rosin C. 1993 Mobilization and redistribution of REEs and thorium in a syenitic lateritic profile: A mass balance study. *Geochimica et Cosmochimica Acta*, **57**, 4419-4434.

Buddemeier R. W. and Hunt J. R. 1988 Transport of colloidal contaminants in groundwater: radionuclide migration at the Nevada Test Site. *Applied Geochemistry*, **3**, 535-548.

Byrne R.H. and Kim K.H. 1993 Rare earth precipitation and coprecipitation behavior: The limiting role of PO_{43}- on dissolved rare earth concentrations in seawater. *Geochimica et Cosmochimica Acta*, **57**, 519-526.

Byrne R.H., Liu X., and Schijf J. 1996 The influence of phosphate coprecipitation on rare earth distributions in natural waters. *Geochimica et Cosmochimica Acta*, **60**, 3341-3346.

Castro M. C., Stute M., and Schlosser P. 2000 Comparison of ^{4}He and ^{14}C ages in simple aquifer systems: implications for groundwater flow and chronologies. *Applied Geochemistry*, **15**, 1137-1167.

Dia A., Grua G., Olivié-Lauquet G., Riou C., Molénat J., and Curmi P. 2000 The distribution of rare earth elements in groundwaters: Assessing the role of source-rock composition, redox changes, and colloidal particles. *Geochimica et Cosmochimica Acta*, **64**, 4131-4152.

Edmunds W.M., Guendouz A.H., Mamou A., Moulla A., Shand P., and Zouarid K. 2003 Groundwater evolution in the Continental Intercalaire aquifer of southern Algeria and Tunisia: trace element and isotopic indicators. *Applied Geochemistry*, **18**, 805-822.

Elderfield H. 1988 The oceanic chemistry of the rare-earth elements. *Philosphical Transactions-Royal Society of London,* **A325**, 105-126.

Elderfield H. and Greaves M.J. 1982 The rare earth elements in seawater. *Nature*, **296**, 214-219.

Elderfield H., Upstill-Goddard, R., and Sholkovitz, E.R. 1990 The rare earth elements in rivers, estuaries, and coastal seas and their significance to the composition of ocean waters. *Geochimica et Cosmochimica Acta*, **54**, 971-991.

Fee J. A., Gaudette H. E., Lyons W. B., and Long D. T. 1992 Rare earth element distribution in the Lake Tyrrell groundwaters, Victoria, Australia. *Chemical Geology*, **96**, 67-93.

Goldstein S. J. and Jacobsen S. B. 1988a REE in the Great Whale River estuary, northwest Quebec. *Earth and Planetary Science Letters*, **88**, 241-252.

Goldstein S.J. and Jacobsen S.B. 1988b Rare earth elements in river waters. *Earth and Planetary Science Letters*, **89**, 35-47.

Gosselin D G., Smith M. R., Lepel E. A., and Laul J. C. 1992 Rare earth elements in chloride-rich groundwater, Palo Duro Basin, Texas, USA. *Geochimica et Cosmochimica Acta*, **56**, 1495-1505.

Hamlin S. H. 1988 Depositional and ground-water flow systems of the Carrizo-upper Wilcox, South Texas. Report of Investigations - Texas, University, Bureau of Economic Geology, Rept. 175, 61 pp.

Johannesson K. H. and Lyons W. B. 1994 The rare earth element geochemistry of Mono Lake water and the importance of carbonate complexing. *Limnology and Oceanography*, **39**, 1141-1154.

Johannesson K. H. and Lyons W. B. 1995 Rare-earth element geochemistry of Colour Lake, an acidic freshwater lake on Axel Heiberg Island, Northwest Territories, Canada. *Chemical Geology*, **119**, 209-223.

Johannesson K. H. and Hendry M. J. 2000 Rare earth element geochemistry of groundwaters from a thick till and clay-rich aquitard sequence, Saskatchewan, Canada. *Geochimica et Cosmochimica Acta*, **64**, 1493-1509.

Johannesson K.H., Lyons W.B., Yelken M.A., Gaudette H.E., and Stetzenbach K.J. 1996a Geochemistry of the rare earth elements in hypersaline and dilute acidic natural terrestrial waters: complexation behavior and middle rare earth enrichments, *Chemical Geology*, **133**, 125-144.

Johannesson K. H., Stetzenbach K. J., Hodge V. F., and Lyons W. B. 1996b Rare earth element comlpexation behavior in circumneutral pH groundwaters: Assessing the role of carbonate and phosphate ions. *Earth and Planetary Science Letters*, **139**, 305-319.

Johannesson K. H., Stetzenbach K. J., Hodge V. F., Kreamer D. K., and Zhou X. 1997a Delineation of ground-water flow systems in the southern Great Basin using aqueous rare earth element distributions. *Ground Water,* **35**, 807-819.

Johannesson K. H., Stetzenbach K. J., and Hodge V. F. 1997b Rare earth elements as geochemical tracers of regional groundwater mixing. *Geochimica et Cosmochimica Acta*, **61**, 3605-3618.

Johannesson K. H., Farnham I. M., Guo C., and Stetzenbach K. J. 1999 Rare earth element fractionation and concentration variations along a groundwater flow path within a shallow, basin-fill aquifer, southern Nevada, USA. *Geochimica et Cosmochimica Acta*, **63**, 2697-2708.

Johnston R.H., Krause R.E., Mayer F.W., Ryder P.D., Tibbals C.H., and Hunn J.D. 1980 Estimated potentiometric surface for the Tertiary limestone aquifer system, southeastern United States, prior to development. *U.S. Geological Survey, Open-file Report*, 80-406.

Jonasson R.G., Bancroft G.M., and Nesbitt H.W. 1985 Solubilities of some hydrous REE phosphates with implications for diagenesis and seawater concentrations. *Geochimica et Cosmochimica Acta*, **49**, 2133-2139.

Jones I.C., Vacher H.L., and Budd D.A. 1993 Transport of calcium, magnesium and SO4 in the Floridan aquifer, west-central Florida: implications to cementation rates. *Journal of Hydrology*, **143**, 455-480.

Kersting A. B., Efurd D. W., Finnegan D. L., Rokop D. J., Smith D. K., and Thompson J. L. 1999 Migration of plutonium in ground water at the Nevada Test Site. *Nature*, **397**, 56-59.

Klungness G.D. and Byrne R.H. 2000 Comparative hydrolysis behavior of the rare earths and yttrium: the influence of temperature and ionic strength. *Polyhedron*, **19**, 99-107.

Koeppenkastrop D. and DeCarlo E. H. 1993 Uptake of rare earth elements from solution by metal oxides. *Environmental Science and Technology*, **27**, 1796-1802.

Lee J. H. and Byrne R.H. 1992 Examination of comparative rare earth element complexation behavior using linear free-energy relationships. *Geochimica et Cosmochimica Acta*, **56**, 1127-1137.

Lee J. H., and Byrne, R. H. 1993 Complexation of trivalent rare earth elements (Ce, Eu, Gd, Tb, Yb) by carbonate ions. *Geochimica et Cosmochimica Acta*, **57**, 295-302.

Leybourne M.I., Goodfellow W.D., Boyle D.R. and Hall G.M. 2000 Rapid development of negative Ce anomalies in surface waters and contrasting REE patterns in groundwaters associated with Zn-Pb massive sulfide deposits. *Applied Geochemistry*, **15**, 695-723.

Luo Y.R. and Byrne 2000 The ionic strength dependence of rare earth and yttrium fluoride complexation at 25°C. *Journal of Solution Chemistry*, **29**, 1089-1099.

Marley N. A., Gaffney J. S., Orlandini K. A., and Cunningham M. M. 1993 Evidence for radionuclide mobilization in a shallow, sandy aquifer. *Environmental Science and Technology*, **27**, 2456-2461.

McCarthy J. F., Sanford W. E., and Stafford P. L. 1998a Lanthanide field tracers demonstrate enhanced transport of transuranic radionuclides by natural organic matter. *Environmental Science and Technology*, **32**, 3901-3906.

McCarthy J. F., Czerwinski K. R., Sanford W. E., Jardine P. M., and Marsh J. D. 1998b Mobilization of transuranic radionuclides from disposal trenches by natural organic matter. *Journal of Contaminant Hydrology*, **30**, 49-77.

Miller J. A. 1986 Hydrogeological framework of the Floridan Aquifer system in Florida and in parts of Georgia, Alabama, and South Carolina. *U.S. Geological Survey Professional Paper*, 1403-B.

Millero F.J. 1992 Stability constants for the formation of rare earth inorganic complexes as a function of ionic strength. *Geochimica et Cosmochimica Acta*, **56**, 3123-3132.

Milliken K.L., Mack L.E., and Land L.S. 1994 Elemental mobility in sandstones during burial: whole-rock chemical and isotopic data, Frio Formation, south Texas. *Journal of Sedimentary Research*, **64**, 788-796.

Möller P., Dulski P., Bau M., Knappe A., Pekdeger A., and Sommer-von Jarmersted C. 2000 Anthropogenic gadolinium as a conservative tracers in hydrology. *Journal of Geochemical Exploration*, **69-70**, 409-414.

Nelson D. M., Penrose W. R., Karttunen J. O., and Mehlhaff P. 1985 Effects of dissolved organic carbon on the adsorption properties of plutonium in natural waters. *Environmental Science and Technology*, **19**, 127-131.

Nesbitt H.W. 1979 Mobility and fractionation of rare earth elements during weathering of a granodiorite. *Nature*, **279**, 206-210.

Parkhurst D.L. and Appelo C.A.J. 1999 User's guide to PHREEQC—A Computer program for speciation, batch-reaction, one-dimensional transport, and inverse geochemical calculations. U.S. Geol. Surv. *USGS Water Resources Investigations Report*, 99-4259.

Pearson F.J., Jr. and White D.E. 1967 Carbon 14 ages and flow rates of water in Carrizo sand, Atascosa County, Texas. *Water Resources Research*, **3**, 251-261.

Penrose W. R., Polzer W. L., Essington E. H., Nelson D. M., and Orlandini K. A. 1990 Mobility of plutonium and americium through a shallow aquifer in a semiarid region. *Environmental Science and Technology*, **24**, 228-234.

Plummer L. N. 1977 Defining reactions and mass transfer in part of the Florida aquifer. *Water Resources Research*, **13**, 801-812.

Plummer L. N. and Sprinkle C. L. 2001 Radiocarbon dating of dissolved inorganic carbon in groundwater from confined parts of the Upper Floridan Aquifer, Florida, USA. *Hydrogeology Journal*, **9**, 127-150.

Plummer L. N., Parkhurst D. L., and Thorstenson D. C. 1983 Development of reaction models for ground-water systems. *Geochimica et Cosmochimica Acta*, **47**, 665-686.

Sacks L.A., Herman J.S., and Kauffman S.J. 1995 Controls on high sulfate concentrations in the Upper Floridan aquifer in southwest Florida. *Water Resources Research*, **31**, 2541-2551.

Schau M. and Henderson J.B. 1983 Archean chemical weathering at three localities on the Canadian Shield. *Precambrian Research*, **20**, 189-224.

Sholkovitz E. R. 1976 Flocculation of dissolved organic and inorganic matter during the mixing of river water and seawater. *Geochimica et Cosmochimica Acta*, **40**, 831-845.

Sholkovitz E. R. 1988 Rare earth elements in the sediments of the North Atlantic Ocean, Amazon Delta, and East China Sea: Reinterpretation of terrigenous input patterns to the oceans. *American Journal of Science*, **288**, 236-281.

Sholkovitz E. R. 1993 The geochemistry of the rare earth elements in the Amazon River estuary. *Geochimica et Cosmochimica Acta*, **57**, 2181-2190.

Sholkovitz E. R. 1995 The aquatic chemistry of the rare earth elements in rivers and estuaries. *Aquatic Geochemistry*, **1**, 1-34.

Sholkovitz E. R., Landing W.M., and Lewis B.L. 1994 Ocean particle chemistry: the fractionation of rare earth elements between suspended particles and seawater. *Geochimica et Cosmochimica Acta*, **58**, 1567-1579.

Smedley P. L. 1991 The geochemistry of rare earth elements in groundwater from the Carnmenellis area, southwest England. *Geochimica et Cosmochimica Acta*, **55**, 2767-2779.

Stanley J.K., Jr. and Byrne R.H. 1990 The influence of solution chemistry on REE uptake by Ulva lactuca L. in seawater. *Geochimica et Cosmochimica Acta*, **54**, 1587-1596.

Stute M., Schlosser P., Clark J. F., and Broecker W. S. 1992 Paleotemperatures in the southwestern United States derived from noble gases in ground water. *Science*, **256**, 1000-1003.

Tang J. and Johannesson K.H. 2003 Speciation of rare earth elements in natural terrestrial waters: assessing the role of dissolved organic matter from the modeling approach. *Geochimica et Cosmochimica Acta*, **67**, 2321-2339.

Tipping E. and Hurley, M.A. 1992 A unifying model of cation binding by humic substances. *Geochimica et Cosmochimica Acta*, **56**, 3627-3641.

Tipping E. 1993 Modelling the binding of Europium and Actinides by humic substances. *Radiochimica Acta*, **62**, 141-152.

Tipping E. 1994 WHAM—A chemical equilibrium model and computer code for waters, sediments, and soils incorporating a discrete site/electrostatic model of ion-binding by humic substances. *Computers & Geosciences*, **20**, 973-1023.

Wicks C.M. and Herman J.S. 1994 The effect of a confining unit on the geochemical evolution of groundwater in the Upper Floridan aquifer system. *Journal of Hydrology*, **153**, 139-155.

Chapter 10

RARE EARTH ELEMENTS (REE) AND Nd AND Sr ISOTOPES IN GROUNDWATER AND SUSPENDED SEDIMENTS FROM THE BATHURST MINING CAMP, NEW BRUNSWICK: WATER-ROCK REACTIONS AND ELEMENTAL FRACTIONATION

MATTHEW I. LEYBOURNE[1] & BRIAN L. COUSENS[2]

[1]*Department of Geosciences, University of Texas at Dallas, Richardson, Texas 75083-0688 USA*

[2]*Department of Earth Sciences, Carleton University, Ottawa, Ontario, Canada K1S 5B6*

1. Introduction

It is important to better understand metal fractionation between suspended sediments and groundwaters to improve understanding of metal mobility during weathering and water-rock reaction. It is also important to make the distinction between metals that are transported hydromorphically from those that are transported in detrital form, as this influences interpretations of metal reactivity and bioavailibility. The REE are important because although they generally display coherent behavior across the series, differences exist in terms of particle reactivity (primarily through adsorption) and speciation as a function of atomic weight (Johannesson et al., 1999; Johannesson and Hendry, 2000). In addition, some of the REE are redox sensitive; with the ability to reflect changes in water properties, e.g., Leybourne et al. (2000) showed that the redox transformation of Ce^{3+} to Ce^{4+} may be rapid. The REE are therefore interesting because although commonly considered to be strongly adsorbed to particles, REE mobility is enhanced at low pH, at high pH for the HREE (Johannesson and Lyons, 1994) and in the presence of organic matter (Andersson et al., 2001; Aubert et al., 2001; Banner et al., 1989; Möller et al., 1998). The REE have also been used as conservative tracers to model groundwater flow and mixing in large aquifer systems (Johannesson et al., 1997).

Several studies have investigated contributions of rivers to oceans with respect to REE budgets and Nd isotopic compositions (Andersson et al., 2001; Goldstein and Jacobsen, 1988), but comparatively few have looked at the relationship between groundwaters and aquifer materials/suspended sediments, in particular, in small catchments. Radiogenic isotopes are useful in hydrogeochemical studies for understanding the styles and extents of water-rock reactions and, by extension, understanding elemental mobilization and dispersion in the surficial environment (Goldstein and Jacobsen, 1988). One of the purposes of this study was to investigate the utility of analyzing the Nd isotopic composition of dissolved REE in groundwaters. Previous studies of river catchments have focused on the analysis of Nd isotopes on

K.H. Johannesson,(ed), Rare Earth Elements in Groundwater Flow Systems , 253-293.

suspended sediments, primarily to estimate global weathering and river contributions to oceanic elemental budgets (e.g. Allègre et al., 1996). There have been relatively few studies on the Nd isotopic composition of the "dissolved" REE load of terrestrial waters (e.g. Andersson et al., 2001; Aubert et al., 2001; Banner et al., 1989; Möller et al., 1998).

In this paper we use the REE contents of host rocks, groundwaters and suspended sediment to improve our understanding of partitioning of REE between aqueous and solid phases. We also report Sr and Nd isotopic compositions on the groundwaters and representative host lithologies in order to help constrain ideas of water-rock interaction and preferential weathering.

2. Geological Setting

The areas investigated are part of the Miramichi terrane of northern New Brunswick, Canada (Figure 1). The Cambro-Ordovician Miramichi Group is the oldest unit in the study area and is interpreted as representing flysch deposited on a continental margin (van Staal et al., 2003). The California Lake and Tetagouche groups conformably to disconformably overlie Miramichi group rocks, are more or less coeval (van Staal et al., 1992), and are composed of a lower predominantly felsic volcanic part and an upper, predominantly mafic volcanic part. California Lake and Tetagouche rocks are interpreted to have formed during development of an ensialic back-arc basin. Volcanogenic massive sulfide (VMS) deposits, hosted by felsic volcanic and volcaniclastic rocks, are common in California Lake and Tetagouche group rocks (in particular the Mount Brittain and Nepisiguit Falls formations, respectively; Figure 1). California Lake group rocks are tectonically overlain by the Fournier Group, which is dominated by mafic back-arc basin rocks. The Ordovician sequence is unconformably overlain by shallow marine sedimentary rocks of the Chaleurs Group.

The VMS deposits typically consist of fine- to medium-grained, layered, stratiform lenses with pyrite, sphalerite, galena, chalcopyrite and pyrrhotite as the dominant sulfide minerals, and lesser arsenopyrite, marcasite, tetrahedrite, and bournonite. Disseminated mineralization, mainly pyrite, underlies most of the VMS lenses but is generally absent in hanging-wall rocks. Some of the stratiform deposits are underlain by discordant stockwork zones containing pyrrhotite, pyrite, and chalcopyrite (McAllister, 1960). Many of the stratiform sulfide deposits are overlain by, or have laterally equivalent oxide (± carbonate and silicate) iron-formation (Peter and Goodfellow, 1996).

The Halfmile Lake deposit is a moderate sized VMS deposit, hosting around 26 Mt of massive sulfides. The deposit is structurally complex, is overturned, and dips steeply to the north and northwest. Felsic quartz-porphyry rocks, felsic pyroclastic rocks and fine-grained volcaniclastic sedimentary rocks (Nepisiguit Falls Formation) host the massive sulfide mineralization. Mafic to intermediate flows and pyroclastic rocks (Flat Landing Brook Formation) form the structural footwall to the deposit. The bulk of the sulfide mineralization (as breccia matrix and laminated sulfides) occurs at depth, although sulfides locally crop out on a local topographic high as a well developed gossan (formed by oxidation in the Late Tertiary; Symons et al., 1996).   Pyrrhotite - quartz -

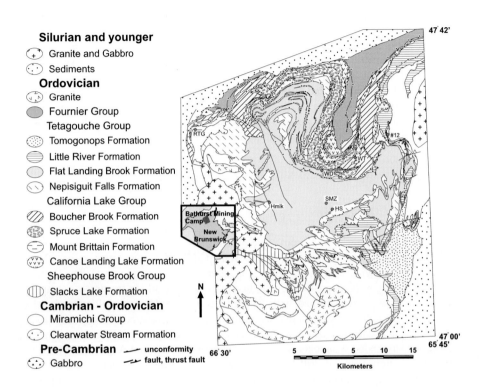

Figure 1. Location and regional geology of the Bathurst Mining Camp. Also shown is the location of the deposits from which groundwaters were sampled in this study. Hmlk = Halfmile Lake deposit; RTG = Restigouche deposit; SMZ = Stratmat Main Zone; WT = Willet deposit; WD = Wedge mine; HS = Heath Steele mine; #12 = Brunswick #12 mine.

chalcopyrite stringers from 3 to 150 m thickness characterize the stockwork zone (Adair, 1992).

VMS-hosting rocks at the Restigouche deposit are dominantly felsic volcanic and associated sedimentary rocks (Mount Brittain Formation; McCutcheon, 1997). The felsic volcanic rocks are underlain by older metasedimentary rocks and conformably overlain by mafic volcanic and associated sedimentary rocks. The Restigouche deposit is cross-cut by a NE-trending mafic dyke swarm (Barrie, 1982). Quartz and carbonate veins are common in both footwall and hanging wall units; feldspar veins are developed locally. Massive sulfides are shallow (<200 m depth) and crop-out in a cross-cutting stream bed (Charlotte Brook) as a small gossan. Massive sulfides are dominantly pyrite, marcasite, sphalerite and galena with minor chalcopyrite (Barrie, 1982).

3. Methods

3.1 SAMPLE COLLECTION

Groundwater was recovered from several undisturbed VMS deposits (Restigouche, Halfmile Lake, Stratmat Main Zone and Willett deposits), two active mines (Heath Steele and Brunswick #12) and a past producing mine (Wedge) (Figure 1). The Halfmile Lake and Restigouche deposits were studied in greatest detail. At the time of sampling, both deposits were undisturbed with the exception of exploration drilling undertaken since the 1950's. Subsequent to our sampling in 1996, the Restigouche deposit was put into open-pit production and has subsequently ceased operation. Groundwaters and associated suspended sediments were collected between June and August (1994 - 1996) from exploration diamond drill holes. Boreholes were drilled between 2 and 40 years prior to sampling, sufficient time for equilibrium groundwater flow conditions to be re-established. Local surface water sources (streams and ponds) were typically used as drilling fluid. Groundwaters were recovered by flow-through bailer and straddle-packer systems. These sampling systems have been described in detail elsewhere (Leybourne et al., 2002; Leybourne et al., 2000). The great advantage of the straddle-packer system is that it permits sampling of groundwater at discrete depth intervals, as opposed to the standing borehole water column sampled by the flow-though bailer. During development of a zone with the straddle-packer, the suspended sediment content of the recovered groundwater typically decreased with time. Groundwater recovered by bailer typically had greater suspended sediment loads than by straddle-packer, most likely due to disturbance during bailer deployment.

Specific conductance, pH, Eh (oxidation-reduction potential), temperature, and dissolved oxygen (DO) were measured in the field. Suspended sediments were collected on 0.45 μm filter papers, air dried and transferred to glass vials prior to dissolution and analysis. Waters (cations and anions) were collected in 60 or 125-mL Nalgene bottles. Cation and anion samples were filtered in situ through 0.45 μm filters and acidified (cations) with ultrapure nitric acid (to 0.4%) at base camp. All samples were stored refrigerated prior to analysis. At the Halfmile Lake deposit, 12 boreholes were sampled, most at multiple depths as deep as 710 m. At the Restigouche deposit, 13 boreholes were sampled, as deep as 550 m.

3.2 GROUNDWATER ANALYSIS

Major and trace ions were measured by inductively coupled plasma optical emission spectrometry (ICP-OES; Si, Al, Fe, Mg, Ca, Na, K, B, Ba, Cu, Li, Sr and Zn) and inductively coupled plasma mass spectrometry (ICP-MS; Co, Cr, Cs, Mn, Pb, Rb, Sb, U, V, Al, As, Ba, Cd, Cu, Mo, Ni, Se and Zn). Anions (SO_4, Cl, F, Br, PO_4, NO_3, and NO_2) were measured by ion chromatography and alkalinity by acid-base titration. Rare earth elements (REE) and selected trace elements (V, Co, Ni, Cu, Y, Mo, Cd, Th and U) were analyzed by ICP-MS following 5x pre-concentration by chelation, using the method of Hall et al. (1995). Pre-concentration was carried out on CC-1 chelating resin of macroporous iminodiacetate that preferentially retains free ionic REE and transition metals relative to the alkali and alkaline-earth elements. As all samples for cation analysis were acidified, this method is quantitative for the REE in these waters. The chelation ICP-MS technique allows determination of all 14 naturally occurring REE to very low detection limits (< 1 ng/L). For the REE, the following isotopes were measured in order to minimize the effects of isobaric interferences: ^{139}La, ^{140}Ce, ^{141}Pr, ^{146}Nd, ^{149}Sm, ^{151}Eu, ^{157}Gd, ^{159}Tb, ^{163}Dy, ^{165}Ho, ^{166}Er, ^{169}Tm, ^{172}Yb, and ^{175}Lu. Bromide and PO_4 were also measured by ICP-MS for improved detection limits (detection limits of 0.54 and 1.1µg/L, respectively).

REE-speciation calculations were performed on all groundwaters (Leybourne et al., 2000). Speciation calculations were based on calculated contents of the major inorganic free (uncomplexed) ligands (OH, F, Cl, HCO_3^-, CO_3^{2-}, $H_2PO_4^-$, HPO_4^{2-}, PO_4^{3-}, SO_4^{2-}), as determined by PHREEQC (using the WATEQ4F thermodynamic database). Stability constants for the REE species were taken from Millero (1992) except those for carbonate species, which were taken from Lee and Byrne (1993) and phosphate, which were taken from Lee and Byrne (1992).

3.3 SUSPENDED SEDIMENT AND WHOLE ROCK ANALYSIS

Thirty-one samples of suspended sediment were analyzed by ICP-MS for the rare earth elements. Details of the mineralogy and major and trace element compositions of the groundwater suspended sediments are described elsewhere (Leybourne, 2001). Major elements in the suspended sediments were analyzed by ICP-OES (TiO_2, Al_2O_3, Fe_2O_3, MnO, MgO, CaO, Na_2O, and K_2O) and trace elements by ICP-MS (Ag, Ba, Be, Bi, Cd, Co, Cr, Cs, Cu, Ga, Hf, In, Mo, Nb, Ni, Pb, Rb, Sc, Sn, Sr, Ta, Th, Tl, U, V, Y, Zn, Zr and the REE). Suspended sediments were taken into solution by multi-acid digestion (HCl-HF-HNO_3-$HClO_4$), with any residue undergoing lithium metaborate fusion and further acid dissolution. Silica was not measured due to low sample volumes.

Representative whole-rock samples (mafic and felsic volcanics and massive sulfides) from boreholes that intersect mineralization at the Restigouche and Halfmile Lake deposits as well as background boreholes more distal from mineralization were also analyzed by ICP-OES, XRF (X-Ray fluorescence; major elements) and ICP-MS (trace elements and REE) following multi-acid digestion as described above.

3.4 Sr AND Nd ISOTOPIC ANALYSES

Groundwaters and whole rocks were analyzed for Sr and Nd isotopes at Carleton University using the procedures described by Cousens (1996). Groundwater was decanted into Teflon bombs and evaporated under clean laboratory conditions to achieve the required Sr and Nd concentrations. For Nd, only a few samples had sufficient Nd and available sample volume to attempt analysis. Samples of representative mafic dykes and felsic metavolcanic rocks, as well as two samples of vein carbonate, were also analyzed. Carbonate was separated from silicate minerals by dissolution in dilute HCl. Rocks were crushed to < 200 mesh in an agate mill, with between 100 and 300 mg of powder dissolved in 50% HF/8 N HNO$_3$, 8 N HNO$_3$, 6 N HCl, and finally, 2.5 N HCl.

Strontium was separated chromatographically in Bio-Rad 15-ml Econocolumns and Dowex AG-50-X8 cation resin. The resin was pre-cleaned with 6 N HCl. Strontium was then eluted using 15 ml of 2.5 N HCl. For samples being analyzed for Nd isotopes, the REEs were eluted after Sr in 6 N HCl.

The REE residue was dissolved in 0.26 N HCl, and loaded into 15-ml Bio-Rad Econocolumns with a 2-cm high bed of HDEHP-coated Teflon® powder capped with a thin layer of anion resin (see Richard et al., 1976). Neodymium was eluted with 0.26N HCl. For mass spectrometry, Sr fractions were loaded with H$_3$PO$_4$ onto Ta filaments and run at temperatures of 1450-1500°C. ^{87}Sr/^{86}Sr ratios were corrected for fractionation using an ^{86}Sr/^{88}Sr = 0.1186. Fifty runs (September 1996-May 1998) of the NBS 987 standard yielded an average ^{87}Sr/^{86}Sr = 0.710251 ± 18 (2-σ). Neodymium was loaded with HNO$_3$ onto one side of a Re double-filament assembly and run as a metal species at temperatures of 1800-1850°C. ^{143}Nd/^{144}Nd ratios are corrected for fractionation using a ^{146}Nd/^{144}Nd = 0.7219. Twenty-two runs (September 1996-May 1998) of the La Jolla Nd standard yielded an average ^{143}Nd/^{144}Nd ratio of 0.511877 ± 18 (2-σ).

It should be noted that all Sr and Nd isotope ratios presented hereafter represent present-day ratios.

4. Results

4.1 REE IN HOST METAVOLCANIC AND METASEDIMENTARY ROCKS

Major lithologies in the study areas are dominantly felsic volcanic and volcaniclastic rocks, with younger mafic volcanics, sills and dykes. Excellent discussions of the geochemistry of the felsic and mafic host rocks are given by Lentz (1996), Rogers (1995) and Rogers et al. (in press). The compositions of the least altered felsic rocks from both deposits were averaged to provide normalizing values for waters and sediments. Most of the rocks analyzed in this study have flat REE patterns, normalized to North American Shale Composite (NASC), with the exception of mafic rocks, massive sulfides, and rocks with extensive carbonate alteration (Figure 2). The

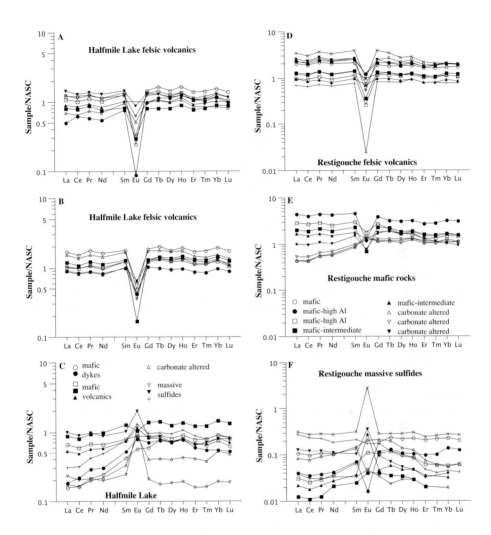

Figure 2. A, B, C; Rare-earth element concentrations of Halfmile Lake lithologies from borehole HN94-65 normalized to North American Shale Composite (NASC). D, E, F; Rare-earth element concentrations of Restigouche deposit lithologies normalized to North American Shale Composite (NASC). Normalizing values are from Haskin et al. (1968) with Dy value from Gromet et al. (1984). Data from Table 1.

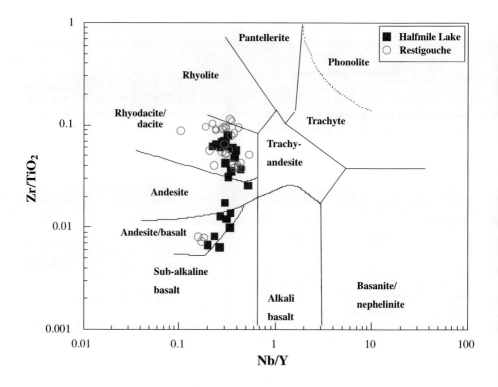

Figure 3. Discrimination diagram showing the variation in composition of the host rocks from the Halfmile Lake and Restigouche deposits. The mafic rocks analyzed in this study are tholeiitic dykes. Boucher Brook mafic volcanic rocks are typically alkalic. Discrimination diagram from (Winchester and Floyd, 1977). Data from Leybourne (1998).

lithologies at both deposits show a bimodal distribution between dacitic to rhyolitic volcanics and mafic tholeiites (Figure 3). Host lithologies from the Halfmile Lake deposit have essentially no Ce anomalies ([Ce/Ce*]$_{NASC}$ varies from 0.88 to 1.14). Felsic lithologies generally exhibit negative Eu anomalies ([Eu/Eu*]$_{NASC}$ = 0.103 -

0.693), whereas mafic dykes ($[Eu/Eu^*]_{NASC}$ = 0.79 to 1.52) and massive sulfides ($[Eu/Eu^*]_{NASC}$ = 2.17 - 4.95) generally have positive Eu anomalies. The positive Eu anomalies for the massive sulfides (Figure 2) are typical of massive sulfides observed in modern settings (hydrothermal deposits at mid-ocean ridges) due to the redox conditions of ore-forming hydrothermal fluids (Klinkhammer et al., 1994a; Klinkhammer et al., 1994b). The most REE-depleted samples are the massive sulfides that also display the greatest depletion in the HREE (Figure 2). Mafic rocks show two different trends. Two mafic dykes have strongly LREE depleted patterns typical of tholeiitic rocks. The other mafic dykes have patterns similar to the felsic rocks (Figure 2).

Least altered felsic metavolcanic rocks at the Restigouche Deposit typically have higher REE contents than those from the Halfmile Lake deposit. For example, least altered felsic volcanic rocks from the Halfmile Lake deposit have average ΣREE = 182 ppm, $[La/Yb]_{NASC}$ = 0.90, $[Gd/Yb]_{NASC}$ = 1.02, $[La/Sm]_{NASC}$ = 0.91 and $[Eu/Eu^*]_{NASC}$ = 0.34. In contrast, least altered felsic units from the Restigouche deposit have average ΣREE = 336 ppm, $[La/Yb]_{NASC}$ = 1.16, $[Gd/Yb]_{NASC}$ = 1.24, $[La/Sm]_{NASC}$ = 0.88 and $[Eu/Eu^*]_{NASC}$ = 0.41. For Restigouche deposit lithologies, negative europium anomalies are ubiquitous (average $[Eu/Eu^*]_{NASC}$= 0.40) with the exception of highly carbonate-altered rocks, mafic sills, and massive sulfides, which tend to have positive to highly positive Eu anomalies (Figure 2). Two stockwork sulfide samples have strong negative Eu anomalies, in contrast to the massive sulfide samples. Mafic rocks have LREE-depleted patterns and lower overall REE abundances compared to the felsic volcanic rocks. Massive sulfide samples have the lowest REE abundances (Figure 2).

4.2 REE IN GROUNDWATERS

Groundwaters at the Halfmile Lake deposit are characterized by low total dissolved solids (TDS < 230 mg/L), and Ca-HCO$_3$ to Na-HCO$_3$ compositions. In contrast, groundwaters at the Restigouche deposit range up to > 20,000 mg/L (TDS), with Ca-HCO$_3$, Na-HCO$_3$, Na-Cl and Na-Ca-SO$_4$ compositions.

Halfmile Lake groundwaters vary from slightly LREE-enriched to strongly LREE depleted compared to NASC. Many of these waters have an unusual concave-up profile between the LREE and the MREE, and many are convex about the MREE (Figure 4). Fe-oxyhydroxide-rich groundwaters from a borehole that intersects the gossan have the lowest TDS contents at the Halfmile Lake deposit, yet have the highest ΣREE concentrations (up to ΣREE = 1816 ng/L). Whereas surface waters in the Bathurst Mining Camp are characterized by strongly developed negative Ce anomalies (average $[Ce/Ce^*]_{NASC}$ = 0.51), Halfmile Lake groundwaters show relatively weak Ce-anomalies ($[Ce/Ce^*]_{NASC}$ varies from 0.55 to 1.41, most ~ 1; Figure 4).

Although groundwaters at the Restigouche deposit are compositionally more variable than at the Halfmile Lake deposit, REE contents display less variation (ΣREE varies from 5.4 ng/L to 240 ng/L, average = 65.9 ng/L). Most Restigouche deposit groundwaters are LREE-depleted with $[La/Sm]_{NASC}$ ranging from 0.23 to 2.08 (average

Table 1. REE contents (in ppm) and ratios for Halfmile Lake and restigouche deposit whole rocks. [La/Yb] is NASC-normalized (i.e., SN). Ce/Ce* = $Ce_{SN}/(La_{SN}*Pr_{SN})^{1/2}$ and Eu/Eu* = $Eu_{SN}/(Sm_{SN}*Gd_{SN})^{1/2}$.

Sample	La	Ce	Pr	Nd	Sm	Eu	Gd	Tb	Dy	Ho	Er	Tm	Yb	Lu	[La/Yb]	Ce/Ce*	Eu/Eu*
Halfmile Lake																	
ML95-320-65	39	89	10	38	7.8	0.30	7.6	1.4	7.9	1.7	4.7	0.71	4.8	0.66	0.787	0.982	0.171
ML95-321-65	16	45	4.6	18	4.3	0.11	5.1	0.97	6.3	1.3	3.6	0.51	3.6	0.47	0.431	1.142	0.103
ML95-322-65	35	74	8.9	34	7.3	0.40	7.1	1.2	7.0	1.4	3.6	0.54	3.6	0.50	0.942	0.913	0.244
ML95-323-65	5.0	12	1.7	8.1	2.4	0.70	3.1	0.65	4.0	0.84	2.3	0.32	2.3	0.30	0.211	0.896	1.127
ML95-324-65	5.9	16	2.3	10	3.0	0.97	3.7	0.64	3.9	0.80	2.0	0.27	1.7	0.25	0.336	0.946	1.278
ML95-325-65	26	57	6.9	25	4.7	0.37	4.2	0.68	4.4	0.93	2.6	0.40	2.8	0.42	0.900	0.927	0.366
ML95-326-65	29	63	7.4	28	5.8		5.2	0.91	5.5	1.2	3.1	0.48	3.2	0.48	0.878	0.937	0.000
ML95-326-65b	22	47	5.8	23	5.4	0.63	5.1	0.88	5.4	1.1	2.9	0.42	2.6	0.39	0.820	0.906	0.527
ML95-328-65	26	57	7.0	28	5.8	0.77	6.0	0.98	5.6	1.1	3.1	0.45	3.1	0.48	0.813	0.920	0.573
ML95-329-65	40	83	9.8	38	7.1	0.77	6.4	1.1	6.6	1.4	3.9	0.59	4.1	0.56	0.945	0.913	0.502
ML95-330-65	10	22	2.6	9.9	2.0		1.9	0.32	1.9	0.39	1.1	0.17	1.3	0.20	0.745	0.940	0.000
ML95-331-65	39	82	10	41	7.9	0.99	6.8	1.1	6.5	1.4	4.1	0.58	4.0	0.55	0.945	0.904	0.593
ML95-332-65	46	94	11	43	8.2	1.1	6.9	1.1	6.4	1.3	3.7	0.54	3.9	0.56	1.143	0.910	0.642
ML95-333-65	54	110	14	53	10	0.60	9.6	1.7	9.7	2.0	5.7	0.87	6.0	0.83	0.872	0.871	0.269
ML95-334-65	21	43	5.3	22	4.5	1.1	4.5	0.76	4.4	0.93	2.6	0.39	2.8	0.39	0.727	0.888	1.073
ML95-335-65	28	59	7.4	32	7.2	1.3	7.2	1.2	6.8	1.4	4.1	0.61	4.5	0.63	0.603	0.893	0.793
ML95-336-65	17	36	4.5	19	4.2	1.4	4.3	0.69	3.9	0.84	2.3	0.34	2.5	0.34	0.659	0.896	1.446
ML95-338-65	29	61	6.9	27	5.6	0.52	5.3	0.83	5.1	1.0	2.9	0.42	3.0	0.42	0.936	0.939	0.419
ML95-339-65	60	120	14	54	8.8	1.3	7.7	1.4	8.8	1.7	4.7	0.66	4.5	0.61	1.292	0.902	0.693
ML95-341-65	32	71	8.4	33	6.9	0.52	6.5	1.2	6.9	1.4	3.9	0.57	3.8	0.50	0.816	0.943	0.341
ML95-342-65	38	78	9.6	37	7.3	0.21	6.8	1.2	7.4	1.5	4.4	0.64	4.4	0.60	0.837	0.889	0.131
ML95-343-65	28	59	6.8	26	5.7	0.81	6.2	1.1	7.2	1.4	4.2	0.59	4.0	0.51	0.678	0.931	0.598
ML95-344-65	48	100	12	47	9.3	0.75	9.0	1.5	9.3	1.8	4.9	0.69	4.7	0.63	0.989	0.908	0.360
ML95-345-65	9.9	23	3.3	16	4.3	1.6	5.0	0.84	5.1	1.1	3.0	0.39	2.8	0.40	0.343	0.876	1.515
ML95-346-65	16	36	4.2	17	3.8	0.25	3.3	0.53	3.1	0.60	1.7	0.27	1.9	0.29	0.816	0.956	0.310
ML95-347-65	33	71	8.2	31	6.9	0.44	6.6	1.1	6.6	1.3	3.7	0.55	3.9	0.54	0.820	0.940	0.286
ML95-348-65	5.6	12	1.6	7.2	1.9	1.1	2.1	0.35	2.2	0.45	1.4	0.19	1.6	0.23	0.339	0.873	2.418
ML95-349-65	7.4	15	1.7	6.6	1.4	1.4	1.1	0.15	1.0	0.20	0.54	0.08	0.59	0.09	1.215	0.921	4.953
ML95-358-65	32	65	7.7	29	5.8	2.5	4.4	0.73	4.0	0.81	2.2	0.35	2.5	0.38	1.240	0.902	2.173

Sample	La	Ce	Pr	Nd	Sm	Eu	Gd	Tb	Dy	Ho	Er	Tm	Yb	Lu	[La/Yb]	Ce/Ce*	Eu/Eu*
Restigouche																	
ML95-235	65	140	17	66	12	0.92	10	1.5	9.7	1.9	5.4	0.80	5.2	0.83	1.211	0.917	0.369
ML95-237	73	150	19	73	14	1.5	11	1.8	10	2.1	5.8	0.89	6.1	0.91	1.159	0.877	0.531
ML95-240	17	39	5.1	23	5.6	1.5	6.7	1.1	7.0	1.4	3.9	0.56	3.5	0.53	0.471	0.912	1.075
ML95-241	71	140	18	71	14	1.6	11	1.7	9.1	1.8	4.8	0.73	4.9	0.73	1.404	0.853	0.566
ML95-250	3.5	7.2	0.84	3.4	0.77	0.05	0.97	0.20	1.2	0.24	0.70	0.11	0.71	0.10	0.478	0.915	0.254
ML95-253	14	31	4.3	19	4.8	1.7	5.9	0.99	5.8	1.3	3.5	0.52	3.3	0.52	0.411	0.870	1.403
ML95-255	1.3	2.6	0.3	1.4	0.4	0.02	0.6	0.11	0.59	0.11	0.33	0.05	0.43	0.06	0.293	0.907	0.175
ML95-256	31	68	8.4	31	6.5	0.32	6.1	1.0	6.0	1.3	3.6	0.53	3.5	0.51	0.858	0.918	0.223
ML95-258	7.6	18	2.2	9.2	2.3	0.21	2.4	0.40	2.1	0.40	1.1	0.16	1.2	0.17	0.614	0.959	0.392
ML95-259	28	67	8.9	35	6.2	0.42	5.2	0.81	4.8	1.0	2.9	0.43	2.9	0.43	0.935	0.924	0.325
ML95-260	41	89	11	40	8.1	0.45	6.9	1.1	6.5	1.3	3.7	0.53	3.9	0.57	1.018	0.913	0.264
ML95-262	31	61	7.0	26	5.4	0.86	5.0	0.82	4.8	0.95	2.7	0.40	2.8	0.40	1.073	0.902	0.727
ML95-263	39	81	11	38	7.7	1.2	6.3	1.1	6.2	1.2	3.4	0.49	3.5	0.51	1.079	0.852	0.756
ML95-264	14	32	4.4	20	5.1	1.8	6.1	1.0	6.5	1.4	3.7	0.53	3.8	0.54	0.357	0.888	1.417
ML95-265	65	140	18	68	14	2.0	13	2.1	13	2.6	7.1	1.1	7.1	1.0	0.887	0.891	0.651
ML95-266	72	160	21	77	15	0.66	13	2.2	12	2.5	6.4	0.95	6.3	0.94	1.107	0.896	0.208
ML95-269	82	170	22	81	15	1.5	13	1.8	11	2.2	5.8	0.91	6.3	0.92	1.261	0.872	0.472
ML95-273a	110	230	29	110	24	1.0	20	3.1	18	3.5	8.9	1.3	8.4	1.2	1.269	0.887	0.200
ML95-273b	22	48	5.6	22	4.4	0.03	4.2	0.70	4.3	0.95	2.7	0.39	2.4	0.36	0.888	0.942	0.031
ML95-274	69	140	17	66	12	0.76	9.8	1.5	8.6	1.7	4.4	0.65	4.4	0.64	1.519	0.890	0.308
ML95-276	32	71	8.7	34	8.6	0.96	9.1	1.6	8.9	1.8	4.4	0.56	3.2	0.44	0.969	0.927	0.476
ML95-277	110	220	29	110	22	1.2	20	2.9	15	2.9	7.4	1.0	6.3	0.91	1.691	0.848	0.251
ML95-278	64	140	17	63	13	0.85	12	1.9	10	2.0	5.4	0.79	5.1	0.74	1.216	0.924	0.299
ML95-279	52	110	14	53	12	0.60	10	1.7	9.1	1.8	4.7	0.68	4.4	0.66	1.145	0.888	0.240
ML95-280	140	290	35	140	26	1.5	20	2.7	17	3.3	9.1	1.4	10.0	1.5	1.356	0.902	0.289
ML95-283	94	200	23	84	17	2.2	14	1.8	9.2	1.7	4.3	0.66	4.5	0.70	2.024	0.937	0.626
ML95-284	53	110	13	50	10	1.6	8.9	1.6	9.3	1.9	5.1	0.74	5.0	0.75	1.027	0.913	0.745
ML95-285	40	91	11	43	9.0	1.2	8.3	1.4	8.1	1.7	4.7	0.71	4.7	0.72	0.824	0.945	0.610
ML95-297	1.0	1.9	0.23	1.2	0.38	0.14	0.54	0.10	0.55	0.09	0.21	0.03	0.18	0.03	0.538	0.863	1.357
ML95-298	0.40	0.80	0.10	0.70	0.14	0.05	0.21	0.03	0.19	0.03	0.07		0.10		0.678	0.871	1.280
ML95-299	0.70	1.3	0.17	0.90	0.21	0.34	0.24	0.05	0.26	0.05	0.12	0.03	0.17	0.03	1.539	0.821	6.649
ML95-300	2.7	5.7	0.73	3.3	0.86	0.26	1.1	0.17	0.77	0.13	0.26		0.06		1.938	0.884	1.174
ML95-321	1.2	2.3	0.24	1.1	0.19	0.06	0.18	0.02	0.16	0.03	0.07					0.933	1.424
ML95-322	4.2	8.9	0.97	3.7	0.59	0.44	0.49	0.06	0.30	0.05	0.11	0.02	0.12	0.02	3.391	0.960	3.593
ML95-324	8.9	17.0	1.8	6.2	1.2	0.22	0.92	0.10	0.50	0.09	0.16		0.14		6.158	0.925	0.919
ML95-371	10	20	2.3	9.2	1.8	3.5	1.5	0.24	1.6	0.32	0.83	0.13	0.85	0.13	1.140	0.908	9.352

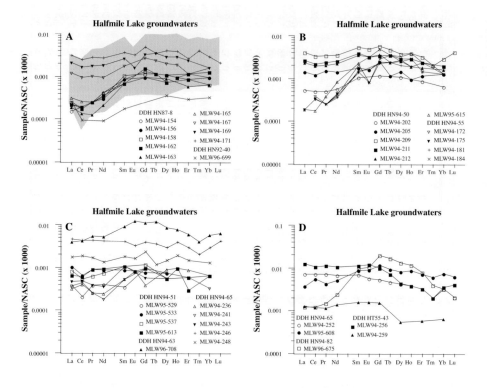

Figure 4. Rare-earth element concentrations of Halfmile Lake borehole waters normalized to NASC. Normalizing values are from Haskin et al. (1968) with Dy value from Gromet et al. (1984). Data are from Table 2. Shaded region represents surface waters from the Halfmile Lake deposit (Leybourne et al., 2000).

= 0.63). In general, REE patterns show some MREE-enrichment with $[Gd/Yb]_{NASC}$ ranging from 0.38 to 3.11 (average = 1.25; Figure 5). In rare cases, groundwaters show a concave-up pattern from La to Gd. In general, negative Ce anomalies are only rarely well developed (average $[Ce/Ce^*]_{NASC}$ = 0.73), and much less so than surface waters (average $[Ce/Ce^*]$ = 0.33; Figure 5).

Groundwaters from the Heath Steele and Wedge deposits are characterized by elevated REE contents, up to 690,000 ng/L (Figure 5). Waters of variable, though generally low, pH from deeper in the Heath Steele mine, from Brunswick #12 mine, and from Fe-rich waters at the Wedge deposit have the highest REE abundances. Mine waters are variable, with $[La/Sm]_{NASC}$ varying from 0.23 to 2.05 with an average of 0.67. The $[Gd/Yb]_{NASC}$ varies from 0.61 to 3.97 with an average of 1.51. Several waters adjacent to the main ore zone at Heath Steele have strong positive Eu anomalies, similar to anomalies for Halfmile Lake massive sulfides (Figures 2, 5).

Speciation calculations for Halfmile Lake groundwaters suggest that in the shallow waters with concave-up LREE patterns, the LREE occur mainly as free ions, whereas the deeper waters show a greater degree of complexation of the LREE by carbonate ions. These data are consistent with the idea that the MREE and HREE form more stable carbonate complexes for the shallow waters with the LREE lost due to adsorption to Mn- and Fe-oxyhydroxides. The abundance of $[CO_3^{2-}]$ increases with increasing pH, resulting in greater carbonate-REE complexation, as shown elsewhere (Lee and Byrne, 1993; Johannesson et al., 1994). At the Halfmile Lake deposit, the groundwater with the highest pH (9.86) and TDS (226 mg/L) is modeled as having essentially 100% of the REE complexed as $Ln(CO_3)_2^-$ species.

Restigouche deposit groundwaters have higher pH and carbonate alkalinity than those at the Halfmile Lake deposit resulting in complexation of REE primarily by carbonate. Although REE-speciation by phosphate is considered important in marine waters (Byrne and Kim, 1993), for the Bathurst Mining Camp waters, REE-phosphate species represent at most 11% for low-carbonate waters from Halfmile Lake. Waters with the most elevated free $[PO_4^{3-}]$ contents also have the highest free $[CO_3^{2-}]$ contents (higher pH) but low REE-phosphate complex formation, consistent with carbonate ions being more effective REE complexers than phosphate ions (Johannesson and Lyons, 1994).

4.3 REE IN SUSPENDED SEDIMENTS

Suspended sediment REE profiles are variable, reflecting the variation in suspended sediment mineralogy and the variation in the REE contents of the groundwaters with which they are associated (Table 4; Figures 6, 7). Suspended sediments from the Halfmile Lake deposit show less variation ($[La/Sm]_{NASC}$ ranges from 0.80 to 1.01, $[La/Yb]_{NASC}$ from 0.91 to 2.29 and $[Gd/Yb]_{NASC}$ from 0.77 to 1.74) in REE profiles and contents than suspended sediment from the Restigouche deposit ($[La/Sm]_{NASC}$ ranges from 0.29 to 1.01, $[La/Yb]_{NASC}$ from 0.15 to 1.94 and $[Gd/Yb]_{NASC}$ from 1.10 to 2.07). Suspended sediments from the Halfmile Lake deposit have only minor Eu and Ce anomalies ($[Eu/Eu^*]_{NASC}$ ranges from 0.90 to 1.11, with one sample at 1.84; $[Ce/Ce^*]_{NASC}$ ranges from 0.91 to 0.99). Suspended sediments from the Restigouche deposit are more variable ($[Eu/Eu^*]_{NASC}$ ranges from 0.19 to 0.96; $[Ce/Ce^*]_{NASC}$ ranges from 0.37 to 0.98).

Groundwaters from the Restigouche deposit have REE patterns that show greater differences compared to the corresponding suspended sediment. For example, groundwaters from two shallow boreholes that intersect massive sulfides close to their surface expression are characterized by convex-up REE profiles from the LREE to the MREE and are generally HREE-enriched, some with small negative Ce anomalies. In contrast, suspended sediments have flat to LREE-depleted profiles, are HREE-depleted compared to the MREE, and lack negative Ce anomalies. The REE patterns displayed by suspended sediments from borehole MM-89-106 are antithetical to the waters; the suspended sediments have LREE/HREE > 1 and have significant negative Eu anomalies

Table 2. REE contents and ratios for Halfmile Lake deposit groundwaters. All ratios are normalized to the North America Shale Composite (NASC) as defined in Table 1.

Sample	La	Ce	Pr	Nd	Sm	Eu	Gd	Tb	Dy	Ho	Er	Tm	Yb	Lu	[La/Yb]	[Ce/Ce*]	Eu/Eu*
MLW94-154	0.005	0.014	0.002	0.014	0.005		0.007	0.001	0.006	0.001	0.003		0.003		0.161	0.964	
MLW94-156	0.007	0.013	0.002	0.010	0.005		0.005		0.004		0.003		0.003		0.226	0.757	
MLW94-158	0.008	0.013	0.002	0.011	0.006	0.001	0.006	0.001	0.004	0.001	0.003	0.001	0.002	0.001	0.388	0.708	0.732
MLW94-162	0.007	0.014	0.002	0.012	0.004		0.008	0.001	0.006	0.001	0.004		0.004		0.170	0.815	
MLW94-163	0.006	0.010	0.001	0.007	0.004		0.005		0.005		0.002		0.002		0.291	0.889	
MLW94-165	0.010	0.019	0.002	0.014	0.004	0.001	0.008	0.001	0.006	0.001	0.004	0.001	0.003	0.001	0.323	0.925	0.776
MLW94-167	0.038	0.069	0.008	0.032	0.010	0.001	0.014	0.002	0.011	0.002	0.004	0.001	0.005	0.001	0.736	0.862	0.371
MLW94-169	0.066	0.123	0.014	0.061	0.015	0.002	0.017	0.003	0.016	0.003	0.008		0.005		1.279	0.881	0.550
MLW94-171	0.099	0.179	0.022	0.091	0.021	0.004	0.025	0.003	0.022	0.004	0.009	0.002	0.009	0.001	1.066	0.835	0.766
MLW96-674	0.005	0.005	0.001	0.001	0.001	0.001	0.002	0.001	0.001	0.001	0.001	0.001	0.001	0.001	0.484	0.487	3.105
MLW96-699	0.008	0.007	0.001	0.003	0.001	0.001	0.002	0.001	0.002	0.001	0.003	0.001	0.001	0.001	0.775	0.539	3.105
MLW94-202	0.017	0.037	0.004	0.019	0.006		0.006		0.006	0.001	0.003		0.002		0.823	0.977	
MLW95-615	0.006	0.013	0.003	0.028	0.013	0.004	0.025	0.003	0.018	0.004	0.008	0.001	0.004	0.001	0.145	0.667	0.974
MLW94-205	0.046	0.088	0.012	0.046	0.010	0.002	0.013	0.001	0.008	0.001	0.004	0.001	0.004	0.001	1.114	0.816	0.770
MLW94-209	0.130	0.238	0.027	0.116	0.031	0.006	0.030	0.004	0.021	0.004	0.011	0.001	0.009	0.002	1.399	0.875	0.864
MLW94-211	0.083	0.156	0.019	0.084	0.022	0.004	0.018	0.003	0.016	0.003	0.008		0.006		1.340	0.856	0.883
MLW94-212	0.074	0.141	0.016	0.073	0.020	0.004	0.019	0.002	0.014	0.002	0.006		0.005		1.434	0.893	0.901
MLW94-172		0.028	0.002	0.013	0.008		0.012	0.002	0.014	0.003	0.005		0.004		0.000		
MLW94-175	0.006	0.025	0.002	0.015	0.008	0.001	0.013	0.002	0.015	0.002	0.008	0.001	0.004	0.001	0.145	1.572	0.431

Table 2. Continued.

Sample	La	Ce	Pr	Nd	Sm	Eu	Gd	Tb	Dy	Ho	Er	Tm	Yb	Lu	[La/Yb]	Ce/Ce*	Eu/Eu*
MLW94-181	0.005	0.024	0.002	0.012	0.009	0.002	0.012	0.002	0.011	0.002	0.006	0.001	0.005	0.001	0.097	1.653	0.845
MLW94-184		0.025	0.002	0.014	0.012	0.001	0.015	0.003	0.014	0.003	0.005	0.001	0.004		0.000		0.327
MLW95-529	0.012	0.015	0.003	0.013	0.002		0.005		0.004		0.002		0.002		0.581	0.545	
MLW95-613	0.021	0.044	0.007	0.031	0.006	0.001	0.005		0.003	0.001	0.001		0.002		1.017	0.790	0.802
MLW95-533	0.033	0.048	0.007	0.029	0.005		0.004		0.004	0.002	0.002		0.002		1.598	0.688	
MLW95-537	0.026	0.039	0.005	0.024	0.006		0.006		0.003		0.002		0.002		1.259	0.745	
MLW96-708	0.126	0.300	0.043	0.171	0.052	0.015	0.058	0.010	0.047	0.008	0.020	0.002	0.018	0.003	0.678	0.888	1.199
MLW94-236	0.010	0.029	0.002	0.008	0.003	0.001	0.002		0.005	0.001	0.003		0.002		0.484	1.412	1.792
MLW94-241	0.012	0.032	0.002	0.006	0.003	0.001	0.005	0.001	0.004	0.001	0.002	0.001	0.001	0.001	1.163	1.423	1.134
MLW95-608	0.120	0.408	0.034	0.169	0.050	0.011	0.060	0.008	0.045	0.009	0.024	0.003	0.023	0.003	0.505	1.391	0.882
MLW94-243	0.015	0.035	0.003	0.012	0.006	0.001	0.003	0.001	0.003	0.001	0.002	0.001	0.002	0.001	0.727	1.136	1.035
MLW94-246	0.146	0.317	0.034	0.137	0.023	0.004	0.020	0.003	0.016	0.004	0.010	0.001	0.009	0.002	1.572	0.980	0.819
MLW94-248	0.055	0.134	0.013	0.044	0.010	0.002	0.012	0.001	0.008	0.002	0.005	0.001	0.004	0.001	1.332	1.091	0.802
MLW94-252	0.231	0.514	0.056	0.218	0.039	0.007	0.028	0.004	0.023	0.004	0.011	0.001	0.010	0.001	2.238	0.984	0.930
MLW96-673	0.005	0.005	0.001	0.001	0.001	0.001	0.002	0.001	0.001	0.001	0.001	0.001	0.001	0.001	0.484	0.487	3.105
MLW96-675	0.039	0.089	0.012	0.081	0.054	0.013	0.101	0.014	0.072	0.012	0.027	0.002	0.010	0.001	0.378	0.896	0.773
MLW94-256	0.404	0.781	0.088	0.352	0.065	0.015	0.051	0.006	0.024	0.004	0.011	0.001	0.011	0.001	3.558	0.902	1.144
MLW94-259	0.040	0.088	0.009	0.046	0.009	0.002	0.008		0.003		0.002		0.002	0.002	1.938	1.010	1.035

All concentrations in µg/L; blank = below detection

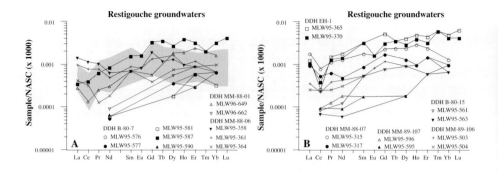

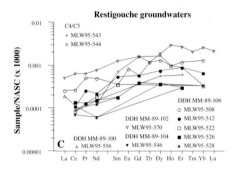

Figure 5. Rare-earth element concentrations of Restigouche deposit borehole waters normalized to NASC. Normalizing values are from Haskin et al. (1968) with Dy value from Gromet et al. (1984). Data are from Table 3. Shaded region represents surface waters from the Restigouche deposit (Leybourne et al., 2000).

and slight negative Ce anomalies, whereas the groundwaters are LREE depleted compared to NASC. The suspended sediments from borehole MM-88-01 are the most fractionated compared to NASC with LREE depleted profiles and HREE contents between 10 and 20 x NASC. In addition, these sediments have significant negative Eu and Ce anomalies (Figure 6).

Flat patterns exist for sediment-normalized groundwater compositions with elevated REE from the Halfmile Lake deposit. At lower total REE contents (lower water/sediment) there tends to be greater fractionation of the LREE, though generally less so for La than for Ce, Pr, and Nd (Figure 7). For the Restigouche deposit groundwater-sediment pairs, most groundwaters are depleted in the LREE (especially Ce, Pr, Nd) relative to the suspended sediments (Figure 7).

Table 3. REE content and ratios for Restigouche deposit groundwaters. All ratios are NASC-normalized as defined in Table 1.

Sample	La	Ce	Pr	Nd	Sm	Eu	Gd	Tb	Dy	Ho	Er	Tm	Yb	Lu	[La/Yb]	Ce/Ce*	Eu/Eu*
MLW95-561	0.005	0.006	0.001	0.004	0.003	0.001	0.004	0.001	0.005	0.001	0.002	0.001	0.003	0.001	0.161	0.584	
MLW95-563	0.005	0.005	0.001	0.002	0.001	0.001	0.002	0.001	0.001	0.001	0.002	0.001	0.002	0.001	0.242	0.487	3.105
MLW95-590				0.002					0.002		0.001		0.001				
MLW95-581				0.002					0.001		0.002		0.002				
MLW95-577				0.002					0.002		0.002						
MLW95-587	0.011	0.029	0.005	0.028	0.009	0.002	0.017	0.003	0.015	0.004	0.011	0.001	0.010	0.002	0.107	0.852	0.710
MLW95-543	0.005	0.006	0.001	0.005	0.001	0.001	0.002	0.001	0.003	0.001	0.002	0.001	0.001	0.001	0.484	0.584	3.105
MLW95-544	0.005	0.007	0.001	0.004	0.004	0.001	0.008	0.001	0.009	0.001	0.003	0.001	0.001	0.001	0.484	0.682	0.776
MLW95-365	0.040	0.019	0.012	0.059	0.018	0.001	0.027	0.003	0.022	0.005	0.015	0.003	0.015	0.003	0.258	0.189	0.199
MLW95-370	0.034	0.028	0.010	0.046	0.018	0.002	0.016	0.003	0.016	0.004	0.011	0.003	0.013	0.002	0.253	0.331	0.517
MLW96-649	0.009	0.010	0.002	0.010	0.004	0.001	0.003	0.001	0.010	0.002	0.008	0.001	0.005	0.001	0.174	0.513	1.267
MLW96-662	0.005	0.005	0.001	0.003	0.001	0.001	0.002	0.001	0.003	0.001	0.003	0.001	0.002	0.001	0.242	0.487	3.105
MLW95-358	0.046	0.082	0.008	0.021	0.004	0.001	0.010	0.001	0.007	0.001	0.003	0.001	0.002	0.001	2.228	0.931	
MLW95-361	0.031	0.056	0.006	0.016	0.005	0.001	0.007	0.001	0.009	0.001	0.004	0.001	0.003	0.001	1.502	0.894	
MLW95-364	0.013	0.023	0.001	0.006	0.004	0.002	0.002	0.002	0.004	0.003	0.003	0.001	0.004	0.001	0.420	1.389	1.552
MLW95-315	0.057	0.058	0.009	0.043	0.010	0.002	0.012	0.001	0.013	0.001	0.008		0.003		1.380	0.558	0.802
MLW95-317	0.030	0.039	0.005	0.016	0.005	0.001	0.009	0.001	0.007	0.001	0.005	0.001	0.003	0.001	0.969	0.694	0.655
MLW95-556	0.006	0.008	0.001	0.002	0.001	0.001	0.002	0.001	0.001	0.001	0.001	0.001	0.001	0.001	0.581	0.711	3.105
MLW95-570	0.016	0.046	0.005	0.025	0.007		0.008	0.001	0.010	0.003	0.009	0.001	0.008	0.001	0.194	1.120	
MLW95-546	0.005	0.005	0.001	0.002	0.001	0.001	0.002	0.001	0.002	0.001	0.001	0.001	0.001	0.001	0.484	0.487	3.105
MLW95-528				0.004	0.001				0.002								
MLW95-526		0.010	0.001	0.005	0.001		0.002				0.001		0.001				
MLW95-522		0.006	0.001	0.006	0.002								0.002				
MLW95-512		0.008	0.002	0.007	0.003		0.003		0.004		0.003		0.004				
MLW95-508	0.008	0.019	0.004	0.016	0.002	0.001	0.003	0.001	0.007	0.001	0.004		0.004		0.194	0.732	1.792
MLW95-504	0.008	0.018	0.002	0.010	0.003	0.001	0.004		0.005	0.001	0.002		0.003		0.258	0.980	
MLW95-503	0.011	0.016	0.003	0.014	0.003		0.006		0.005	0.001	0.005		0.002		0.533	0.607	1.035
MLW95-596		0.007	0.001	0.006	0.001												
MLW95-595		0.007		0.003	0.001				0.001								

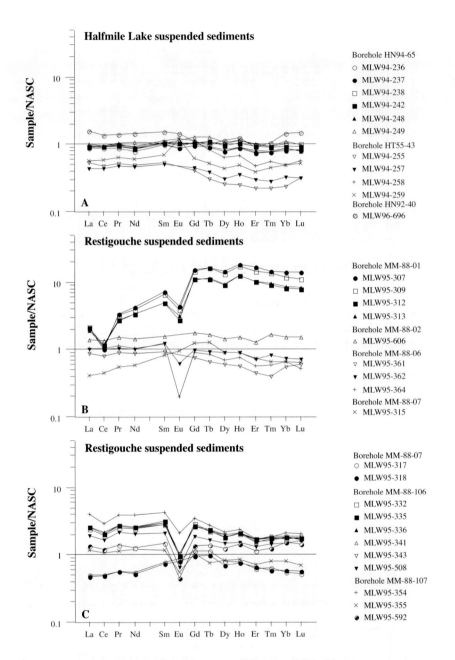

Figure 6. Rare-earth element concentrations of Halfmile Lake and Restigouche deposit groundwater suspended sediments normalized to NASC. Normalizing values are from Haskin et al. (1968) with Dy value from Gromet et al. (1984). Data are from Table 4.

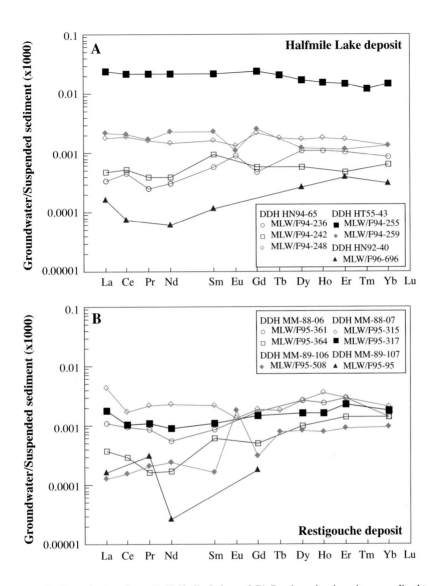

Figure 7. Groundwaters from A) Halfmile Lake and B) Restigouche deposits normalized to the corresponding suspended sediment. Data are from Tables 2-4.

The suspended sediments from boreholes MM-88-06 and MM-88-07 at the Restigouche deposit have different REE patterns despite many similarities in the chemistries of the waters from these boreholes. The suspended sediments from borehole MM-88-07 have lower La/Yb and La/Sm, higher Gd/Yb, and lower REE abundances than the borehole MM-88-06 sediments. The MM-88-07 sediments also have higher Fe and lower Al, K, Ca, Mg, and base metal contents (Leybourne, 2001).

Table 4. REE concentrations and ratios of groundwater suspended sediments. All ratios are NASC-normalized as defined in Table 1.

Sample	Borehole	Mineralogy	La	Ce	Pr	Nd	Sm	Eu	Gd	Tb	Dy	Ho	Er	Tm	Yb	Lu	[La/Yb]	Ce/Ce*	Eu/Eu*
							Halfmile Lake deposit												
F94-236	HN94-65		28	63	7.0	26	5.6	1.1	5.0	0.77	4.3	0.95	2.6	0.38	2.6	0.39	1.043	0.980	0.913
F94-237	HN94-65	scpa	29	67	7.5	28	5.8	1.1	5.0	0.76	4.3	0.90	2.5	0.38	2.5	0.40	1.124	0.989	0.897
F94-238	HN94-65	scpa	30	68	7.4	28	6.1	1.2	5.5	0.92	5.5	1.2	3.3	0.46	3.1	0.48	0.938	0.994	0.910
F94-242	HN94-65		30	68	7.8	29	5.9	1.3	5.5	0.94	5.3	1.1	3.2	0.45	2.9	0.45	1.002	0.968	1.002
F94-248	HN94-65		31	69	8.0	30	6.1	1.3	5.2	0.79	4.5	0.94	2.7	0.39	2.7	0.37	1.112	0.954	1.013
F94-249	HN94-65		32	70	8.3	35	6.0	1.4	6.6	1.1	5.9	1.2	3.4	0.47	3.4	0.47	0.912	0.936	0.977
F94-255	HT55-43		17	35	4.1	16	3.0		2.1	0.26	1.4	0.26	0.75	0.11	0.72	0.15	2.287	0.913	
F94-257	HT55-43	csf	14	32	3.7	15	2.9		2.3	0.33	1.7	0.37	1.0	0.14	1.0	0.15	1.356	0.968	
F94-258	HT55-43		30	65	7.7	31	6.5	1.5	5.9	0.70	3.5	0.70	1.6	0.27	1.5	0.27	1.938	0.931	1.063
F94-259	HT55-43		18	42	5.0	20	4.0	1.5	3.2	0.44	2.4	0.50	1.3	0.22	1.5	0.25	1.163	0.964	1.841
F96-696	HN92-40	cskfp	50	99	11	48	8.8	1.8	5.8	0.84	6.3	1.3	3.1	0.52	4.5	0.72	1.076	0.919	1.106
							Restigouche deposit (MM boreholes)												
F95-307	88-01		67	73	27	140	41	5.4	81	14.0	81	19.0	57.0	7.4	44.0	6.8	0.148	0.374	0.411
F95-309	88-01		69	79	26	130	38	4.8	75	14.0	74	18.0	50.0	6.8	37.0	5.4	0.181	0.406	0.395
F95-312	88-01	sckp	67	86	22	110	28	3.4	58	9.5	50	13.0	35.0	4.5	25.0	3.8	0.260	0.488	0.370
F95-313	88-01		61	83	21	110	28	3.9	56	9.9	52	13.0	35.0	4.8	27.0	4.0	0.219	0.505	0.432
F95-606	88-02		45	99	12	48	9.0		9.1	1.4	7.9	1.6	4.3	0.83	4.7	0.74	0.928	0.928	
F95-361	88-06		28	59	7.1	29	5.3		3.9	0.55	3.3	0.58	1.5	0.20	1.7	0.29	1.596	0.911	
F95-362	88-06	skaf-sz-sp-s	32	73	8.3	33	7.0	0.76	5.1	0.79	4.9	0.94	2.4	0.41	2.3	0.34	1.348	0.976	0.558
F95-364	88-06	skapf-sp-s	33	77	9.0	34	6.8	0.24	4.7	0.72	3.8	0.78	1.9	0.29	2.0	0.25	1.598	0.973	0.186
F95-315	88-07		13	33	4.3	19	4.7	1.2	6.4	1.1	5.0	0.95	2.5	0.33	2.0	0.30	0.630	0.961	0.961
F95-317	88-07	kaspp-s	16	37	4.4	18	4.4	1.1	5.9	0.97	4.4	0.82	2.1	0.32	1.7	0.25	0.912	0.960	0.948
F95-318	88-07		15	36	4.4	17	4.2	0.98	5.0	0.83	3.8	0.78	2.2	0.29	1.8	0.27	0.807	0.965	0.939
F95-332	89-106	sckak-f	79	140	20	80	16	1.2	15	1.9	10	2.1	5.6	0.89	5.3	0.84	1.444	0.767	0.340
F95-335	89-106		84	150	22	85	18	1.2	14	2.0	10	2.2	5.9	0.89	5.6	0.86	1.453	0.760	0.332
F95-336	89-106		83	150	21	83	16	1.2	14	1.9	11	2.2	5.3	0.81	5.6	0.80	1.436	0.783	0.352
F95-341	89-106	sckak-f	84	160	22	86	17	1.2	14	1.9	10	2.1	5.2	0.85	5.5	0.83	1.480	0.811	0.342
F95-343	89-106		82	150	21	83	17	1.3	14	2.0	11	2.2	5.7	0.93	5.8	0.90	1.370	0.787	0.370
F95-508	89-106	kREE-p	62	120	17	68	12	0.80	9.8	1.4	8.1	1.6	4.5	0.73	4.6	0.77	1.306	0.805	0.324
F95-354	89-107		130	210	31	130	24	2.6	18	2.3	12	2.5	5.9	0.92	6.5	0.97	1.938	0.720	0.549
F95-355	89-107		38	78	9.1	40	6.7	0.67	5.5	0.65	4.6	0.88	2.4	0.40	2.5	0.33	1.473	0.914	0.485
F95-592	89-107	skcak-f	44	89	11	42	8.4	0.55	7.1	1.2	6.8	1.5	3.9	0.63	4.7	0.69	0.907	0.881	0.313

Concentrations in ppm.

Mineralogy (SEM, XRD; see Leybourne, 2001 for analytical details)

s = sericite; c = chamosite; p = pyrite; a = albite; k = kaolinite; f = Fe(+Mn) oxides; f-s = Fe-sulfate; z-s = Zn-sulfate; p-s = Pb-sulfate; k-f = K-feldspar; REE-p = REE phosphates

4.4 Sr ISOTOPIC COMPOSITION OF BATHURST MINING CAMP GROUNDWATERS

Groundwaters at the Restigouche deposit with high base metal loads recovered from the main ore zone at the Restigouche deposit have $^{87}Sr/^{86}Sr$ values ranging from 0.71339 to 0.71404 (Table 5; Figures 8, 9). These shallow groundwaters have moderately low Sr contents (43 to 100 μg/L), and low Rb/Sr and Sr/Ca ratios. Shallow groundwaters distal from the deposit have less radiogenic Sr than groundwaters from the ore zone, with two samples having $^{87}Sr/^{86}Sr$ values of 0.71111 and 0.71249. These waters have similar Sr/Ca and Sr contents to the ore zone groundwaters (95 - 100 μg/L) but somewhat lower Rb/Sr (Figures 8, 9). The more saline waters from the Restigouche deposit are more variable in Sr isotopic composition. Groundwater from below the ore zone (TDS = 510 mg/L) and groundwater northeast of the ore zone (TDS = 693 mg/L) have non-radiogenic Sr isotopic compositions ($^{87}Sr/^{86}Sr$ = 0.71156 and 0.71097, respectively), with the latter having the lowest Sr isotopic composition in this study (Table 5). In contrast, the more saline water from borehole MM-89-106 (TDS = 2510 mg/L) is more radiogenic, with a $^{87}Sr/^{86}Sr$ value of 0.71519. Finally, the saline waters from borehole MM-88-02 are moderately radiogenic ($^{87}Sr/^{86}Sr$ = 0.71223 - 0.71237) and fall within the range of the brackish waters. The saline waters have much higher Sr contents (180 - 89000 μg/L) and Sr/Ca and lower Rb/Sr than the other groundwaters (Figures 8, 9).

Five groundwaters were analyzed from other deposits for their Sr isotopic composition, three from the Heath Steele deposit, and one each from the Brunswick #12 and Wedge deposits. The water from the Wedge deposit is moderately radiogenic and is similar to shallow groundwaters from the Restigouche deposit with $^{87}Sr/^{86}Sr$ = 0.7149. This groundwater represents shallow groundwater flow through the waste pile at the Wedge deposit and has very low pH (< 3) and high metal loads. All three waters analyzed from the Heath Steele deposit are more radiogenic than any of the groundwaters from the Restigouche deposit with $^{87}Sr/^{86}Sr$ values ranging from 0.71826 to 0.71932. Despite similar Sr isotopic values, these waters are variable in other chemical characteristics although all are high-SO_4 waters. The groundwater from the Brunswick #12 deposit has the highest salinity of all the waters collected (TDS = 45,800 mg/L) and also has the most radiogenic Sr isotope signature ($^{87}Sr/^{86}Sr$ = 0.7241) with the highest Rb/Sr and lowest Sr/Ca.

Overall, there are two discernible trends between Sr isotopic composition and Rb/Sr. For shallow groundwaters from the Restigouche deposit, the Sr isotopic composition decreases with increasing Sr content, whereas for the mine waters the Sr isotopic composition generally increases with increasing Sr content (Figure 8). Saline waters from borehole MM-88-02 differ in that they have the highest Sr contents (88,000 – 89,000 μg/L) but have generally low $^{87}Sr/^{86}Sr$ values and very low Rb/Sr (Figure 8). Brackish and saline waters from the Restigouche deposit have higher Sr/Ca and lower Ca/Na than shallow groundwaters and mine waters.

Table 5. Sr isotope and abundance results for waters and rocks from the Restigouche, Wedge, Heath Steele, and Brunswick # 12 deposits.

Sample#	Location	$^{87}Sr/^{86}Sr$	Sr	Rb	Rb/Sr
		Waters			
MLW95-300	Seep	0.71599	6.40	0.70	0.1094
MLW95-586	B-80-7	0.71156	180	1.20	0.0067
MLW95-306	MM-88-01	0.71404	100	2.50	0.0250
MLW95-344	MM-88-02	0.71237	88000	28	0.0003
MLW95-601	MM-88-02	0.71225	89000	27	0.0003
MLW95-606	MM-88-02	0.71223	89000	25	0.0003
MLW95-361	MM-88-06	0.71339	43	0.60	0.0140
MLW95-314	MM-88-07	0.71376	69	0.90	0.0130
MLW95-571	MM-89-102	0.71097	360	1.20	0.0033
MLW95-546	MM-89-104	0.71111	100	1.00	0.0100
MLW95-504	MM-89-106	0.71519	580	3.10	0.0053
MLW95-594	MM-89-107	0.71249	95	0.70	0.0074
MLW95-482	Wedge	0.71490	320	14	0.0438
MLW95-486	Heath Steele	0.71826	2000	35	0.0175
MLW95-610	Heath Steele	0.71932	5000	14	0.0028
MLW95-485	Heath Steele	0.71826	400	4.50	0.0113
MLW95-492	Brunswick #12	0.72410	880	200	0.2273
		Rocks			
ML95-252	Carbonate	0.708889			
ML95-259	Carbonate	0.714467			
ML95-256	Felsic volcanic	0.745397	56	130	2.3214
ML95-262	Dacitic volcanic	0.730763	99	100	1.0101
ML95-264	Mafic dyke	0.708429	170	3.90	0.0229
ML95-277	Felsic volcanic	0.745655	45	120	2.6667
ML95-285	Felsic volcanic	0.739896	50	89	1.7800
ML95-297	Massive sulphide	0.713662	35	0.50	0.0143
MLS96-664	Stream sediment	0.714138	37	20	0.5405

Water elemental abundances in µg/L; Rock abundances in mg kg^{-1}.

4.4 Nd ISOTOPIC COMPOSITION OF BATHURST MINING CAMP GROUNDWATERS

Four groundwaters were successfully analyzed for Nd isotopic composition (Table 6, Figure 9) and all have high REE contents, with ΣREE ranging from 3.1 to 690 µg/L. All four waters are characterized by Ca-SO$_4$ compositions with elevated Fe (4400 to 650,000 µg/L) and base metal (e.g., Zn ranges from 3600 to 340,000 µg/L) contents. All the waters have low pH (2.33 to 6.56). However, the four waters differ in REE

Table 6. Nd isotope and abundance results for BMC waters and rocks.

Groundwater	Deposit/Type	^{143}Nd/^{144}Nd	error	Nd	Sm	Sm/Nd
		Waters				
MLW95-300	Restigouche	0.5123	0.000030	0.75	0.41	0.55
MLW95-482	Wedge	0.5123	0.000018	45.61	11.55	0.25
MLW95-485	Heath Steele	0.5122	0.000014	70.59	16.39	0.23
MLW95-486	Heath Steele	0.51216	0.000023	112.09	15.91	0.14
		Rocks				
ML95-252	Carbonate	0.51258	0.000020	11.2	4.62	0.41
ML95-259	Carbonate	0.51535	0.000016	0.045	0.033	0.73
ML95-256	Felsic volcanic	0.51217	0.000023	31	6.5	0.21
ML95-262	Dacitic volcanic	0.51219	0.000016	26	5.4	0.21
ML95-264	Mafic dyke	0.51275	0.000014	20	5.1	0.26
ML95-277	Felsic volcanic	0.51217	0.000015	110	22	0.20
ML95-285	Felsic volcanic	0.51210	0.000011	43	9	0.21
MLS96-664	Stream sediment	0.51244	0.000012	15	3.7	0.25

Water elemental abundances in µg/L (ICP-MS data); Rock elemental abundances in ppm (ICP-MS data).

abundances and REE profiles. The groundwater seep from the Restigouche deposit has the lowest REE abundances of the four waters and is also characterized by a moderately strong negative Ce anomaly. In contrast, the other three groundwaters from the Wedge and Heath Steele deposits have elevated REE contents. Groundwaters from the Wedge and Heath Steele deposits have convex-up, NASC-normalized REE patterns and are strongly LREE enriched with $[La/Sm]_{chon} > 1$. The groundwater from the Heath Steele deposit has a strong positive Eu anomaly, similar to massive sulfides from the Halfmile Lake and Restigouche deposits (Figure 2).

As a result of these differences in REE patterns, the waters have different Sm/Nd ratios, ranging from 0.55 to 0.14 (Table 6). Similarly, the waters vary in their Nd isotopic composition with ^{143}Nd/^{144}Nd ratios ranging from 0.51216 to 0.5123 (Figure 9). There is a moderate positive correlation between ^{147}Sm/^{144}Nd and ^{143}Nd/^{144}Nd and a negative correlation between ^{143}Nd/^{144}Nd and ^{87}Sr/^{86}Sr (Figure 9). Assuming an age of the host rocks of 465 Ma and a chondritic uniform reservoir (CHUR) value of 0.51264, the groundwaters have ε_{Nd} values in the range +2.66 to -14.49.

4.5 Sr AND Nd ISOTOPIC COMPOSITION OF HOST LITHOLOGIES

Host lithologies (all from the Restigouche deposit) analyzed for Sr and Nd isotopic composition include two vein carbonates, a mafic dyke, four felsic volcanic rocks, a

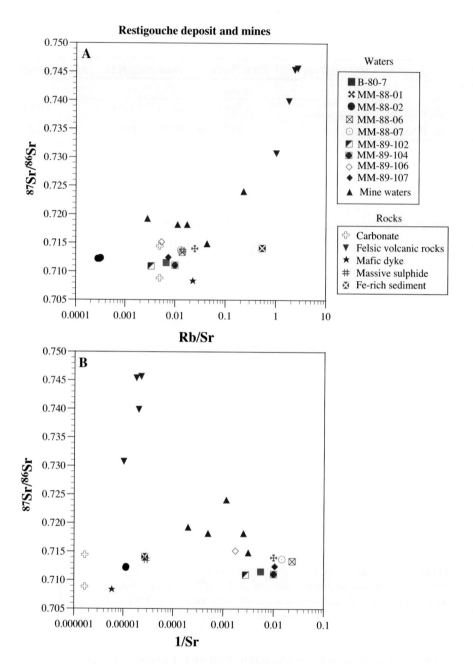

Figure 8. Plots of ^{87}Sr/^{86}Sr vs. A) Rb/Sr and B) 1/Sr for Restigouche deposit groundwaters, host lithologies and mine waters from Heath Steele, Brunswick #12 and Wedge deposits. Data from Table 5.

massive sulfide and an Fe-oxyhydroxide-rich stream sediment. There has been little previous work on Sr isotope analyses of whole rocks or mineral separates for the Bathurst Mining Camp. Goodfellow and Peter (unpublished data) analyzed several siderite iron formation samples from the Brunswick #12 deposit. The ^{87}Sr/^{86}Sr values for the siderite are moderately radiogenic, ranging from 0.71189 to 0.71508.

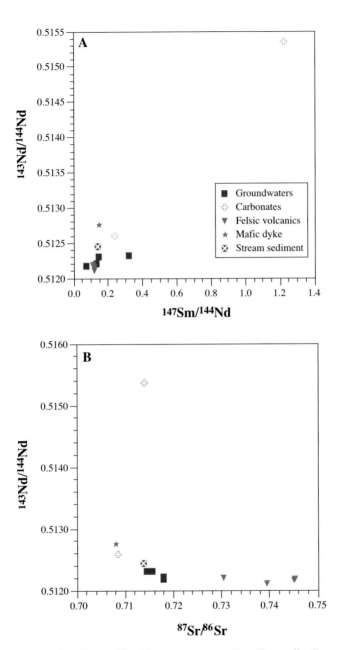

Figure 9. Plots of ^{143}Nd/^{144}Nd vs ^{147}Sm/^{144}Nd (panel A) and ^{143}Nd/^{144}Nd vs ^{87}Sr/^{86}Sr (panel B). Nd and Sr isotope data are from Tables 5 and 6.

The two vein carbonates are similar to those analyzed by Goodfellow and Peter (unpublished data) in that they range from unradiogenic (^{87}Sr/^{86}Sr = 0.70889) to moderately radiogenic (^{87}Sr/^{86}Sr = 0.71447). The felsic host rocks are the most radiogenic, with ^{87}Sr/^{86}Sr values ranging from 0.73076 for the dacite to a high of 0.74566 for the felsic volcanic rocks (Table 5). In contrast, the cross-cutting tholeiitic dyke has the lowest $^{87}Sr/^{86}Sr$ value (0.70843). The massive sulfide (dominantly galena, sphalerite and pyrite) has a moderately radiogenic Sr isotopic composition ($^{87}Sr/^{86}Sr$ = 0.71366), which is similar to the stream sediment ($^{87}Sr/^{86}Sr$ = 0.71414).

The four felsic rocks have similar Nd isotopic compositions and are non-radiogenic with $^{143}Nd/^{144}Nd$ ranging from 0.5121 to 0.51219 (Figure 9). The mafic dyke is the most radiogenic of the whole rocks, with a $^{143}Nd/^{144}Nd$ value of 0.51275. The stream sediment is intermediate between the mafic and felsic rocks, suggesting a contribution from both mafic and felsic rocks (Figure 9). The carbonate veins span a wide range in Nd isotopic composition. Sample ML95-252 is slightly less radiogenic than the mafic dyke, whereas carbonate ML95-259 is highly radiogenic, with a $^{143}Nd/^{144}Nd$ value of 0.51535.

With the exception of carbonate from a vein (ML95-259), the other rocks and the four waters form an array between the mafic dyke with radiogenic Nd and non-radiogenic Sr isotopic compositions and the felsic volcanic rocks with radiogenic Sr and non-radiogenic Nd (Figure 9). The stream sediment and groundwaters fall between these two end-members.

5. Discussion

5.1 FRACTIONATION BETWEEN HOST LITHOLOGIES, GROUNDWATERS, AND SUSPENDED SEDIMENTS

Both the Halfmile Lake and Restigouche deposits occur in relatively small catchments (< 15 km^2), yet appear to have very different hydrologies. Groundwaters at the Halfmile Lake deposit have generally low salinities (< 250 mg/L). The low salinities, moderately low pH values, major ion chemistry and stable (O, H) isotopes were previously interpreted to have been produced by rapid recharge of modern meteoric water along steep fracture systems (Leybourne et al., 1998). In contrast, groundwaters at the Restigouche deposit range to much higher salinities (up to 22,000 mg/L), more evolved major ion compositions and much larger variation in O and H isotopes compared to surface waters, consistent with presence of much older, saline groundwaters relatively close to surface (Leybourne et al., 1998). Thus, groundwaters at the two deposits have different flow paths and age distributions, with implications for relative fractionation of the REE compared to host lithologies and also to the shape of REE patterns.

Groundwater REE abundances and patterns are ultimately controlled by the composition of the rocks (and minerals) through which the waters flow, modified to varying degrees by complexation, differential mineral solubility, and adsorption and precipitation reactions (Johannesson and Lyons, 1994; Johannesson et al., 1994; Johannesson and Zhou, 1999). For example, deeper groundwaters at the Halfmile Lake deposit have REE patterns that are very similar to the patterns of the host metavolcanic rocks, suggesting that for these waters the bulk host-rock exerts a major influence over the relative abundances of the REE. In contrast, Restigouche groundwaters generally have LREE-depleted profiles, unlike typical metavolcanic and metasedimentary rocks in the Bathurst Mining Camp so that there is little correlation between the relative groundwater and rock REE abundances. Halfmile Lake deposit groundwaters typically have lower pH, resulting in enhanced leaching of host rocks and minerals and likely also influences the degree of Mn- and Fe-oxyhydroxide formation with the result that differential REE adsorption is less important for the Halfmile Lake groundwaters. Thus, the REE profiles of the Halfmile Lake deposit groundwaters are less fractionated than those from the Restigouche deposit and, therefore, mimic the host rocks to a greater extent.

There are three primary sources of suspended sediment (and inorganic colloidal material) in groundwater systems (Degueldre et al., 1996); 1) suspension following alteration of host rock minerals by groundwaters, 2) mechanical abrasion or disaggregation of non-cemented particles of aquifer material, and 3) precipitation of minerals directly from the groundwater. Hydrothermal alteration about the Halfmile Lake and Restigouche deposits is characterized by quartz-white mica-chlorite, with albite and K-feldspar more common with increasing distance from the ore zones (Lentz and Goodfellow, 1993b; Lentz and Goodfellow, 1996). This alteration assemblage is typically also observed in fracture zones (Leybourne, 1998) and in the groundwater suspended sediments (see Mineralogy column in Table 4).

The REE contents of the suspended sediment may represent only disaggregated fracture zone material, or they may be controlled, in part at least, by the chemistry of the present groundwaters. Most suspended sediments have REE patterns similar to the host lithologies (compared to local least-altered volcanic rocks), with some notable exceptions for Eu (Figures 10, 11). At the Restigouche deposit, suspended sediments distal from the ore zone (boreholes MM-88-106, 107) have REE patterns that are relatively flat compared to unaltered felsic volcanic host rocks (Figure 11). In contrast, suspended sediments from boreholes that intersect massive sulfide mineralization (HN94-65 at the Halfmile Lake deposit and MM-88-06 and MM-88-07 at the Restigouche deposit) have positive Eu anomalies relative to unaltered felsic volcanics (Figures 10B, 11).

Although groundwaters and sediments typically possess negative NASC-normalized Eu anomalies, these Eu anomalies are not as large as those that exist in the host lithologies. Thus, water or suspended sediment normalized to least-altered rocks for both deposits typically possess strong positive Eu anomalies (Figures 10, 11). In addition, groundwaters with positive Eu anomalies have negative suspended sediment Eu anomalies (Figure 7), also suggesting that Eu is partitioned into the aqueous phase. The difference in Eu anomalies indicates that Eu is preferentially removed from the host lithologies relative to the other REE. Hopf (1993) noted that, in general, alteration of fresh rhyolite by acidic or alkaline fluids resulted in positive Eu anomalies if the altered rocks were normalized to the unaltered precursor. However, Lentz and Goodfellow (1993b) have shown that moderately hydrothermally altered rocks from the Bathurst Mining Camp show depletion of Eu, except for the most intensely altered rocks and massive sulfides. With the exception of carbonate-rich samples and massive sulfides, the more altered rocks in this study typically have negative Eu anomalies when normalized to least altered equivalents. Note that the alteration at 465 Ma that led to the formation of massive sulfide mineralization was by hydrothermal fluids that likely had large positive Eu anomalies, as shown in modern hydrothermal systems (Klinkhammer et al., 1994b).

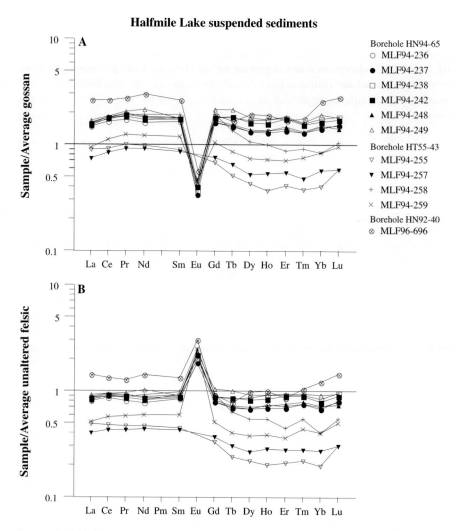

Figure 10. Halfmile Lake deposit suspended sediments normalized to A) average gossan, and B) average unaltered felsic volcanic rocks. Data from Tables 1, 4 with Halfmile Lake deposit gossan data from Volesky et al. (2000).

There appear to be two possible explanations to account for the relative positive Eu anomalies in suspended sediments from the mineralized zone; 1) Eu is relatively mobile and is being mass-transferred into solution and subsequently to the suspended sediment, or 2) rocks proximal to mineralization also have relative enrichment in Eu so that sediment from these rocks would also be expected to have positive anomalies relative to unaltered rocks. Compared to unaltered felsic rocks, Eu is preferentially enriched in the groundwaters suggesting that feldspar is a weathering contributor (Middelburg et al., 1988) and/or that Eu^{2+} is preferentially mobilized during water-mineral reaction compared to the trivalent REE (Bau, 1991; Sverjensky, 1984). Preferential mobilization of Eu^{2+} implies dissolution of minerals that formed at temperatures greater than 250°C at

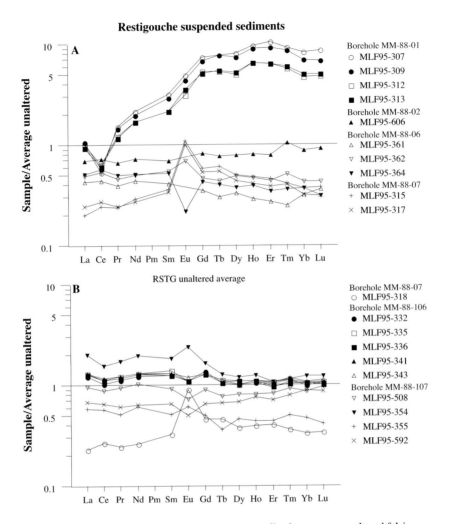

Figure 11. Restigouche deposit suspended sediments normalized to average unaltered felsic volcanic rocks. Data from Tables 1, 4.

which temperature Eu is typically divalent (Möller et al., 1998). Feldspar in Bathurst Mining Camp host rocks is dominated by albite and K-feldspar, which should possess positive Eu anomalies.

Suspended sediments from borehole HT55-43 at the Halfmile Lake deposit are unusual. This borehole penetrates the gossan at the surface expression of the deposit. Suspended sediments from this borehole have slight to large positive Eu anomalies, normalized to NASC and unaltered felsic volcanics, respectively (Figures 6, 10B). However, normalized to average Halfmile Lake gossan, these suspended sediments instead have large negative Eu anomalies (Figure 10A). The REE profiles of these suspended sediments are fractionated compared to the gossan from Halfmile Lake (Figure 10) with a convex-up pattern from La to Sm and concave-up pattern from Gd to Lu.

Boreholes MM-88-06 and MM-88-07 at the Restigouche deposit intersect massive sulfide mineralization close to surface. Groundwaters and suspended sediment from these boreholes are characterized by elevated base metal contents, Pb- and Fe-sulfate minerals (Table 4) and compared to unaltered felsic volcanic rocks show fractionated patterns (Figure 11). Both suspended sediment and massive sulfides show less depletion in Eu compared to the felsic volcanics. The enrichment in Eu in the suspended sediment relative to host rocks is greater than for the massive sulfides, suggesting that Eu is preferentially being mobilized during rock and sulfide weathering, but is strongly adsorbed to the suspended phases.

In general, suspended sediments are more LREE-enriched than corresponding groundwater (Figure 7), implying that LREE are preferentially transferred to the suspended fraction, presumably through adsorption. Several sediment-groundwater pairs also display concave-up patterns from La to Sm, similar to many shallow groundwaters at the Halfmile Lake deposit (c.f., Figures 4 and 7), suggesting preferential adsorption of the LREE to suspended sediment.

Suspended sediments from borehole MM-88-01 at the Restigouche deposit are strongly enriched in Al relative to host rocks and other suspended sediments. Co-existing groundwaters have Al contents (20 – 300 µg/L) generally more elevated than most groundwaters at the Restigouche deposit. The suspended sediments in this borehole shows the strong influence of the aqueous phase compared to that of the host rocks (Figure 11A) as shown by the REE patterns with strong depletion in the LREE and strong enrichment in the HREE compared to NASC and host rocks (Figures 6, 11). In addition, these suspended sediments have strong negative Ce anomalies unlike any of the other suspended sediments, and are similar only to local surface waters (Figures 5, 12). Condie (1991) noted that moderate correlation between the REE and Al_2O_3 in shale suggests that clays are important in hosting the REE's. SEM studies of the suspended material in Bathurst Mining Camp groundwaters shows that this fraction is dominated by silicate phases, suggesting that for some of the groundwaters, the REE patterns are also controlled by clay minerals. This control of the REE by clay minerals is also suggested by generally strong correlation between Al_2O_3 and the REE for most of the suspended sediments. For the Al-rich sediments from borehole MM-88-01, however, the control by aqueous phase processes is shown by the Al_2O_3-REE relationships (Figure 12). That the elevated Al_2O_3 in these sediments is not simply a detrital control is also shown by the strong correlation between Al_2O_3 and Ga for most sediment, except those from borehole MM-88-01 with relatively low Ga contents (Figure 12), indicating that Al is more mobile in these groundwaters than Ga.

5.2 Sr ISOTOPES

The primary control over the Sr isotopic composition of a water is the lithology and mineralogy through which the water has flowed, and the time and length of the flow path. In addition, the age and Rb/Sr of the host rock mineralogy is important given that older rocks will have greater time to produce ^{87}Sr through ^{87}Rb decay. All of the waters are more radiogenic than modern seawater, which has $^{87}Sr/^{86}Sr$ ranging from 0.70921 to 0.70924 (Burke et al., 1982; Veizer, 1989). The Bathurst Mining Camp waters are also more radiogenic than middle Ordovician seawater ($^{87}Sr/^{86}Sr$ = 0.7086 at 465 - 470 Ma; Veizer, 1989). The $^{87}Sr/^{86}Sr$ ratios of the Bathurst Mining Camp saline waters fall within the range for brines from the Canadian Shield, which vary between 0.707 and 0.755 (Franklyn et al., 1991; McNutt et al., 1987).

There are several trends that can be discerned from the $^{87}Sr/^{86}Sr$ vs Rb/Sr and 1/Sr relationships (Figure 8). The shallow groundwaters and saline waters from the Restigouche deposit may define a mixing/evolutionary trend such that with increasing water-rock interaction, Sr content increases and the isotopic ratio is increasingly controlled by the isotopic composition of the most soluble phases in the host rocks. This type of trend has been observed at the Eye-Dashwa Lakes Pluton (Franklyn et al., 1991), where the deeper saline waters had a common value essentially identical to plagioclase phenocrysts in the host-rock. However, given the quite different trace element ratios, major cation chemistry and stable isotopes between the shallow groundwaters and the saline waters at the Restigouche deposit, it is unlikely that these different waters are in any way related by simple mixing. Indeed, it was shown previously (Leybourne et al., 1998) that major and trace element and stable isotopic considerations preclude simple mixing between shallow groundwaters and a saline end member to produce the range of groundwater compositions recovered from the Restigouche deposit. In particular, the $^{87}Sr/^{86}Sr$ values for the brackish water from the bottom of borehole MM-89-106 are inconsistent with simple mixing between shallow waters and deep saline waters at the Restigouche deposit; the Sr isotopic composition of this water is too radiogenic compared to the other Restigouche groundwaters (Figure 8). Rather, the decreasing $^{87}Sr/^{86}Sr$ ratios with increasing Sr concentration may represent increasing degrees of water-rock interaction, indicating an increasing component from plagioclase (low Rb/Sr silicate) weathering. In addition, the groundwaters with the lowest $^{87}Sr/^{86}Sr$ ratios are those that are more distal from the ore body; the more radiogenic groundwaters associated with the ore body may be influenced to a greater degree by dissolution of carbonate and/or K-silicate minerals associated with hydrothermal alteration.

The high $^{87}Sr/^{86}Sr$ but low Sr contents in the shallow groundwaters close to the Restigouche deposit most likely represent easily dissolved, weathered materials in the soil/till zone and/or the shallowest portion of the bedrock. At the Restigouche deposit, examination of drill core showed that the uppermost 20-50 meters of bedrock is highly weathered and broken. In addition, although vein and vug carbonate is common at greater depth in the stratigraphy, much of the carbonate has been weathered out of the upper part of the section, suggesting that carbonate mineral dissolution dominates the isotopic signature of the shallow groundwaters. Carbonate minerals reported here and those analyzed by Peter and Goodfellow (unpublished data) have $^{87}Sr/^{86}Sr$ values between 0.70889 and 0.71508.

Strontium isotope ratios decrease in Restigouche deposit groundwaters with increasing water-rock reaction, indicating that the waters are increasingly influenced by low-K (and therefore low Rb) minerals. The trend shown by the shallow groundwaters is similar to Andean rivers and other river waters (Edmond et al., 1996). Edmond et al. (1996) suggested that this trend reflected mixing between waters derived from weathering of carbonate units with high Sr and non-radiogenic $^{87}Sr/^{86}Sr$ with waters derived from silicate weathering with low Sr and high $^{87}Sr/^{86}Sr$. However, for the

Restigouche deposit, carbonate close to the deposit is more radiogenic than the groundwaters further away from the ore deposit, with the exception of carbonate ML95-252. This carbonate may not be representative of most of the hydrothermal carbonate associated with the ore deposit; although the calcite component is within the range of other vein calcite, dolomite and siderite (with $\delta^{13}C_{VPDB}$ = -4 to –8 ‰ and $\delta^{18}O_{VPDB}$ = -12 to –18 ‰), the dolomite fraction is substantially lighter (with $\delta^{13}C_{VPDB}$ = -11.6 ‰ and $\delta^{18}O_{VPDB}$ = -23.5 ‰) (Leybourne, unpublished data).

There is a moderate positive correlation between $^{87}Sr/^{86}Sr$ and Rb/Sr (Figure 8). The most saline waters (Restigouche and Brunswick #12 deposits, with salinities of 22,000 and 45,000 mg/L, respectively) are at opposite ends of this Sr isotope spectrum, indicating that, at least for Sr, they have interacted with very different minerals and/or have mixed with very different fluids. Thus, it is most likely that there is a trend between the shallow groundwaters and the deeper brackish waters at the Restigouche deposit, and that the most saline waters are unrelated to this trend. The low $^{87}Sr/^{86}Sr$ ratios for saline groundwaters at the Restigouche deposit suggest that they may have equilibrated with more mafic host rocks than the other waters, consistent with elevated Cr, Co, Ni, and Cu values in these groundwaters compared to other groundwaters from ore zones at either the Restigouche or Halfmile Lake deposits (Leybourne and Goodfellow, 2003). Influence from the mafic sills/dykes would imply that these waters are locally derived, or at least that the Sr isotopic composition is locally derived, or that the sills represent preferential pathways for the saline groundwaters. In fact, the very low $^{87}Sr/^{86}Sr$ coupled with very high Sr concentrations in the Restigouche saline waters indicates that their Sr isotopic composition was derived from dissolution of a low Rb/Sr mineral. The mafic dyke has the least radiogenic Sr isotopic composition of all the rock samples (Table 5). Although there is extensive development of carbonate veins and fractures in the footwall of the Restigouche deposit, the $\delta^{13}C_{VPDB}$ data is not consistent with carbonate dissolution as the source of the non-radiogenic Sr (Leybourne, unpublished data).

For the Brunswick #12 groundwater, the low Sr contents but radiogenic Sr isotopic signature is most consistent with dissolution of a high Rb/Sr mineral such as K-feldspar or mica. This interpretation agrees with much higher K/Na for this groundwater compared to other saline waters in the Bathurst Mining Camp. The low Sr/Ca ratio for this groundwater suggests that mica dissolution exerted a greater influence than K-feldspar dissolution. The preferential dissolution of micas could be a reflection of the more intense and extensive alteration halo associated with the Brunswick #12 deposit (there is extensive sericitic alteration in the footwall; Lentz and Goodfellow, 1993a; Lentz and Goodfellow, 1996) and/or longer, more closed-system, water-rock reactions. In contrast, the brackish water from borehole MM-89-106 at the Restigouche deposit is moderately radiogenic with higher Sr/Ca and very low Ca/Na, suggesting that this water may have equilibrated with a high Rb/Sr, moderate Sr content mineral, such as K-feldspar.

The Heath Steele and Wedge deposit groundwaters show increasingly radiogenic Sr with increasing Sr contents and lower Rb/Sr. All four of these groundwaters are Ca-SO$_4$ to Na-Ca-SO$_4$-type waters. The trend to higher $^{87}Sr/^{86}Sr$ is consistent with increasing water-rock reactions (producing higher Na/Ca and Sr/Ca; not shown). The Wedge deposit groundwater has the shortest flow path of the four groundwaters and has the

$^{87}Sr/^{86}Sr$ value closest to that of the shallow groundwaters near the Restigouche deposit. The Heath Steele groundwaters are Ca-SO$_4$ waters with low pH and high base metal loads from moderate depth at the Heath Steele deposit (400-450 m). These groundwaters have had longer flow paths than the Wedge deposit groundwater and have interacted with minerals with higher Sr contents and higher $^{87}Sr/^{86}Sr$. The most radiogenic of the Heath Steele groundwaters is probably the most evolved (in terms of chemical evolution; Na-SO$_4$ water) and was recovered from greater depth (840 m).

The Heath Steele and Wedge deposit groundwaters plot along a similar Sr/Ca trend to the shallow groundwaters from the Restigouche deposit. Unlike the saline and brackish groundwaters from the Restigouche deposit, these groundwaters are undersaturated with respect to carbonate minerals. Therefore, it is possible that the isotopic composition of these groundwaters is controlled by dissolution of relatively radiogenic carbonate. The Sr isotopic compositions of the Heath Steele and Wedge groundwaters may indicate dissolution of carbonate with elevated Rb/Sr or carbonate formed from fluids which had acquired more radiogenic Sr during or subsequent to hydrothermal alteration of the host rocks. For example, one of the carbonate veins analyzed in this study is relatively radiogenic (Table 5). Siderite from the Brunswick Belt in the Bathurst Mining Camp ranges in Sr isotopic composition from 0.71189 to 0.71508, which was interpreted by Peter and Goodfellow (pers. comm.) to reflect precipitation from hydrothermal fluids with Sr derived from Ordovician seawater ($^{87}Sr/^{86}Sr$ = 0.7086; Veizer, 1989) and Sr derived from more radiogenic continental crust. Note that these $^{87}Sr/^{86}Sr$ values for Bathurst Mining Camp siderite are more radiogenic than some of the groundwaters at the Restigouche deposit. The less radiogenic groundwaters have equilibrated with minerals with lower time integrated Rb/Sr than the continental crustal end-member of the siderites. Given that most of the host lithologies at the Restigouche deposit are felsic volcanic rocks (presumably produced by anatexis of continental crust) or sediments derived from the felsic volcanic rocks, presence of minerals with less radiogenic Sr than the Bathurst Mining Camp siderite implies that either the continental crust was isotopically heterogeneous during Ordovician times or that the siderite has become more radiogenic with time due to accumulation of ^{87}Sr from ^{87}Rb decay. Clow et al. (1997) reported calcite with a $^{87}Sr/^{86}Sr$ value of 0.73215 compared to plagioclase in the same rocks with a $^{87}Sr/^{86}Sr$ value of 0.71956. Thus, the trend shown by waters associated with the massive sulfide deposits at Wedge, Restigouche and Heath Steele may reflect weathering of sericite, radiogenic carbonate, or both.

5.3 Nd ISOTOPES

The four waters analyzed do not correspond to an age consistent with the host rocks. Regression of all four waters yields an "age" of 77 Ma, based on $^{147}Sm/^{144}Nd$ ratios. Regression of the three groundwaters with the highest REE contents (and with REE patterns most similar to the whole rocks) yields an "age" of 244 Ma. Therefore, the Nd isotopic compositions of the different groundwaters reflect the time-integrated signature of the mineral or minerals controlling the REE composition of the water. The closer the REE compositions reflect the bulk rock, the closer the regression will be to the true age.

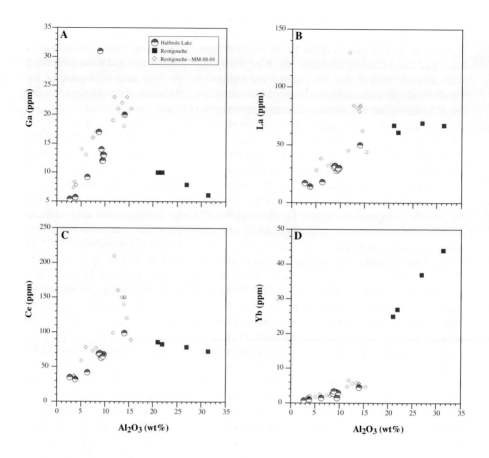

Figure 12. Geochemistry of suspended sediments, highlighting the unusual composition of the sediments from borehole MM-88-01 at the Restigouche deposit. Plots of Al_2O_3 vs A) Ga; B) La; C) Ce; and D) Yb. Data from Table 4 and Leybourne (2001).

Several whole rocks were also analyzed for their Nd isotopic composition. However, the dacitic and felsic volcanic rocks do not have enough variation in Sm/Nd to yield an age. The mafic dyke is tholeiitic in composition and clearly unrelated to the felsic volcanic rocks and cannot be used to derive an age. Although Nd model ages are difficult to interpret, especially for rocks with complex histories, model ages were calculated for the waters and felsic volcanic rocks. The four felsic rocks yield model ages between 940 Ma and 1.2 Ga, suggesting derivation from older (Grenville) basement. Two of the waters yield similar model ages (1.2 Ga) and have REE patterns very similar to the felsic volcanic rocks (Figure 9).

Although there are only a limited number of analyses, the Nd isotope results of the groundwaters highlight an important point with respect to the solubility of the REE in natural waters. The REE patterns of the surface waters and the majority of the groundwaters from the Restigouche deposit are controlled primarily by REE solubility, whereas many of the groundwaters from the Halfmile Lake deposit have REE profiles which appear to be mainly controlled by water-rock reactions (Leybourne et al., 2000). Regardless of the control over the REE distribution in water, if the water was sampling the whole-rock, the Nd isotopic composition should be similar to the host lithology and consistent with its age. If a given groundwater is preferentially interacting with one or more mineral phases, its derived Nd isotopic composition will reflect the time-integrated value of that mineral(s). For groundwaters MLW95-482 and MLW95-485, the similarity in REE patterns and model ages compared to felsic volcanic rocks suggests that the REE budgets of these groundwaters is bulk-rock dominated. Both of these waters have low pH values (3.03 and 2.33, respectively) consistent with this interpretation.

The Sr isotopes indicate that preferential weathering plays a significant role and this is also the preferred interpretation for the Nd isotope systematics of the BMC groundwaters. An alternative hypothesis that the Nd isotopic compositions of the waters reflect altered source rocks would require fairly rapid changes in the Nd values of the water, given the short spatial extent of potential whole rock materials. This seems unlikely. For example, Banner et al. (1989) reported Nd isotopic compositions of saline waters recovered from springs in Mississippian carbonates and Ordovician sandstone and carbonates from central Missouri. Due to differences of Nd isotopes between waters and host carbonates, Banner et al. (1989) concluded that the groundwaters had attained their Nd (and Sr) isotopes from crustal sources prior to migration through the carbonate host rocks. This implies preservation of the Nd isotopic signature over long time periods and flow distances. Furthermore, the Nd contents of groundwaters MLW95-482, MLW95-485 and MLW95-486 are high for natural waters (Leybourne et al., 2000) making it difficult to rapidly perturb the Nd isotopic compositions.

Nd isotope results for surface waters, soil waters and stream suspended sediments are also consistent with preferential weathering controlling dissolved REE contents (Andersson et al., 2001; Aubert et al., 2001). Aubert et al. (2001) concluded, on the basis of Sr and Nd isotopes of surface waters, soil waters and suspended sediments that leaching of apatite controlled the dissolved REE, whereas the Nd and Sr budget of the suspended load was mainly accounted for by feldspar weathering. Möller et al. (1998) also looked at Nd and Sr isotopic composition of groundwaters and suggested that the location of the waters in Nd-Sr isotope space between basalt and granite aquifer end members represented mixing of waters between the two aquifers. However, an alternative interpretation would be that there are different preferentially soluble minerals in the two different aquifer materials so that in the granite aquifer, less radiogenic Sr and more radiogenic Nd minerals are leached preferentially and vice versa in the mafic rocks. These studies support the conclusions drawn above.

6. Conclusions

Although there are broad similarities between host rock and suspended sediment REE contents and patterns, many of the suspended sediments display REE fractionation compared to NASC and least-altered host lithologies. In particular, Eu appears to be preferentially mobilized into groundwaters and suspended sediments, either because of preferential weathering of Eu-rich minerals (plagioclase, vein carbonate) or because divalent Eu is more readily mobilized from the host rocks compared to the trivalent REE.

Suspended sediments from one borehole at the Restigouche deposit are highly fractionated compared to local host rocks and are characterized by elevated Al_2O_3 contents, enriched HREE and negative Ce anomalies. These sediments are clearly influenced by mass transport of REE from host rocks to suspended sediment via the aqueous phase.

Most of the waters appear to follow two trends with respect to Sr isotopic compositions. Groundwaters from the Restigouche deposit trend to less radiogenic Sr compositions with increasing water-rock reaction, probably reflecting increasing contributions of Sr from plagioclase weathering. Groundwaters interacting with the Restigouche massive sulfides, and the Wedge and Heath Steele deposits appear to trend to more radiogenic Sr isotopic compositions and higher Rb/Sr with increasing depth and water-rock reaction. These relationships are most consistent with increasing contribution of ^{87}Sr from white-mica associated with hydrothermal alteration proximal to the massive sulfides, and/or from radiogenic carbonate in the stockwork zones to the deposits. The saline waters from borehole MM-88-02 at the Restigouche deposit have very high Sr and relatively non-radiogenic Sr isotopes, most likely indicating equilibration with plagioclase, perhaps associated with mafic dykes that cross-cut the deposit. In contrast, saline waters from the Brunswick #12 deposit have the most radiogenic Sr compositions in the study, suggesting interaction with K-feldspar and/or micas.

Sr and Nd isotopes all indicate that for natural waters, the dissolved loads of these elements have isotopic compositions that reflect differential mineral weathering rather than the bulk rock compositions. This interpretation is consistent with the fact that the groundwater $^{143}Nd/^{144}Nd$ ratios yield incorrect (young) ages, indicating preferential leaching of Nd from mineral(s) with low Sm/Nd ratios.

Acknowledgements

The fieldwork was funded by the Geological Survey of Canada, as part of the EXTECH-II (EXploration TECHnology) program. We thank Wayne Goodfellow in particular for providing the funding for this project. Peter Belanger is thanked for overseeing the geochemical analyses and Judy Vaive for the chelation ICP-MS analyses. R. Forconi and K. Lalonde of the GSC Instrument Development Section are thanked for their efforts in constructing the straddle packer system. Jan Peter, Wayne Goodfellow and Dave Lentz are thanked for discussions on the geology of the Bathurst Mining Camp during the course of the project. Very special thanks to our friend and colleague Dan Boyle, who died in 2000, for designing the straddle-packer system, for advice and encouragement throughout the project, and for improving MIL's golf swing. Comments from Dr. Karen Johannesson and an anonymous reviewer helped improve the manuscript. This is University of Texas at Dallas Geosciences Department contribution #1017.

References

Adair R. N. (1992) Stratigraphy, structure, and geochemistry of the Halfmile Lake massive-sulfide deposit, New Brunswick. *Exploration and Mining Geology* **1**, 151-166.

Allègre C. J., Dupré B., Négrel P., and Gaillardet J. (1996) Sr-Nd-Pb isotope systematics in Amazon and Congo River systems: Constraints on erosion processes. *Chemical Geology* **131**, 93-112.

Andersson P. S., Dahlqvist R., Ingri J., and Gustafsson O. (2001) The isotopic composition of Nd in a boreal river: A reflection of selective weathering and colloidal transport. *Geochimica et Cosmochimica Acta* **65**, 521-527.

Aubert D., Stille P., and Probst A. (2001) REE fractionation during granite weathering and removal by waters and suspended loads: Sr and Nd isotopic evidence. *Geochimica et Cosmochimica Acta* **65**, 387-406.

Banner J. L., Wasserburg G. J., Dobson P. F., Carpenter A. B., and Moore C. H. (1989) Isotopic and trace element constraints on the origin and evolution of saline groundwaters from central Missouri. *Geochimica et Cosmochimica Acta* **53**, 383-398.

Barrie C. Q. (1982) Summary Report Restigouche Property N.T.S. 21O/7E and 10E. Billiton Canada Limited.

Bau M. (1991) Rare-earth element mobility during hydrothermal and metamorphic fluid-rock interaction and the significance of the oxidation state of europium. *Chem. Geol.* **93**, 219-230.

Burke W. H., Denison R. E., Hetherington E. A., Koepnick R. B., Nelson N. F., and Otto J. B. (1982) Variation of seawater 87Sr/86Sr throughout Phanerozoic time. *Geology* **10**, 516-519.

Byrne R. H. and Kim K.-H. (1993) Rare earth precipitation and coprecipitation behaviour: The limiting role of PO_4^{3-} on dissolved rare earth concentrations in seawater. *Geochimica et Cosmochimica Acta* **57**, 519-526.

Clow D. W., Mast M. A., Bullen T. D., and Turk J. T. (1997) Strontium 87/strontium 86 as a tracer of mineral weathering reactions and calcium sources in an alpine/subalpine watershed, Loch Vale, Colorado. *Water Resources Research* **33**, 1335-1351.

Condie K. C. (1991) Another look at rare earth elements in shales. *Geochim. Cosmochim. Acta* **55**, 2527-2531.

Cousens B. L. (1996) Magmatic evolution of Quaternary mafic magmas at Long Valley Caldera and the Devils Postpile, California: Effects of crustal contamination on lithospheric mantle-derived melts. *Journal of Geophysical Research* **101**, 27,673-27,689.

Degueldre C., Grauer R., laube A., Oess A., and Silby H. (1996) Colloid properties in granitic groundwater systems. II: Stability and transport study. *Applied Geochemistry* **11**, 697-710.

Dupré B., Gaillerdet J., Rousseau D., and Allègre C. J. (1996) Major and trace elements of river-borne material: the Congo Basin. *Geochimica et Cosmochimica Acta* **60**, 1301-1321.

Edmond J. M., Palmer M. R., Measures C. I., Brown E. T., and Huh Y. (1996) Fluvial geochemistry of the eastern slope of the northeastern Andes and its foredeep in the drainage of the Orinoco in Columbia and Venezuela. *Geochimica et Cosmochimica Acta* **60**, 2949-2976.

Elderfield H., Upstill-Goddard R., and Sholkovitz E. R. (1990) The rare earth elements in rivers, estuaries, and coastal seas and their significance to the composition of ocean waters. *Geochimica et Cosmochimica Acta* **54**, 971-991.

Franklyn M. T., McNutt R. H., Kamineni D. C., Gascoyne M., and Frape S. K. (1991) Groundwater $^{87}Sr/^{86}Sr$ values in the Eye-Dashwa Lakes pluton, Canada: evidence for plagioclase-water reaction. *Chemical Geology* **86**, 111-122.

Goldstein S. J. and Jacobsen S. B. (1988) Nd and Sr isotopic systematics of river water suspended material: implications for crustal evolution. *Earth and Planetary Science Letters* **87**, 249-265.

Hall G. E. M., Vaive J. E., and McConnell J. W. (1995) Development and application of a sensitive and rapid analytical method to determine the rare-earth elements in surface waters. *Chemical Geology* **120**, 91-109.

Hopf S. (1993) Behaviour of rare earth elements in geothermal systems of New Zealand. *Journal of Geochemical Exploration* **47**, 333-357.

Johannesson K. H., Farnham I. M., Guo C., and Stetzenbach K. J. (1999) Rare earth element fractionation and concentration variations along a groundwater flow path within a shallow, basin-fill aquifer, southern Nevada, USA. *Geochimica et Cosmochimica Acta* **63**, 2697-2708.

Johannesson K. H. and Hendry M. J. (2000) Rare earth element geochemistry of groundwaters from a thick till and clay-rich aquitard sequence, Saskatchewan, Canada. *Geochimica et Cosmochimica Acta* **64**, 1493-1509.

Johannesson K. H. and Lyons W. B. (1994) The rare earth element geochemistry of Mono Lake water and the importance of carbonate complexing. *Limnology and Oceanography* **39**, 1141-1154.

Johannesson K. H., Lyons W. B., and Bird D. A. (1994) Rare earth element concentrations and speciation in alkaline lakes from the western U.S.A. *Geophysical Researh Letters* **21**, 773-776.

Johannesson K. H., Stetzenbach K. J., and Hodge V. F. (1997) Rare earth elements as geochemical tracers of regional groundwater mixing. *Geochimica et Cosmochimica Acta* **61**, 3605-3618.

Johannesson K. H. and Zhou X. (1999) Origin of middle rare earth element enrichments in acid waters of a Canadian High Arctic lake. *Geochimica et Cosmochimica Acta* **63**, 153-165.

Klinkhammer G., German C. R., Elderfield H., Greaves M. J., and Mitra A. (1994a) Rare earth elements in hydrothermal fluids and plume particulates by inductively coupled plasma mass spectrometry. *Marine Chemistry* **45**, 179-186.

Klinkhammer G. P., Elderfield H., Edmond J. M., and Mitra A. (1994b) Geochemical implications of rare earth element patterns in hydrothermal fluids from mid-ocean ridges. *Geochimica et Cosmochimica Acta* **58**, 5105-5113.

Lee J. H. and Byrne R. H. (1992) Examination of comparative rare earth element complexation behaviour using free-energy relationships. *Geochimica et Cosmochimica Acta* **56**, 1127-1137.

Lee J. H. and Byrne R. H. (1993) Complexation of trivalent rare earth elements (Ce, Eu, Gd, Tb, Yb) by carbonate ions. *Geochimica et Cosmochimica Acta* **57**, 295-302.

Lentz D. R. (1996) Recent advances in lithogeochemical exploration for massive-sulphide deposits in volcano-sedimentary environments: petrogenetic, chemostratigraphic, and alteration aspects with examples from the Bathurst camp, New Brunswick. In *Current Research 1996*, Vol. Mineral; Resource Report 96-1 (ed. B. M. W. Carroll), pp. 73-119. New Brunswick Department of Natural Resources and Energy, Minerals and Energy Division.

Lentz D. R. and Goodfellow W. D. (1993a) Mineralogy and petrology of the stringer sulphide zone in the Discovery Hole at the Brunswick No. 12 massive sulphide deposit, Bathurst, New Brunswick. *Current Research, Part E; Geological Survey of Canada* **Paper 93-1E**, 249-258.

Lentz D. R. and Goodfellow W. D. (1993b) Petrology and mass-balance constraints on the origin of quartz-augen schist associated with the Brunswick massive sulfide deposits, Bathurst, New Brunswick. *Canadian Mineralogist* **31**, 877-903.

Lentz D. R. and Goodfellow W. D. (1996) Intense silicification of footwall sedimentary rocks in the stockwork alteration zone beneath the Brunswick no. 12 massive sulphide deposit, Bathurst, New Brunswick. *Canadian Journal of Earth Sciences* **33**, 284-302.

Leybourne M. I. (1998) Hydrochemistry of ground and surface waters associated with massive sulphide deposits, Bathurst Mining Camp, New Brunswick: Halfmile Lake and Restigouche deposits. Ph.D., University of Ottawa.

Leybourne M. I. (2001) Mineralogy and geochemistry of suspended sediments from groundwaters associated with undisturbed Zn-Pb massive sulfide deposits. *Canadian Mineralogist* **39**, 1597-1616.

Leybourne M. I. and Goodfellow W. D. (2003) Processes of metal solution and transport in ground and surface waters draining massive sulfide deposits, Bathurst Mining Camp, New Brunswick. In *Massive Sulphide Deposits of the Bathurst Mining Camp, New Brunswick, and Northern Maine* (ed. W. D. Goodfellow, S. R. McCutcheon, and J. M. Peter). Economic Geology Monograph 11, 723-740.

Leybourne M. I., Goodfellow W. D., and Boyle D. R. (1998) Hydrogeochemical, isotopic, and rare earth element evidence for contrasting water-rock interactions at two undisturbed Zn-Pb massive sulphide deposits, Bathurst Mining Camp, N.B., Canada. *Journal of Geochemical Exploration* **64**, 237-261.

Leybourne M. I., Goodfellow W. D., and Boyle D. R. (2002) Sulfide oxidation and groundwater transport of base metals at the Halfmile Lake and Restigouche Zn-Pb massive sulfide deposits, Bathurst Mining Camp, New Brunswick. *Geochemistry: Exploration, Environment, Analysis* **2**, 37-44.

Leybourne M. I., Goodfellow W. D., Boyle D. R., and Hall G. M. (2000) Rapid development of negative Ce anomalies in surface waters and contrasting REE patterns in groundwaters associated with Zn-Pb massive sulphide deposits. *Applied Geochemistry* **15**, 695-723.

McAllister A. L. (1960) Massive sulphide deposits in New Brunswick. In *Symposium on the occurrence of massive sulphide deposits in Canada*. Canadian Institute of Mining and Metallurgy. Montreal, PQ, Canada.

McCutcheon S. R. C. (1997) *Geology and massive sulphides of the Bathurst Camp, New Brunswick*. Geological Association of Canada – Mineralogical Association of Canada, Joint Annual Meeting, Ottawa '97, Field Trip B7, Guidebook.

McNutt R. H., Gascoyne M., and Kamineni D. C. (1987) $^{87}Sr/^{86}Sr$ values in groundwaters of the East Bull Lake pluton, Superior Province, Ontario, Canada. *Applied Geochemistry* 2, 93-101.

Middelburg J. L., Van Der Weijden C. H., and Woittiez J. R. W. (1988) Chemical processes affecting the mobility of major, minor and trace elements during weathering of granitic rocks. *Chemical Geology* 68, 253-273.

Millero F. J. (1992) Stability constants for the formation of rare earth inorganic complexes as a function of ionic strength. *Geochimica et Cosmochimica Acta* 56, 3123-3132.

Möller P., Dulski P., Gerstenberger H., Morteani G., and Fuganti A. (1998) Rare earth elements, yttrium and H, O, C, Sr, Nd and Pb isotope studies in mineral waters and corresponding rocks from NW-Bohemia, Czech Republic. *Applied Geochemistry* 13, 975-994.

Peter J. M. and Goodfellow W. D. (1996) Mineralogy, bulk and rare earth element geochemistry of massive sulphide-associated hydrothermal sediments of the Brunswick Horizon, Bathurst Mining Camp, New Brunswick. *Canadian Journal of Earth Sciences* 33, 252-283.

Richard P., Shimizu N., and Allègre C. J. (1976) $^{143}Nd/^{146}Nd$, a natural tracer: an application to oceanic basalts, *Earth and Planetary Science Letters* 31, 269-278.

Rogers N. and van Staal C. R. (1995) Distribution and origin of felsic volcanic rocks, Tetagouche Group, Bathurst Mining Camp, New Brunswick. *Abstracts/Résumés - 1995, twentieth Annual review of Activities.*, 7.

Rogers N., Wodicka N., McNicoll V., and van Staal C. R. (in press) U-Pb zircon ages of Tetagouche Group felsic volcanic rocks, northern New Brunswick. In *Radiogenic Age and Isotopic Studies: Report 11*, Vol. Bulletin 72. Geological Survey of Canada.

Smedley P. L. (1991) The geochemistry of rare earth elements in groundwater from the Carnmenellis area, southwest England. *Geochim. Cosmochim. Acta* 55, 2767-2779.

Sverjensky D. A. (1984) Europium redox equilibria in aqueous solution. *Earth Planet. Sci. Lett.* 67, 70-78.

Symons D. T. A., Lewchuk M. T., and Boyle D. R. (1996) Pliocene-Pleistocene genesis for the Murray Brook and Heath Steele Au-Ag gossan ore deposits, New Brunswick, from paleomagnetism. *Canadian Journal of Earth Sciences* 33(1), 1-11.

van Staal C. R., Fyffe L. R., Langton J. P., and McCutcheon S. R. (1992) The Ordovician Tetagouche Group, Bathurst Camp, northern New Brunswick, Canada: history, tectonic setting, and distribution of massive-sulfide deposits. *Exploration and Mining Geology* 1, 93-103.

van Staal C. R., Wilson R. A., Fyffe I. R., Langton J. P., McCutcheon S. R., Rogers N., McNicoll V., and Ravenhurst C. E. (2003) Geology and tectonic history of the Bathurst Mining Camp and its relationship to coeval rocks in southwestern New Brunswick and adjacent Maine - a synthesis. In *Massive Sulphide Deposits of the Bathurst Mining Camp, New Brunswick, and Northern Maine*, Vol. 11, Economic Geology Monograph (ed. W. D. Goodfellow, S. R. McCutcheon, and J. M. Peter), pp. 37-60.

Veizer J. (1989) Strontium isotopes in seawater through time. *Annual Reviews Earth and Planetary Science Letters* **17**, 141-167.

Winchester J. A. and Floyd P. A. (1977) Geochemical discrimination of different magma series and their differentiation products using immobile elements. *Chemical Geology* **20**, 325-343.

Water Science and Technology Library

1. A.S. Eikum and R.W. Seabloom (eds.): *Alternative Wastewater Treatment.* Low-Cost Small Systems, Research and Development. Proceedings of the Conference held in Oslo, Norway (7–10 September 1981). 1982   ISBN 90-277-1430-4

2. W. Brutsaert and G.H. Jirka (eds.): *Gas Transfer at Water Surfaces.* 1984
ISBN 90-277-1697-8

3. D.A. Kraijenhoff and J.R. Moll (eds.): *River Flow Modelling and Forecasting.* 1986
ISBN 90-277-2082-7

4. World Meteorological Organization (ed.): *Microprocessors in Operational Hydrology.* Proceedings of a Conference held in Geneva (4–5 September 1984). 1986
ISBN 90-277-2156-4

5. J. Němec: *Hydrological Forecasting.* Design and Operation of Hydrological Forecasting Systems. 1986   ISBN 90-277-2259-5

6. V.K. Gupta, I. Rodríguez-Iturbe and E.F. Wood (eds.): *Scale Problems in Hydrology.* Runoff Generation and Basin Response. 1986   ISBN 90-277-2258-7

7. D.C. Major and H.E. Schwarz: *Large-Scale Regional Water Resources Planning.* The North Atlantic Regional Study. 1990   ISBN 0-7923-0711-9

8. W.H. Hager: *Energy Dissipators and Hydraulic Jump.* 1992   ISBN 0-7923-1508-1

9. V.P. Singh and M. Fiorentino (eds.): *Entropy and Energy Dissipation in Water Resources.* 1992   ISBN 0-7923-1696-7

10. K.W. Hipel (ed.): *Stochastic and Statistical Methods in Hydrology and Environmental Engineering.* A Four Volume Work Resulting from the International Conference in Honour of Professor T. E. Unny (21–23 June 1993). 1994
10/1: Extreme values: floods and droughts   ISBN 0-7923-2756-X
10/2: Stochastic and statistical modelling with groundwater and surface water applications   ISBN 0-7923-2757-8
10/3: Time series analysis in hydrology and environmental engineering
ISBN 0-7923-2758-6
10/4: Effective environmental management for sustainable development
ISBN 0-7923-2759-4
Set 10/1–10/4: ISBN 0-7923-2760-8

11. S.N. Rodionov: *Global and Regional Climate Interaction: The Caspian Sea Experience.* 1994   ISBN 0-7923-2784-5

12. A. Peters, G. Wittum, B. Herrling, U. Meissner, C.A. Brebbia, W.G. Gray and G.F. Pinder (eds.): *Computational Methods in Water Resources X.* 1994
Set 12/1–12/2: ISBN 0-7923-2937-6

13. C.B. Vreugdenhil: *Numerical Methods for Shallow-Water Flow.* 1994
ISBN 0-7923-3164-8

14. E. Cabrera and A.F. Vela (eds.): *Improving Efficiency and Reliability in Water Distribution Systems.* 1995   ISBN 0-7923-3536-8

15. V.P. Singh (ed.): *Environmental Hydrology.* 1995   ISBN 0-7923-3549-X

16. V.P. Singh and B. Kumar (eds.): *Proceedings of the International Conference on Hydrology and Water Resources* (New Delhi, 1993). 1996
16/1: Surface-water hydrology   ISBN 0-7923-3650-X
16/2: Subsurface-water hydrology   ISBN 0-7923-3651-8

Water Science and Technology Library

16/3: Water-quality hydrology   ISBN 0-7923-3652-6
16/4: Water resources planning and management   ISBN 0-7923-3653-4
Set 16/1–16/4 ISBN 0-7923-3654-2

17. V.P. Singh: *Dam Breach Modeling Technology*. 1996   ISBN 0-7923-3925-8
18. Z. Kaczmarek, K.M. Strzepek, L. Somlyódy and V. Priazhinskaya (eds.): *Water Resources Management in the Face of Climatic/Hydrologic Uncertainties*. 1996
                                                                 ISBN 0-7923-3927-4
19. V.P. Singh and W.H. Hager (eds.): *Environmental Hydraulics*. 1996
                                                                 ISBN 0-7923-3983-5
20. G.B. Engelen and F.H. Kloosterman: *Hydrological Systems Analysis*. Methods and Applications. 1996   ISBN 0-7923-3986-X
21. A.S. Issar and S.D. Resnick (eds.): *Runoff, Infiltration and Subsurface Flow of Water in Arid and Semi-Arid Regions*. 1996   ISBN 0-7923-4034-5
22. M.B. Abbott and J.C. Refsgaard (eds.): *Distributed Hydrological Modelling*. 1996
                                                                 ISBN 0-7923-4042-6
23. J. Gottlieb and P. DuChateau (eds.): *Parameter Identification and Inverse Problems in Hydrology, Geology and Ecology*. 1996   ISBN 0-7923-4089-2
24. V.P. Singh (ed.): *Hydrology of Disasters*. 1996   ISBN 0-7923-4092-2
25. A. Gianguzza, E. Pelizzetti and S. Sammartano (eds.): *Marine Chemistry*. An Environmental Analytical Chemistry Approach. 1997   ISBN 0-7923-4622-X
26. V.P. Singh and M. Fiorentino (eds.): *Geographical Information Systems in Hydrology*. 1996   ISBN 0-7923-4226-7
27. N.B. Harmancioglu, V.P. Singh and M.N. Alpaslan (eds.): *Environmental Data Management*. 1998   ISBN 0-7923-4857-5
28. G. Gambolati (ed.): *CENAS. Coastline Evolution of the Upper Adriatic Sea Due to Sea Level Rise and Natural and Anthropogenic Land Subsidence*. 1998
                                                                 ISBN 0-7923-5119-3
29. D. Stephenson: *Water Supply Management*. 1998   ISBN 0-7923-5136-3
30. V.P. Singh: *Entropy-Based Parameter Estimation in Hydrology*. 1998
                                                                 ISBN 0-7923-5224-6
31. A.S. Issar and N. Brown (eds.): *Water, Environment and Society in Times of Climatic Change*. 1998   ISBN 0-7923-5282-3
32. E. Cabrera and J. García-Serra (eds.): *Drought Management Planning in Water Supply Systems*. 1999   ISBN 0-7923-5294-7
33. N.B. Harmancioglu, O. Fistikoglu, S.D. Ozkul, V.P. Singh and M.N. Alpaslan: *Water Quality Monitoring Network Design*. 1999   ISBN 0-7923-5506-7
34. I. Stober and K. Bucher (eds): *Hydrogeology of Crystalline Rocks*. 2000
                                                                 ISBN 0-7923-6082-6
35. J.S. Whitmore: *Drought Management on Farmland*. 2000   ISBN 0-7923-5998-4
36. R.S. Govindaraju and A. Ramachandra Rao (eds.): *Artificial Neural Networks in Hydrology*. 2000   ISBN 0-7923-6226-8
37. P. Singh and V.P. Singh: *Snow and Glacier Hydrology*. 2001   ISBN 0-7923-6767-7
38. B.E. Vieux: *Distributed Hydrologic Modeling Using GIS*. 2001 ISBN 0-7923-7002-3

Water Science and Technology Library